This is the first book in a new series that will publish the very best work in the philosophy of biology. The series will be nonsectarian in character, will extend across the broadest range of topics, and will be genuinely interdisciplinary.

The Immune Self is a critical study of immunology from its origins at the end of the nineteenth century to its contemporary formulation. The book offers the first extended philosophical critique of immunology, in which the function of the term *self,* which underlies the structure of current immune theory, is analyzed. However, this analysis is carefully integrated into a broad survey of the major scientific developments in immunology, a discussion of their historical context, and a review of the conceptual arguments that have molded this sophisticated modern science.

In its highly distinctive blending of conceptual and historical narrative this book will prove invaluable to philosophers and historians of science, historians of ideas, immunologists, and biologists.

The immune self: Theory or metaphor?

CAMBRIDGE STUDIES IN PHILOSOPHY AND BIOLOGY

General Editor
Michael Ruse *University of Guelph*

Advisory Board
Michael Donoghue *Harvard University*
Jonathan Hodge *University of Leeds*
Jane Maienschein *Arizona State University*
Jesus Mosterin *University of Barcelona*
Elliott Sober *University of Wisconsin*

This major new series will publish the very best work in the philosophy of biology. Nonsectarian in character, the series will extend across the boadest range of topics: evolutionary theory, population genetics, molecular biology, ecology, human biology, systematics, and more. A special welcome will be given to contributions treating significnt advances in biological theory and practice, such as those emerging from the Human Genome Project. At the same time, due emphasis will be given to the historical context of the subject, and there will be an important place for projects that support philosophical claims with historical case studies.

Books in the series will be genuinely interdisciplinary, aimed at a broad cross-section of philosophers and biologists, as well as interested scholars in related disciplines. They will include specialist monographs, collaborative volumes, and – in a few instances – selected papers by a single author.

Forthcoming
Elliot Sober *From a Biological Point of View*
Peter Godfrey-Smith *Teleonomy and the Philosophy of Mind*

The immune self:
Theory or metaphor?

ALFRED I. TAUBER
Boston University

CAMBRIDGE
UNIVERSITY PRESS

PUBLISHED BY THE PRESS SYNDICATE OF THE UNIVERSITY OF CAMBRIDGE
The Pitt Building, Trumpington Street, Cambridge CB2 1RP

CAMBRIDGE UNIVERSITY PRESS
The Edinburgh Building, Cambridge CB2 2RU, United Kingdom
40 West 20th Street, New York, NY 10011-4211, USA
10 Stamford Road, Oakleigh, Melbourne 3166, Australia

First published 1994
Reprinted 1996
First paperback edition 1997

Printed in the United States of America

Library of Congress Cataloging-in-Publication Data is available.

A catalog record for this book is available from the British Library.

ISBN 0-521-46188-X hardback
ISBN 0-521-57443-9 paperback

To my father,
Laszlo Nandor Tauber,
with love and admiration

Contents

Acknowledgments

The Immune Self attempts to address a complex agenda, no doubt because it was written in response to a complicated challenge. The project took its final form in June 1992, in Ischia, where Gus Nossal, under the auspices of the Stazione Zoologica di Napoli, had invited scientists and their critics to discuss the history of immunology. At that meeting, it became apparent to me that the way "selfhood" framed post–World War II immune theory and experimentation deserved close scrutiny. Although stimulated by discussions with various scientists (particularly Gus Nossal, Ian Mackay, Lee Hood, Ron Zinkernagel, David Talmage, and Leslie Brent), and historians (especially Thomas Söderqvist, Barbara Rosenkrantz, Peter Keating, Alberto Cambrosio, Ann Marie Moulin, Horace Judson, and Arthur Silverstein), it became obvious that this key concept of contemporary immunology had somehow escaped notice, or at least had not received the attention it deserved. I was, however, well on the path of studying the immune self, from previous work with Leon Chernyak. This book began within a week of sending off the manuscript for *Metchnikoff and the Origins of Immunology: From Metaphor to Theory* (Tauber and Chernyak, 1991). The excitement of the first book ignited the sequel. That earlier work was always viewed as the beginning of an extended interpretation of immunology, and I knew there were untended issues that would require attention. Although certain obvious themes were easy to follow, I was haunted by the challenge to explain the modern resonances of Metchnikoff's theory. The dialogue at Ischia between many of the scientists responsible for our modern theory and the historians assembled to hear their stories and offer their own interpretations magically focused my own project.

   The Immune Self: Theory or Metaphor? explores some of the questions posed by *Metchnikoff* and begins where the previous work ended.

Acknowledgments

But *The Immune Self* is a very different kind of work. The sketch of immunology's history is drawn with broad strokes here (except for the detailed treatment of Burnet's use of *self*), and its philosophical analyses are both explicit and central to its theme. In beginning this endeavor, I was encouraged by numerous friends and colleagues as well as institutional support.[1] Splitting my time between my immunology laboratory at the Boston University School of Medicine and the Philosophy Department at the Boston University College of Liberal Arts has been an invigorating and challenging opportunity to examine my critique from diverse perspectives. The confidence to thus engage the problem is very much the product of that interaction, which is embodied both in spirit and in fact at the Center for Philosophy and History of Science, where Bob Cohen has been a mentor and confidant as I have slowly undergone my intellectual metamorphosis. Congenial colleagues, especially Burton Dreben, Charles Griswold, Erazim Kohák, Alan Olson, Lee Rouner, Roger Scruton, Roger Shattuck, and my collaborator Sahotra Sarkar, have enriched my philosophical excursions and tutored me with patience and tact. To each of them I express heartfelt gratitude.

The Immune Self truly profited from the editorial advice of the anonymous critics of several papers, published as preambles to this study, that have been incorporated in various guises through the courtesy of the editors of their respective sources.[2] I appreciate their cooperation and the opportunity now to integrate my various themes into a more comprehensive and coherent work. The project profited enormously through discussions with Henri Atlan, Leon Chernyak, Irun Cohen, Antonio Coutinho, Anne Dubitzky, Scott Gilbert, George Klein, Bob Schwartz, John Stewart, Francisco Varela, and Judah Weinberger, and to them I offer my sincere gratitude. The editorial advice of Michael Ruse and the assistance of Eileen Crist have been invaluable, as have the research assistance and critical comments of Scott Podolsky, whose careful reading of the Burnet papers and books helped to buoy my theme. The preparation of the initial manuscript by Ann Marie Happnie and its final production by Katharita Lamoza are gratefully acknowledged. As always, the support of my family and friends is most deeply appreciated, especially that of Alice and the children, from whom I borrowed the time to write this book.

Introduction

Alteration, movement without rest,
Flowing through the six empty places;
Rising and sinking without fixed law,
Firm and yielding transform each other.
They cannot be confined within a rule;
It is only change that is at work here.
– *The I Ching* 1967, p. 348

The first lecture I give my undergraduates in courses in philosophy of science or philosophy of medicine is designed to jolt them into recognizing how profoundly what we *know* determines what we *see*. The case I use is Leonardo da Vinci's anatomic drawings, which were the unpublished product of his autopsy studies. Shortly after Pope Sixtus IV (in 1482) granted "permission to take the bodies of legally executed criminals from the place of execution, and dissect them according to medical rules and practice," Leonardo began a series of studies in which his Galenic prejudice distorted the accuracy of some of his drawings (Clayton 1992). A striking example is the manner in which Leonardo depicted the cerebral ventricles. As taught by Aristotle and faithfully transmitted to European anatomists and philosophers, the mind was supposed to be composed of sensory, cognitive, and memory functions; thus, Leonardo's cerebral ventricles were drawn in 1489 as three connected in-line bullous structures (ibid., pp. 26–9). When he innovatively made a wax cast of the ventricular system twenty years later, he configured it correctly (ibid., p. 74); the circulatory system, however, he never got right (nor did anyone else prior to Harvey), and when my incredulous students demand why Leonardo did not simply "draw what was there," I show them his exquisite and accurate rendition of the

1

musculoskeletal system and thus demonstrate that persuasive preconceptions, suffered even by a Leonardo, can guide cognition. The study of the body as an object of scrutiny by new scientific criteria was but part of those events leading to a radically revised formulation of ourselves. The anatomic consciousness of the sixteenth century may justly be viewed as setting the stage of scientific medicine, but, more important, it was the underlying orientation toward Man as subject of scientific examination that spawned these inquiries. And underlying this self-consciousness was a metaphysical shift concerning the nature of reality itself. Such events are rare and have far-reaching repercussions, spreading into venues of thought and behavior well beyond the immediate concern of apparent interest. We justly speak of "revolutions," in the full sense of the term, for such profound reorientations. One of these, the Darwinian paradigm, serves as the core of a reformulation in the biological sciences that still reverberates both in the direct area of its purview and more subtly, albeit profoundly, in seemingly distant biomedical disciplines.

With the publication of Darwin's *On the Origin of Species,* a prescient observation was made by a contemporary: "Dykes have been burst; boundaries removed; we hardly know the old landmarks" (Masson 1867). Only six generations later, we are still adjusting to a reorientation of Man in nature, jostled from our anthropocentric primacy to a more circumspect view of our species. Like Leonardo, we have scrutinized and carefully examined our anatomy in light of a new perspective. Slowly we continue to uncover ourselves, reduced to universal biological mechanisms *and* enlarged by new mysteries. The very conception of our nature, composed of so many social, historical, psychological, and biological tributaries, each entwined with its neighboring threads such that none can be isolated and depicted alone, is still undergoing adjustment to the profound challenge of Darwinism. In the process this evolutionary self-awareness markedly altered how Man was self-conceived as a biological entity – beyond his evolutionary past, it would include his current biological nature. I need not argue whether *Origins* was the beginning or the culmination of this evolutionary ethos, for in either case, it clearly was its watershed. In this broadest Darwinian context *The Immune Self* is an attempt to chart the underlying philosophical expression of these forces in immunology.

Immunology was born in the controversies of that fresh announcement that no species, including our own, was a static entity; all were subject to change as a result of the vicissitudes of time and happenstance. Each life form was challenged to respond in endless competition and collectively adapt. Although not explicitly designated by Darwin, a crucial ingredient in this conception is a model of immunity that must be constructed in light of competitive dynamics. That evolutionary theory revealed an ever-changing species, defined by historical exigencies, immediately raises questions about the nature of the organism. In this schema the organism is not *given,* but *evolves;* it is always adapting, always changing. Thus, the very core issue of *identity* is for the first time raised as a problem. With the Darwinian revolution, a physiological ability to differentiate self from nonself must be postulated. This demand arises from two cardinal problems: (1) If the self is not given, it must be defined in process, which in turn requires a mechanism to identify *self,* and (2) in a dynamic interaction of self and other, as an articulated problem, self-identifying processes must in turn recognize the other. This latter concern suggests that the mechanism is fundamentally cognitive. The immune system assumes the role of discerning host from foreigner, a cognitive function, and like the nervous system, it also has a second function of response, that is, effector mechanisms by which to defend the host. The first problem of self-identification is addressed by an immune system that defines the host. The linked destructive mechanisms protect that identity. The first issue, then, is host *identity;* the second is host *integrity.* Together the establishment of identity and subsequent integrity-preservation functions to a large extent determine the organism's capacity to survive and compete with others. And to compete is to interact, and to interact is to cognize. Darwinism, then, implicitly demanded an elaboration of an auxillary science to deal with this aspect of dynamic interaction. Consequently one aim of this essay is an examination of the revolutionary insight into the very nature of the organism in terms that account for this cognitive function.

Immunology is one of those daughter sciences that in terms of both its origins and its guiding principles owes its birth directly to the response to Darwin's challenge. The argument already detailed in *Metchnikoff and the Origins of Immunology* (Tauber and Chernyak 1991) is that the discipline of immunology emerged as a direct product of the evolutionary problematics engendered by *On the Origin of Species,* and it incorporated both explicitly and implicitly the central tenets of that theory. The case seems unassailable that the immune concepts that defined the nascent field arose from the debates concerning evolutionary biology in the late nineteenth century; moreover, despite its divergence into the particular concerns of serology and immunochemistry in the first half of the twentieth century, immunology has returned to its original agenda, both in the specific nature of selectionist theories serving as the common dogma and in the deeper questions governing its scientific program. In brief, the case is as follows: Immunology might be viewed as the sibling of microbiology. As the etiologic agents of infectious diseases were defined in the 1870s and 1880s, with the corresponding principle that these maladies were the expression of a conflict between species (viz. human and pathogen), immunology and microbiology concurrently assumed their respective roles in explaining infectious diseases. Commonly histories have focused upon human disease as their dominant theme. But diphtheria, cholera, and tuberculosis are only particular cases of the more general problem of species competition in the context of evolutionary concerns. For some this was an explicit issue, but for most investigators this broad view remained a hidden agenda. And in this regard the popular narratives accurately reflected the dominant intellectual concerns of the period. I have attempted to take a more critical stance.

In respect to the underlying biological drama of host defense as an evolutionarily determined mechanism to fight pathogens, a comprehensive theory of inflammation was required that would be grounded in observations of simple animals and extrapolated to the human organism. Such a phylogenetically based theme of what we now call immunity was not dealt with by Darwin, but ironically it was articulated as an explicit problem in the year of his death, 1882. The theoretical and experimental origins of immunology were first enunciated along those principles by Elie Metchnikoff (1845–1916), a Russian zoologist who began his research career in the mid-1860s, shortly after the publication of *Origin*. He was the first to recognize immunity as an active response of the host to infection. The immune reaction was comprised of specialized and directed inflammatory processes that normally occurred upon any insult to the integrity of the organism. Thus, immunity was a specialized expression of inflammation that included common repair mechanisms stimulated by injury of any type, as well as surveillance and destruction of effete or dying cells. The common problem regarding the pathobiology of the host response, namely, the inflammatory reaction, was discerned as a complex of both specific and nonspecific reactions: Those classically described symptoms and signs of inflammation (e.g., arising from wounds or aseptic fevers) were differentiated from the specific immune responses elicited by a particular infecting organism. Tauber and Chernyak (1991) traces in detail this sweeping theory of pathology, documenting how Metchnikoff developed the basic foundation of immunology within this novel theoretical framework.

The genesis of Metchnikoff's hypothesis arose through his embryological studies. Careful examination of his published record reveals a fundamental concern to elucidate genealogical relationships in phylogeny and to refute the theory of ontogenetic recapitulation. Metchnikoff's entire research effort was a response to Darwinism and to those who wished to use developmental biology to sustain that argument. A central feature of Metchnikoff's early theory was the rejection of the von Baerian notion of a simple (predicted) developmental plan whose realization was the adult individual; instead, he sought the integrity of the organism in some activity or "mechanism" that was responsible for integrating (and constructing) intermediate designs and their final adoption. Such a mechanism eventuated in the notion of the phagocyte serving as the "harmonizing" element, at first seen as purveyor of what he called physiological inflammation in development and normal physiological function, only later seen as defender of the organism. The evolution of his thinking, specifically with respect to attitudes regarding Darwinism and embryological theory, underwent marked changes during his career, but by 1882, when he performed the famous Messina experiments (showing that phagocytes attacked rose thorns injected into starfish larvae), he was committed to a theory of the organism that sought to explain organismal development and its relation to evolutionary history. That theory, "physiological inflammation," was readily adapted to specific explanations of infectious diseases and thereafter became simply the "phagocytosis theory" (ibid., 1991). This odyssey in the turbulent seas of evolutionary biology was the origin of the first modern theory of immunity. Although Metchnikoff's organismic vision, based on a keen sensitivity of the evolutionary episteme, formed the foundation of early immunology, it was subsumed beneath the powerful reductionist position of his humoral detractors.

The Immune Self is not another book about Metchnikoff, but I must begin with his innovative theory to explain the origins of immunology's conceptual infrastructure. I begin by examining the development of immunology's conflicting theoretical orientations, which originated in the cellularist–humoralist debate, and attempt to show that the Metchnikovian question, which addresses the general issue of how immune function defines organismal identity, has finally achieved its rightful position in immunology. Imre Lakatos's conception of the scientific research program comprised of central and auxiliary hypotheses would be a suitable vehicle to relate immunology to Darwin's evolutionary paradigm (Lakatos 1978). For example, in Newton's program the three laws of dynamics and the principle of attraction compose the stable core principles, and the belt of auxiliary hypotheses includes such issues as planetary paths and viscosity of fluids, which are derived from the guiding and encompassing core doctrine (ibid. pp. 48–52). In this scheme, for immunology, natural selection serves as the core principle; in the Metchnikovian context, the belt of surrounding auxillary hypotheses must include the embryological model of competing cell lineages, the notion of physiological inflammation as a general view of physiological regulation, and immunity as the specific mode of engendering organismic harmony, namely, identity and its essential by-product, integrity. One of the crucial features of Lakatos's description of scientific growth is the active struggle that occurs at the edges of the core scientific program in auxillary disciplines. At the level of these auxillary theories we discern the development of immunology as the outcome of a struggle between an explicit evolutionary framework and one based on what must be regarded as a pre-Darwinian model.

It is important to recognize that the research strategy of Metchnikoff's microbiological detractors was based on the reductionist program enunciated by Hermann Helmholtz, Emil du Bois-Reymond, and other German physiologists in the 1840s (and enthusiastically endorsed by Claude Bernard in France during the next decade). As discussed in Chapter 2, this program addressed biological processes from an orientation divorced from concerns of evolution and development, since it viewed physiology as arising from physics and chemistry, that is, as an applied science of those venerable disciplines. In large measure the resistance Metchnikoff encountered may be understood in terms of the disparate research traditions from which the cellularists (evolutionary biology) and the humoralists (physiology) descended. These research traditions diverged not only in terms of methodology, criteria of explanation, language of discourse, professionalization, and the like, but in the very object of their inquiry: The humoralists were concerned primarily in defining those processes, or mechanisms, by which the organism *maintains* its integrity, whereas Metchnikoff asked how identity *developed.* For Metchnikoff the dynamic features thus became paramount. The question became nothing less than "What *is* the organism?" This essay explores this question from the perspective of the modern expression of this issue in immunology: What is the self?

The problem of selfhood, largely ignored after World War I, emerged again after World War II, when immunology turned from its primary immunochemical concerns to address biological issues revolving around tolerance and autoimmunity. Thus, immunology matured to the point of returning to the Metchnikovian problem of how organismal identity was defined, as opposed to its previous preoccupation with

6

describing immune integrity, that is, with identifying those mechanisms that defend an *already* well-demarcated host. In 1949 the explicit statement that the concern of immunology is to decipher the discrimination between self and nonself was formally introduced by Frank Macfarlane Burnet (1899–1985). This scientific agenda has in fact dominated immunology for the past forty years, and Burnet is a key author in the development of our current theory. He became so by introducing the concept of the self; thus, a careful delineation of his theoretical development is offered (Chapter 3). Burnet, originally a virologist, first used *self* in an offhand fashion in a semipopular book on infectious diseases, where his thinking was heavily dominated by an ecological orientation. In the 1940s he became more concerned with genetics and immune tolerance. To trace the emergence of Burnet's self theory in the 1940s is to clarify the underlying conceptual concerns of integrating immunology into its full biological context. Burnet's appropriateness as heir to Metchnikoff is documented in his sensitivity to the evolutionary perspective and the organismically based approach that governed his theory of immunological tolerance. This concern for accounting for organismal identity provides the link to Metchnikoff.

This essay explicitly argues that identity is an evolving and dialectical process of an organism engaged in challenges from both its internal and external environments. The model most closely approximating that activity is our own behavior – both in our encounters with the world and in our own personal inner space. And in the same sense that our personal identity remains a philosophical problem, so our current theories of immune identity orient themselves on certain philosophical "answers" to this ancient question. To trace the development of *self* serves as the orienting structure of this study. The use of the *self* terminology was initially hesitant, but it has now become part of immunology's language. Over the past forty years *self* has taken various meanings, the most difficult to elucidate being the nebulous concern with defining the "source" of immunity. By this I am referring to what might be perceived as an intuition that identity, the core boundary of the organism, the organismic foundation that immune activity is committed to defend, is not "given," but remains elusive. I maintain that this is the first sense in which *self* was employed in immunology and that it remains at the conceptual core of the discipline. This issue is explicated throughout the book; suffice it to note here that the term *self* accurately reflects our conceptual quandary. *Self* as a metaphor has achieved an unassailable status in immunology, not because it is a precise scientific term, but because it resonates with our understanding of core identity, which in actuality is a most nebulous concept. I maintain that there is no such entity as *the* immune self as a definable scientific construct. *Self* is used as a convenient mode to allude to identity – a code for an experimental approach, or perhaps better, an operational construct. I believe it is no coincidence that *self* is employed in many of the same senses as the term is used in philosophy, where it has provoked much debate since Descartes. Does the notion of self, then, retain any useful meaning, when it is used so expansively and with so many implications? I wish to suggest that *self* resonates between theoretical aspirations and the metaphoric latitude offered by its elusive character. Theory must grope for its footing in common experience and language. By its very nature the metaphor evokes and suggests but cannot precisely detail the phenomena of concern. Metaphorical language points the inquiry toward its strategy and object but cannot truly formulate them. The self appeared in the literature as immunology's theoretical leaders discerned the necessity to readdress poorly or vaguely articulated questions concerning a broadened conceptual framework for their discipline. Self, then, serves to orient the science of immunology and is crucial in erecting its theories.

The organismic view of dynamic selfhood, a historic and changing entity, altered by each immune encounter and in a sense renewed, if not recreated, echoes our own being. Challenged by constant engagements between self and nonself, the immune self has come to be viewed analogously to a living entity, continually redefined, reasserted, and redetermined. Echoes of Metchnikoff's self-defining concept of the organism may well again emerge in modern formulations, where models based on complex systems analysis, hierarchical design, cognitive theory, and mathematical logic are beckoning to incorporate the molecular vocabulary into a more meaningful grammar. If the immune system in fact functions as one of cognition, a view gaining wide acceptance, then we have yet to harvest the full implications of immune function in its normal, that is, unprovoked, setting, which posits an ever-challenged and changing self, responding to its internal and external environments in a highly dynamic fashion.

The power and limitations of the immune self serve as a focus of this study. This narrative begins by summarizing the conceptual growth of immunology and attempts to synthesize the main currents of the discipline's more recent development in the context of this broad intellectual problem. The course of the field in the first half of the twentieth century followed the agenda established by Metchnikoff's opponents. Immunochemistry was the prime focus of interest. Serologic studies and various attempts to establish the mechanisms of antibody formation, that is, the principles of immunochemistry, matured in conjunction with the spectacular successes of molecular biology, which triumphantly discerned the mechanism of the generation of antibody diversity.

The molecularization of immunology has been an exciting example of the power of applied molecular biology. In fact, it could be argued that one of the earliest and richest products of the genetic methodology of the 1970s was dramatically harvested in the determination of the basis of immunoglobulin assembly from constituent gene components to yield diverse (and specific) antibodies. The problem of how antibodies are produced against a highly varied universe of antigens dates from the earliest formulation concerning the nature of immunity; to appreciate the historical context of the contemporary findings, it is necessary to delve into the formative arguments of the discipline. A different tributary, arising from the biological problems of transplantation and immune stimulation refocused immunology on the central concern of immune identity.

Following this scientific narration, I explore how the self metaphor has been employed in immunology. To do so, I first examine in the scientific language the metaphorical content of its exposition and then attempt to correlate and probe the philosophical foundation of the selfhood concept (Chapter 4). This analysis is in turn used to reassess the underlying conceptual structure of immunological theory. The second half of the book, then, treats immunology as a philosophical subject. Simply, my intent is to place the science explicitly in its philosophical context. The philosophy of biology has almost entirely been devoted to the problematics surrounding evolutionary biology proper. Yet developmental biology and immunology are rich sources for studying the contours of the organism – the definition of the individual, or more metaphorically of the self. In this regard immunology is particularly appealing because its subject is the discernment of the self in the context of its encounter with the other; that is, it is a science that discriminates the organism from its environment, conceiving the immune system as mediating that interaction analogously to the nervous system. Just as cognition and response are characteristic of neural functions, immunity has perceptive and effector behavior that both identifies and endeavors to preserve the individual in its environment. Placing these concepts within our broadest philosophical understanding should shed light on how we model such function. In Chapter 5 I ask what conceptual assumptions are made when we posit immune function as cognitive. What do we gain by understanding "the self" as the source of immunity? How might such discussion expose assumptions that the science utilizes? To what extent do these issues reflect concerns of modern philosophical discourse? A philosophical inquiry into immunology must address issues that originated with the mind/body problem and now revolve around phenomenological identity. In turn, the science of immunology is likely to profit from scrutiny regarding its own assumptions from the perspectives that philosophy of language and cognition offer, in addition to the more general concerns of philosophy of science. With this book I hope to stimulate that dialogue.

The subtitle of Tauber and Chernyak (1991), *From Metaphor to Theory,* emphasizes the concretization of a vague metaphor for the body's warfare against disease, resulting in an explicit scientific research program based on a firm theoretical argument. The subordinate title here, *Theory or Metaphor?,* is a sequel, illustrating that the science has begun to outgrow its theoretical boundaries and is again using metaphor to approach conceptual issues as yet poorly articulated. In addition, metaphor serves to project immunology beyond the confines of its narrow research concerns to reach the broader culture, offering its contribution toward understanding ourselves. The extent to which contemporary immunology has extended its domain beyond its particular scientific discourse by bestowing a metaphor upon our culture at large should provoke wide interest. What immunology contributes to our concepts of identity and selfhood resonates well beyond the confines of this scientific discipline. Clearly there are powerful cultural determinants associated with the self concept, and to the extent we expose implicit and attendant meanings through a philosophical analysis of the origins of selfhood, we might discover revealing insights concerning our current definitions and their limits. It is at this interface between science and philosophy that the "triumph of immunology" must be critical with respect to how an imperialistic discipline naturally insinuates itself into the public domain and, more covertly, how cultural assumptions might infiltrate and drive science. Cases in other fields are well described, and similar attention to immunology has recently emerged.

This book, then, is not a history of immunology so much as an examination of the underlying ideas that shaped its growth and an initial exploration of the contributions immunology might offer with respect to broad philosophical themes. This dialogue is testimony to the emergence of a new legitimacy in such discourse, for immunology has arrived at a level of sophistication, where it may be included among the fields contributing to philosophy of science. Although there is a growing body of historians who have recognized the importance of examining the conceptual growth of immunology, philosophical analysis of immunology has been scant. It is this latter task I have assigned myself. This book, as a philosophical critique of immunology, explicitly explores the problems of essence, individuality, and identity. These questions reach back to ancient thought, and despite the technicalities of language, method, and theory, the core problem of selfhood posed by modern immunology is at its very base a philosophical matter.

The epistemological approaches of immunologists should be of interest to the wider audience of philosophers and historians of science, and reciprocally the scientists might profit from a more circumspect view of their own endeavor. This study was thus written in an attempt to bridge a gap heretofore separating these two domains. To do so, basic scientific and philosophical orientations are offered, but these must serve only as introductions. The book may be read by the neophyte in either discipline, but I have made no attempt to simplify the critique; there is strong reliance on the general accessibility of introductory immunology texts, current reviews, and monographs. The literature of immunology is rich and multilayered in sophistication. Regarding the history of immunology, the best general study in English is Arthur Silverstein's *A History of Immunology* (1989), which fairly chronicles the field's development. Regarding the philosophy of immunology as a general topic, there is no apparent precursor. It is this lacuna that I have attempted to fill.

The strategy employed here rests on two approaches, which reflect responses to different questions. The first concerns the nature of immune cognition and thus represents an epistemological problem. How immunology might become the object of philosophical inquiry seems to me to begin with that fundamental issue. I have chosen to adapt the phenomenological analyses of William James and Edmund Husserl, whose examination of the perception of experience offer a useful foundation for our own critique of immune recognition (Chapter 6). Sensory perception and the simultaneous awareness of meaning are intrinsically related; when we experience objects, their meaning is derived from the universe of their context. I believe this position is directly applicable to how we might envision immune perception and cognition.

The self metaphor ultimately reflects current attempts both to model immunity as a cognitive function and to define it as identity. To place the cognitive issue in its appropriate philosophical domain is to clarify immunology's dominant questions and recognize its most basic underlying intellectual query. This particular issue is only sketched as a tentative program of inquiry. Theoretical modeling of immunity as a cogni-

11

tive or a linguistic function, for instance, as analogical to neural net theory or to cellular automata, deserve critical assessment beyond their direct scientific utility. Although representing potentially rich lines of inquiry, these topics are not pursued here. My discussion is oriented to a more general analysis: On what basis might the immune self be understood from a phenomenological analysis? This modern school of philosophy endeavors to define the knowing subject by identifying and characterizing the objects of its thought. Such an inquiry seeks to highlight the resonance of this philosophical approach for a broadened view of the scientific problem of immune cognition.

The second philosophical question addressed in this essay concerns immunology's metaphysical foundation. This analysis reflects the origins of my inquiry into the post-Darwinian saga of Metchnikoff and reflects what I believe to be the foundational concerns of his theory and its persistant influence today. From this perspective, I turn to the episteme of evolution as explicitly formulated in the philosophy of Friedrich Nietzsche. There are cogent reasons to view the biologicism of Nietzsche, with its derived notion of will to power, as similar in orientation to the postulated conceptual foundations of Metchnikoff's organism. To plumb these waters is to identify both assimilated philosophical views of the period and to discern their extended influence. In contrasting the 1882–1908 period with our own time, we note that immunologists referred to the organism, not the self. When Metchnikoff proposed the phagocytosis theory, there was a shared view of the "organism" (viz. Man, viz. the self) in Nietzsche's philosophy. Nietzsche's biologicism is shown to utilize views of the organism as a striving, evolving, self-aggrandizing entity as envisioned by Metchnikoff. The philosopher is used to *illustrate* the scientist's formulation. We would not attempt to reduce one to the other, but the ethos of Nietzsche's organicism does offer insight into the Metchnikovian project and explains, in part, the hostile reaction to it. After explaining how Metchnikoff and Nietzsche share a common metaphysical understanding of the organism, I trace the philosophical antecedents of Nietzsche's will to power and the notion of selfhood from Kant and Schopenhauer. This exercise places the *self* problem in its fuller philosophical history, which in turn prepares for the concluding argument, that the self issue is but part of the modernist–postmodernist debate raging in virtually all intellectual domains today (Chapter 8). The discussion of Nietzsche is central: Is he the last modernist or the first postmodernist? The polemics begin with him.

How to situate the self in current philosophical debate hinges largely on the very possibility of *self.* Of the various postmodern theories, I have chosen Foucault's analysis of a historical and socially contingent Man, which like other deconstructionist critiques, leaves the self as highly problematic. Foucault is representative of the view that the self is an artifice, erected by the various forms of analysis, reflecting cultural power structures of particular periods. His analysis consists in deciphering Man as so many expressions of his social, psychological, and historical influences. More than *exterior* to or *separate* from any core identity, the self is forever in question, that is, in doubt. It is a crucial response to Nietzsche, and thus I attempted to place a modernist (i.e., Kantian) reading of Nietzsche in opposition to a postmodern (i.e., deconstructionist) interpretation of his philosophy. The consequences are obviously relevant to what value we might assign selfhood in the context of viewing the immune question as a philosophical problem. Fundamentally I believe any resolution revolves around our recognition of the meaning *evolution* has conferred to our understanding of our very identity. Thus, I conclude with a reexamination of the initial problematic, how Darwinian evolutionary theory has challenged and eventually invoked a new conception of organism, of Man, of identity. Nietzsche, in my view, holds the key to grappling with this fundamental question; therefore, except for the phenomenological analysis, which provides the most direct philosophical model to structure our current understanding of immune cognition, it is with Nietzsche that I perceive the clearest enunciation of what *identity* means in a post-Darwinian world.

The underlying goal of this essay is to decipher how evolution has become part of our metaphysical construct. In a sense the narrative comes full circle and reexplores how our understanding of evolution has insinuated itself into the structure of how we conceive of organism. Immunology as a post-Darwinian science first posed this question directly in Metchnikoff's research program, and it underlay the very conceptual basis of what became the foundation of the matured science. The same issue, submerged for almost fifty years, reappeared in somewhat different guise as self/nonself discrimination. But the nature of the self is fundamentally elusive. What was immune identity if identity was ever-changing, dialectical, and dynamic? What was the *essence* of the organism? That is a metaphysical question, dating to Heraclitus (ca. 500 B.C.): "As they step into the same rivers, other and still other waters flow upon them" (Kahn 1979, p. 53). Immunology is in this view only the latest participant in seeking to define essence, *qua* immune identity, in a cosmos of ceaseless change. The organism, no less than the world in which it lives, is subject to this Heraclitean flux. Thus, I view defining immune selfhood as a specific aspect of the philosophical problem of evolution. I am following the tradition of Metchnikoff in examining immunology philosophically, but I also invoke Burnet, who wrote, "immunology has always seemed to me more a problem in philosophy than a practical science" (1965, p. 17). This book has been written to substantiate that opinion and explain its importance. But the underlying legitimacy of this project rests more generally upon Whitehead's testament:

> If science is not to degenerate into a medley of *ad hoc* hypotheses, it must become philosophical and must enter upon a thorough criticism of its own foundations. (Whitehead, 1925, p. 24)

Read first when I was an aspiring biology student in college more than twenty-five years ago, Whitehead's view has sustained my effort and remains the basic ethos of this endeavor.

1

The phagocytosis theory

A. ELIE METCHNIKOFF

Fin-de-siècle Europe was difficult for its citizens to define: "Fin-de-siècle! Everywhere it stands for all that you might care to name" (quoted by Weber 1986, p. 9; see also Schorske 1979; Kern 1983; Dowling 1986). In the midst of social and cultural upheaval, the subject of Man was being redefined. This period witnessed the publication of William James's *Principles of Psychology* (1890), Ernst Mach's *Contribution to the Analysis of the Sensations* (1886), Freud's *Interpretation of Dreams* (1900), and Nietzsche's *Beyond Good and Evil* (1886). Ibsen produced *Hedda Gabler* (1890), and Cezanne and van Gogh were radicalizing the visual arts. New artistic freedom, in what Oscar Wilde epitomized as the full expression of personality (with a curious brand of individualism, antibourgeois socialism, and exploration – or better, exposure – of forbidden areas of thought and behavior), has been dubbed an era of decadence or, more kindly, *la belle époque* (Shattuck 1969; Ellman 1987, p. 305) It was a period of rapid imperialistic colonization of Africa and Asia and of French–German rivalry in Europe, while America was quietly looming as the economic giant, already the source of nearly a third of the world's production by 1888. Around that same year, the first beauty contest was held in Spa, Belgium, Barnum and Bailey's circus opened, cocaine began to be used regularly as an anesthetic, fructose was synthesized, and the term *chromosome* was first used.

The Tenth International Congress of Medicine opened on August 4, 1890, in Berlin's Circus Renz, the only hall large enough to hold the seven thousand participants. The new excitement over the possibility of understanding immune phenomena following Pasteur's successful treatment of rabies (1885) and earlier successes in immunization cham-

15

Figure 1. Elie Metchnikoff.

pioned by him and his archrival, Robert Koch, had started a revolution in laboratory-based research to combat newly discovered infectious diseases caused by bacteria: gonorrhea (1879), malaria (1880), tuberculosis (1882), cholera (1883), and diphtheria and tetanus (1884). Rudolf Virchow, who chaired the second-day session of the Congress noted that there were 623 visitors from the United States, 421 from Russia, 352 from Great Britain, and 171 from France (Mazumdar 1972). Among them was the Russian expatriate, Elie Metchnikoff (Figure 1), who had joined Pasteur's newly formed Institute two years earlier, where he had found refuge from inhospitable czarist reactionaries and nonwelcoming German compatriots of Koch (Tauber 1991d). Metchnikoff had presented a novel theory of immunity in 1883 that created a great deal of controversy. He argued that immunity arose by active defensive mechanisms of the host and that the principle mode of host resistance resided in the action of the phagocyte. Reception of the so-called phagocytosis theory was mixed. His supporters included Virchow and Pasteur, but the detractors were equally prestigious and notably were led by Koch. By 1890 the debate was polarized not so much over the notion of active defense as a principle of immunity, although that issue was still unresolved, but most clearly over the mechanism by which the host dealt with pathogens. The celebrated debate between the immunological cellularists, led by Metchnikoff, and the German humoralists was in full swing at the meeting (Mazumdar 1972; Silverstein 1989). The polemics were aggressive, with each side stubbornly defending its rigid position.

Metchnikoff's intense persona is perhaps most vividly presented in Paul de Kruif's popular narrative *The Microbe Hunters* (de Kruif 1926, 1954). Although Metchnikoff is depicted there as one of the early gallant fighters against infectious diseases, who "in a manner of speaking founded" immunology, he is also described as being "like some hysterical character out of one of Dostoevski's novels" (ibid., p. 194). Neither aspect of this description is an overstatement, considering Metchnikoff's suicide attempts and disabling psychosomatic headaches and eye problems, his infamous tantrum against his first scientific mentor, Rudolf Leukart, his squabbles with his friend and collaborator, Alexander Kowalevsky, his celebrated polemic against Ernst Haeckel over the evolutionary nature of the first metazoan, his academic and political intrigues at Odessa as a young professor, and, of course, the celebrated, highly emotional debates with his German detractors, who criticized his adamant and highly defensive position concerning the phagocytosis theory (Tauber and Chernyak 1991). As to de Kruif's grudging testament that Metchnikoff founded immunology, there is strong evidence that he was indeed its first author, as the entire discipline still rests on his formulation of immunity (ibid.; Stillwell 1991, pp. 305–14). More fundamentally, de Kruif misrepresents the development of Metchnikoff's theoretical edifice, which resulted in our modern conception of immunity. He is not alone, for virtually every commentator has essentially distorted Metchnikoff's approach along similar lines, namely, that "without one single bit of evidence, without any research at all, Metchnikoff jumped from the digestions of starfish to the ills of men. . . ." (de Kruif 1926, 1954, p. 195). The starfish myth, partly fostered by Metchnikoff himself (Metchnikoff 1908), originates in the following story: Having studied wandering amoeboid cells as an invertebrate embryologist for almost twenty years, Metchnikoff suddenly (in 1882) became a "pathologist" when he had the cathartic insight that these cells ate microbes and thus protected their host.

Many elements of this account are highly romanticized, and the claims of a sudden flash of insight – the spontaneous birth of the concept of immunity and the revolution in his conceptualization of the action of phagocytic cells – may be treated with reservation, if not skepticism (Chernyak and Tauber 1990; Tauber and Chernyak 1991). It is clear that this celebrated tale was Metchnikoff's popular version of a deliberate scientific program. For almost twenty years, his embryologic studies to define the mesoderm in invertebrates, assign genealogical relationships, establish a model of the first metazoan, and contribute to the exciting debate on the nature of evolutionary mechanisms initiated by Darwin's *On the Origin of Species* may be traced as logically culminating in the phagocytosis theory. Metchnikoff's theory of host defense applied to the newly discovered germ theory was rigorously denied by the German microbiologists (Silverstein 1989; Tauber and Chernyak 1989). As de Kruif observed (with only a hint of exaggeration), they viewed Metchnikoff with disdain: "Now Koch, precise microbe hunter that he was, would hardly have trusted Metchnikoff with the wiping of his microscope, but his ignorance of germs was nothing to this wild Russian" (de Kruif 1926, 1954, p. 195). Not surprisingly Metchnikoff was snubbed in Germany but ultimately found a position at the Pasteur Institute in 1888, where he spent the rest of his career defending his phagocytosis theory (Tauber 1991d).

His experience as an embryologist, concerned with relating developmental biology to the newly presented Darwinian theory of common descent, and the broad, often ingenious, application of his research to seemingly unrelated issues of pathology and inflammation placed him in a unique position to contribute seminal insight into the nature of the immune reaction. Metchnikoff's work emanated from his relationship to Darwinism. His earliest reaction to the *Origin of Species* was highly critical, but concurrent with the development of his immune theory, Metchnikoff became an enthusiastic supporter of Darwinism. (Interestingly, it was in the 1880s and 1890s that Darwin was viewed most skeptically in France [Bowler 1983; Corsi and Weindling 1985].) Metchnikoff came to embrace fully Darwin's proposed mechanism of evolutionary change, namely, natural selection; the notion of competitive struggle and the survival of organisms better adapted to their environment was uniquely applied by Metchnikoff to his own embryological research. He viewed the developing organism as composed of competing cell lineages, thus applying the idea of Darwinian struggle to the internal environment of the organism itself.[1] Metchnikoff's line of approach in this early period of immunology has served as an interesting, if not landmark, example of the decline of a descriptive, teleological, and sometimes even vitalistic vision of biology, and the gradual ascent of the physicochemical reductionist program.

Metchnikoff's arguments regarding interspecies forms of biological struggle was tangential to his principal evolutionary position, which turned that struggle into conflict *within* the individual organism. From this point of view, it is understandable why Metchnikoff stressed so often that the struggle of phagocytes with intruders (that is, interspecies competition) was only a secondary effect of their normal activity. The defense of the organism from parasitic attack was only one of several functions, including phagocytic repair of damaged tissue and surveillance of malignant or senile cells. Thus, the phagocyte (the immune system as a whole), in Metchnikoff's system, defined organismic identity; that is, it determined what was not to be destroyed or eaten, whether foreign or native. He placed the newly described active host response in the broad category of inflammation, thereby directly challenging the current theories of Julius Cohnheim and Virchow; he called the process "physiological inflammation," including in it all matters pertaining to both preserving host integrity and defining organismal identity. The latter concept was directly derived from his concerns as an embryologist; the former soon preoccupied him in the polemical debate with the microbiologists, who were only concerned with the mechanisms dealing with pathogens and had little interest in Metchnikoff's hypothesis in its broadest implications. In fact, Metchnikoff had created a new concept: immunity in the modern sense. Immunity for Metchnikoff was but a specialized part of cellular physiology, in which certain general properties have been accentuated and amplified for a particular function.

> All cells are able, by modifying their function under the direction of susceptibility, to adapt themselves to changes in the surrounding conditions. All living elements are able, therefore, to acquire a certain degree of immunity. But amongst all the cells of the animal body, the elements which have retained most independence – the phagocytes – most easily and first acquire immunity to infective diseases. These are the cells which be-

take themselves to situations where micro-organisms and their poisons
make their appearance, and which manifest a reaction against them.
(Metchnikoff 1901, 1905, p. 566)

Thus, the essence of inflammation (and immunity) was seen by Metchnikoff not primarily as an expression of interspecies conflict, but as
arising from the conflicting (sometimes harmonized) relations between
cell lineages of the same body. By the same reasoning, the "host–
parasite" relation did not play a paradigmatic role for Metchnikoff's
phagocytosis theory, but served as a subordinate theme.

Today we understand the phagocyte as a cell capable of engulfing
particles and killing the ingesta. This insight has been attributed to the
celebrated experiment of 1882, where Metchnikoff witnessed phagocytes attack rose thorns that he placed in transparent starfish larvae;
this model of engulfment is analogous to the action of amoeboid
phagocytes at an inflammatory locus to which they have migrated. Certainly this class of cells is crucial for normal host defense against a
variety of pathogens, but it can also be the principal effector cell of
chronic inflammatory diseases, where the tissue of the host itself becomes the pathological target of the phagocyte. Metchnikoff was the
first to recognize the true significance of the phagocyte as an inflammatory cell, define its function in all its important particulars, and construct a modern concept of the immune process as a special case of
inflammation. He understood that the combat between pathogen and
phagocytes represented the specific expression of the body's effort to
maintain its integrity.

But integrity from foreign insult was only one of several kinds of
deleterious effects the body had to "cleanse" in order to maintain its
very identity. The broader rubric of *identity* – the property that *defines*
organismal boundaries and what constituents its own – is a more fundamental assignment of immune function than simple host defense,
that is, to maintain host integrity. Metchnikoff, as will be discussed in
detail, regarded inflammation not only as a general mechanism for
combating microorganisms, but also one responsible for destroying
cancerous cells, repairing damaged tissue, removing senile or effete elements, and monitoring growth processes in general. It was as a developmental biologist that he first recognized the basis of these diverse functions, as he observed the role of phagocytes in embryonic growth.

As mentioned, Metchnikoff began his research career by describing
embryonic layers in primitive invertebrates. He sought to define the
mesoderm, whose origin and character were still problematic in the
1870s. To determine how the mesoderm was formed between the two

primordial germ layers (endoderm and ectoderm), Metchnikoff required a marker (a characteristic cell) to distinguish these layers from its antecedents. He chose the amoeboid phagocyte as that signal of the mesoderm, and in those seminal studies of mesodermal development, he sought to establish genealogical relations between simple animals where such cells first appeared in evolutionary history (Tauber and Chernyak 1991). Through a phylogenetic approach, he discovered that in organisms with a cavitary digestive apparatus, the phagocyte no longer had the nutritive function assigned to it in simpler animals with no gut. Preserving its primordial generic function in more complex organisms, the phagocyte continued to eat, but now for newly acquired purposes.

Metchnikoff first examined the role of these cells not by modeling an infectious process, but by studying their function in the transformation of the larvae of echinoderms and in the metamorphosis of the tadpole. We can appreciate this work as a natural extension of his embryological investigations, in which genealogical reconstruction of simple invertebrates was based on an expeditious role of phagocytosis in development. He viewed ontogenetic development as a struggle of embryonic components, where the organism appears as a foreigner to itself. Its identity must be established during ontogenesis partly as a result of that struggle. Thus, in relation to amoeboid mesodermic cells the question arises about their role in embryonic transformations and constructions, which became the problem of "physiological inflammation" in the adult. In short, the fact that phagocytes preserved their ancient function and actively continued to participate in establishing organismic identity, made it possible to consider the function of intracellular digestion as a step toward more complex inflammatory reactions. In relation to an already established individuality, the ability of these cells to destroy "the other" is phylogenetically transformed into the ability to destroy "a true other," for example, another organism.

In attempting to prove his conjectures, Metchnikoff proceeded step by step from descendants of older life forms to more recent ones in the study of vertebrate inflammation (Metchnikoff 1892). The central theme became how to extrapolate the ancient activity of amoeboid cells in ontogenetic development to diverse processes found in mature adults. He structured his thinking on his novel understanding of the opposition of harmony and disharmony; this opposition assumed the role of a new concept for formulating the connection between pathology and biology. For Metchnikoff, embryological development proceeds with cell lineages that are potentially in competition. The organism cannot be viewed as growing as a harmonious whole; rather, it arises and develops in a state of disharmony. In genealogical reconstructions this opposition played a methodological role: "Disharmony" was only a designation of genealogy as a problem; "harmony" is the stage that was methodologically chosen as the goal or ideal of this genealogy. But another meaning and a more profound issue was revealed by the opposition in respect to ontogenetic development: From the traditional point of view, evolution provided the scheme of organismic wholeness, its harmony as health. Disease is then conceived as a deviation from that balance.

Metchnikoff's new concept of the organism as a disharmonious entity that must strive for harmony was the basis of his entire research program and the intellectual foundation of the phagocytosis theory. If the organism was essentially disharmonious, the outcome of evolution's bequest of organs and functions that required adaptation to a new setting (time, circumstance, and organismal environment/ecology), then, he postulated, there must be a mechanism that mediates harmonization of potentially disparate or atavistic activities. Metchnikoff assigned the phagocyte this harmonizing function, and he called the harmonizing process "physiological inflammation," a natural integrative phenomenon (Chernyak and Tauber 1990; Tauber and Chernyak 1991). He viewed metamorphosis as a key example of phagocytic function, for example, phagocytes eating the tadpole tail and the reparative scavenging of damaged tissues. Immune defense against microbes was but a subordinate theme of the general phenomenon. In this context, phagocytes are not the gendarmes of the organism. Beyond this function, albeit novel and of immense impact on the development of a new biological discipline, immunology, Metchnikoff focused his conceptual orientation around the more general property of these cells. When he turned to the problem of inflammation, he was in the intellectual position to ignore the confusing complexity in higher vertebrates, since it obscured the primary role of the phagocyte. The phagocyte was not a metaphorical gendarme, subject to the "commands" of the "police state" (à la Virchow) of the organism, but an independent center of activity, representing its own evolutionary origin in that function. This phylogenetic and comparative pathological perspective structured his future immunological thinking.

In addition to Metchnikoff's achievement as a crucial founder of immunology, his contribution must be placed in the broader arena of modern pathology. Most obviously the phagocytosis theory deviates from the thrust of nineteenth-century conceptions concerning the nature of pathology and its dynamic role in defining health. Metchnikoff *began* with disharmony. Restoration of organic balance implies an initial harmony, but for Metchnikoff this state remained an ever-elusive ideal. The general process by which the organism strove for health Metchnikoff dubbed *orthobiosis,* and it remained a focus of his general biological interests during the last fifteen years of his life (Metchnikoff 1903; 1906; 1907). In this scheme the phagocyte became the mediator of the process that defined host identity. The process was teleological and also viewed as incipiently vitalistic, in the sense that the phagocytes determined organismic integrity simply by striving for their own aggrandizement. Their activity of competing with other cell lines was Darwin's struggle between individuals of the same or different species turned inward into a struggle within the organism. Here, then, is a profoundly novel concept of health: No longer are the ancient humors in balance; rather, life's cellular components are in conflict. Health is not given; it is actively pursued. The potential disharmonious assembly of evolved constituents must attain harmony. For Metchnikoff this was a dynamic process, one not given, but attained in conflict.[2]

Thus, Metchnikoff saw health as mediated by an essentially pathological process, the expression of the activity of phagocytes (viz. immunity) defining the normative. The norm arose as an outcome of evolutionary history. It is at this interface of physiology (the present) and evolution (the historical) that a conceptual integration must be made with a view to understanding the organism fully. In large measure Metchnikoff's end point was the same, in that this program aimed for practical applications of scientific insight.[3] He endeavored to harness his immune concepts to establish the ideal (harmony) and annul disease.

Metchnikoff's thinking places him outside the thrust of nineteenth-century conceptions concerning the nature of pathology (Canguilhelm 1989). His program clearly falls closest to dynamic models; the orientation is completely reversed from the approach of other nineteenth-century physiologists of the period. Claude Bernard's homeostasis, for instance, is a very different concept from that held by Metchnikoff. (The contrasting reductionism of Bernard is discussed in Chapter 2.) Both agreed on the organism's striving for health and on disease as its disturbance, but their respective orientations were reversed. Bernard assumed a quantitatively and qualitatively idealized physiology. Metchnikoff allowed no such initial harmony, regarding the organism as forever striving toward some ideal perfection. Pathology then became potentially restorative; "physiological inflammation" was his code for that curative process. Metchnikoff's theory of physiological inflammation arose from a novel notion of pathology. The extension of the normal to the pathological, their placement on an uninterrupted continuum, became a formalized problem for nineteenth-century pathology.

Various schools converged on the central concept that the abnormal was subject to the same laws as the normal, that the phenomena of disease differed from those of health only in terms of intensity. French critics would argue that this theme was established by François Broussais and developed by August Comte and Bernard, whereas German historians would trace the same conceptual development from Ferdinand Jahn to Rudolf Virchow (ibid.). The problem was addressed formally by applying mathematical criteria to biological phenomena, which heretofore had been viewed strictly descriptively. The prospect of a quantitative approach was equivocal: While it offered the objectification that a physicochemical mechanistic biology sought, it was resisted by formally intractable phenomena of instability and irregularity, which were understood as essential characteristics of life. To what degree, then, were arithmetic norms descriptive (of individual behavior), or in another format, to what degree were organic phenomena reducible to chemomechanical laws? As this issue was a major axis of dispute, a second crucial debate centered on how to incorporate the profound impact of Darwinism, that is, how to articulate an interface between evolutionary theory and a nonvitalistic physiology. The momentous intellectual reorientation ushered in by Darwin's *On the Origin of Species* not only offered a new paradigm for species transformation, but had enormous influence on general concepts concerning the nature of the organism. It complemented an aspiring reductionism in physiology by offering a severe materialistic orientation, where blind struggle for survival was the historical basis of form and function.

In this setting Metchnikoff's novelty of insight is apparent. He interpreted evolution as destroying the notion that an organism was composed of a stable, balanced mixture of ancient humors, for in his view the potentially disharmonious assembly of evolved constituents must strive for harmony. He was clearly a child of Darwinism in the sense that he exploded the notion of a circumscribable core identity. Endorsing the concept that species were not fixed, but subject to dynamic variation as the defining feature of life (not an accidental divergence from some fixed ideal), Metchnikoff similarly adopted an approach that *in principle* defined organisms as ever-changing. In such a context, health cannot be defined absolutely. Metchnikoff made his theoretical position quite explicit. It was well described in his 1891 public lectures, published as *Lectures in the Comparative Pathology of Inflammation* (1892),

24

in his magnus opus, *Immunity in Infective Diseases* (1901), and in several articles and books extrapolating these ideas beyond pathology to comments on Man's nature and general health and longevity (e.g., Metchnikoff 1903; 1906; 1907). But it was in the debate over the mechanisms of immunity that these ideas had their greatest impact, leading ultimately to his Nobel Prize. As discussed below, the originality of his underlying immune concept was either misunderstood or ignored, and Metchnikoff lost the scientific argument in the short term.[4] The major thrust of immunology in his day followed the agenda of the immunochemists. Simply, the humoral position regarded phagocytes at best as scavengers of a littered battlefield. True defensive (i.e., killing) factors were noncellular and resided as antitoxins in the fluid phase of blood or bodily cavities. Such agents were as yet poorly characterized, but they putatively fulfilled the expectation of being chemically definable. In large measure, it was the general ideal of establishing a physicochemical analysis that drove immunology away from Metchnikoff's largely descriptive biology and prioritized antitoxins. This is not to imply that Metchnikoff lacked a scientific following, nor that his contributions were ignored. In fact, certain scientists made a case for the complementarity of the humoral and cellular positions (for instance, Ernst 1903), and even the strongest supporters of the chemical position acknowledged Metchnikoff's contribution. Grünbaum, for example, wrote:

> The two chief theories of the present day are those promulgated and supported by Metchnikoff and Ehrlich respectively. Originally opposed to one another . . . they are really only rays of light refracted at different angles from the same source of light and will ultimately be made to fuse into one brilliant cellulo-humoral theory of immunity. . . . The spirited contest between the exponents and their partisans of these two rival theories has led to the most ingenious and interesting experiments of modern biological investigation. (1903, p. 776)

Because the research questions quickly became issues of specificity and chemical identification, the more theoretical and far-reaching questions that Metchnikoff posed were of little interest to the central thrust of immunological research before the mid-twentieth century. The chemical nature of antibody and the problem of host defense were questions more easily addressed and therefore first examined. But the fundamental animosity between Metchnikoff and his detractors was not simply a matter of disagreement on methodological priorities; rather, it centered on the very nature of immunity. Metchnikoff's profound conception of organism posited a dynamic, teleologically based notion of selfhood that was vociferously attacked. It could not be aligned with the new image of a biochemically oriented science of organism reduced to the simplest physicochemical elements (Tauber and Cernyak 1991, pp. 135–74; Stillwell 1991, pp. 219–71). The phagocyte as possessor of its own destiny and mediator of the organism's selfhood was received as too vitalistic a conception. Rather than argue the issue, however, his competitors simply pursued their own concerns. It appears that Metchnikoff's concept of immunity, to the extent that it was appreciated at all was generally understood solely as defensive action against pathogenic agencies (phagocytes as "gendarmes"). Microbiologists of the period were not concerned with host defense; the more fundamental, encompassing issues with which Metchnikoff approached phagocyte function, namely, the understanding of developmental processes in a phylogenetic context and the problematics of organismal identity, were never a matter for discussion.[5]

B. THE HUMORALISTS

Metchnikoff's central thesis of organismal identity, as a manifestation of his embryological interests concerning the primacy of form, were not problems of immediate interest to the immunological community. To put it simply, immune processes in the first half of the twentieth century were generally regarded not as establishing identity, but as serving to protect the organism, that is, as ensuring integrity. In this view the immune reaction does not establish the organism's identity, but at best discriminates between the host and the other. But for Metchnikoff, immunological processes are above all those activities that establish and constitute organismic identity; only as a consequence of secondary phenomena do they also *protect.* What, then, was the fate of Metchnikoff's vision? In tracing the reception of Metchnikoff's phagocytosis hypothesis, it is interesting to note that the earliest critics fastened on two issues: (1) the question whether the cells did in fact kill bacteria or simply participated in "necrophagocytosis" (a scavenging or clean-up function), and (2) the apparent vitalistic nature of the theory. The first question, which truly was an argument over mechanism, persisted throughout Metchnikoff's life. The early criticism against Metchnikoff was not immunologic in the sense we now accept. Paul Baumgarten, Ernst Ziegler, Carl Weigert, and others (Tauber and Chernyak 1989; 1991) contended that "immunity" (the state of not being infected by the same pathogen twice) was a passive phenomenon. For instance, in 1880 Pasteur, extrapolating from test tube experiments, assumed that the invading microorganism exhausted an essential nutrient during the first infection and was thus unable to survive in a host depleted of that substance. Such passive theories were the model of immunity and rested upon an ancient metaphysical understanding of health and disease, the balance of humors and the organism's ability to restore its wholeness.

Metchnikoff was responding to a deliberate attempt to disprove the phagocytosis theory initiated in several German laboratories that, almost exclusively along nationalistic lines, had mounted an aggressive attack on Metchnikoff at the Pasteur Institute. (He had emigrated to Paris in 1888 to escape the political unrest in Russia following the assasination of Alexander II.) George Nuttal, an American studying in Carl Fluegge's laboratory, was given the initial assignment. In a series of experiments with a variety of animals, he demonstrated that bacteria were killed independently of the presence or absence of phagocytes (Nuttal 1888, 1988). Following his lead, others would soon proclaim a humoral theory of immunity to replace Metchnikoff's cellular mechanism. Metchnikoff rigorously answered the challenge and obtained results conflicting with Nuttall's, believing that the humoral factors were in fact "ferments" elaborated from the leukocyte (Tauber and Chernyak 1991, pp. 156–7). (Of course, he was correct if we think of the antibody as a lymphocyte product, but he was not hypothesizing such an entity.) Metchnikoff never acknowledged that the action of his immune cells was either subordinate or even complementary to humoral factors (Metchnikoff 1901, 1905). As already mentioned above, in natural immunity, that is, the presensitized state, cellular mechanisms may play a more crucial role, but even in "special" cases illustrated by Emil von Behring's antitoxin generation (1890) or Richard Pfeiffer's phenomenon (1894), Metchnikoff took an adamently opposed position. In studies performed with Shibasaburo Kitasato, von Behring demonstrated that animals injected with either tetanus or diphtheria toxin generated an antitoxin (antibody) response that neutralized the toxin and protected animals against the lethal effects of infection with these microbes. This was a dramatic demonstration of humoral immunity, but because it was a "special" case of sensitization, that is, prechallenge, Metchnikoff viewed the results as not generally applicable. Pfeiffer championed the next experimental assault on Metchnikoff's theory in demonstrating the apparent cell-free killing of cholera vibrios in presensitized guinea pigs. Again, Metchnikoff argued that the humoral factors were in fact escaped ferments from phagocytes (Metchnikoff 1901, 1905).

The humoralist position into the mid-1890s has been reviewed[6] with particular attention to the critical findings of Pfeiffer, who demon-

The immune self

strated bacterial killing in the peritoneum independent of phagocyte participation, a conclusion Metchnikoff vigorously contested (ibid.). Pfeiffer's phenomenon, against which Metchnikoff felt compelled to argue, was obviously considered pivotal, and it instigated a barrage of highly contentious experiments. Although acknowledging the possibility that bacteria might be killed by humoral factors, Metchnikoff always maintained that the cholera vibrios used in Pfeiffer's experiments were unusually fragile and that the soluble (bactericidal) factors were products of the phagocytes, in effect acting as agents of the cells (ibid.; Metchnikoff 1908). These bacterial substances were recognized both as having several targets (e.g., Jules Bordet's studies revealed that bacteriolysis and hemolysis were due to the same factors) and as being heterogeneous in composition. Paul Ehrlich (1897) and his numerous followers, viewed the *amboceptor* (antibody) as a "sensitizing" substance preparing microbes for destruction, and *complement* as the serum factor that was directly bactericidal. Such interactions based on specificity and affinity for antigen are discussed below, but note that by 1890 the lines were already clearly drawn: On the one hand, the phagocyte was the architect of self-definition and host defense; on the other hand, the opposing humoralists argued for the independent presence of soluble bactericidal substances. In the Berlin Medical Congress of 1890 Charles Bouchard suggested that each limb participated in the immune reaction, but this was a compromise position few others shared. The following year at a Hygiene Congress in London Metchnikoff received a cheering ovation from the audience (Lister 1891).

The nascent immunological community was polarized. The seminal studies of Behring and Kitasato, showing that serum factors neutralized the toxins of tetanus and diphtheria and were transferable between animals, were countered by Metchnikoff as explicating a special case rather than a general phenomenon. The debate persisted, especially as the issues were further obfuscated by differences of experimental design, discrepancies in the specific immune reaction to various pathogens in different test animals, and the nonspecificity of many immunogens used for immunization. The specifics of the debate are less important than the recognition of the fundamental difference between the concepts of immunity held by Metchnikoff, on the one hand, and his detractors, on the other. By 1908, when he shared the Nobel Prize with Ehrlich, Metchnikoff's scientific achievements were well recognized: The phagocyte was generally accepted as an important element in immune effector function. But Metchnikoff maintained a more radical position, that the host, through the vehicle of the phagocyte, asserted itself and strived for (in his terms) internal harmony, by active mechanisms, among which the immunologic reaction against a pathogen was but one incidental example. That Metchnikoff failed to define the direction of immunological research for the next half-century is clearly documented; our interest here is to understand why. In the attempt to arrive at an answer, let us focus on one of the key members of the opposition to Metchnikoff, namely, Paul Ehrlich.

Ehrlich came to the field of immunology in 1890 by working in the newly established Institute for Infectious Diseases, led by one of Metchnikoff's arch antagonists, Robert Koch (Baumler 1984). Ehrlich was already an accomplished scientist; his doctoral dissertation on histochemical staining had established his central notion of differential chemical affinities of cells, which later became the core concept of his side chain theory (discussed below). He showed that basic, acid, or neutral dyes showed differential affinities for blood cells (1878). He thus established important principles for histochemistry that were soon applied to visual identification of the tubercle bacillus and later on modified for the identification of other bacteria in the gram stain (Baumler 1984). Ehrlich's immunological research began with studies of antibody formation to the plant toxin ricin, demonstrating that in its dependence on temperature conditions, the toxin–antitoxin reaction behaved like a chemical reaction. However, further characterization of the reaction was stymied, because the content of antitoxin in the immune sera was highly variable and no standardization had been made; this problem was particularly relevant to Behring's antidiphtheritic serum, a new therapy anticipated as having broad potential. It is difficult, if not unfair, to designate a particular investigator or set of experiments as the fulcrum that supported the redirection of investigation from immunopathology to the biochemistry of immunity. Many of the figures of the period reflected the ambivalence of immunology's conceptual focus, but Ehrlich attained a special status by establishing the first practical method for standardizing the antibody–antigen reaction that quantitated diphtheria toxin and antitoxin (Ehrlich 1897, 1957). In that seminal paper of 1897 he also first proposed the side chain theory of antibody formation. Here, then, in the same manuscript, were crucial contributions to quantitative immunochemistry and the biology of the immune response.

"It is beyond question that, in dealing with the problem of the diphtheria curative serum, whether from the practical therapeutic, or the purely scientific point of view, it is necessary to use sera of *accurately known value*" (ibid., p. 107, English edition). So began Paul Ehrlich's seminal paper to quantify antibody and place the humoral theory on a firm immunochemical basis. It is not surprising that Ehrlich successfully applied chemical principles to immunity. In the case of the immune reaction, Ehrlich understood the effect of toxins as a consequence of their differential affinities with target tissues, and he conceived of antitoxins as competing for binding in a highly selective fashion. His first experiments of the immune response to ricin were conceptualized in terms of "toxin and its antitoxin influenc[ing] one another by a direct chemical interaction" possibly by the "formation of double salts" (ibid., p. 114, English edition). He directly applied models from organic chemistry:

> It must be assumed that this ability to combine with antitoxin is attributable to the presence in the toxin complex of a specific group of atoms with a maximum specific affinity to another group of atoms in the antitoxin complex, the first fitting the second easily, as a key does a lock, to quote Emil Fisher's well known simile. (Ibid.)

It was generally regarded that Ehrlich left it to others to determine the effectiveness of leukocytes in the host response to infection and reserved for himself the establishment of the chemical principles of the immune response. He successfully quantified the diphtheria antibody, providing

> . . . the universally desired basis for a really scientific test. *With this information, the immunity-unit is no longer an arbitrary concept but is an exactly determinable quantity, and one, therefore, which can be reproduced afresh at any time.* (Ibid., p. 123, English edition)

To account for the antibody response, Ehrlich postulated that cells projected "side chains" from their surface that served as the receptors for what we now call antigens, and upon challenge, the host reacted by shedding excess side chains to combine with the offending substance by chemical affinity. The side chain theory was an elaboration of Ehrlich's central thesis that various affinities were exhibited by chemicals, nutrients, or in this case antitoxins for target biologicals. Basically Ehrlich employed the same notion of affinity in three diverse areas of research, with extraordinary effectiveness. Early in his career, he demonstrated that the differing affinities of dyes would exhibit the heterogeneity of blood cells. His immune theory was based on the same notion of chemical affinity as applied to the binding of antitoxin with a targeted toxin (or microorganisms, as the case might be). Later, he applied the same general principles to the use of chemicals as therapeutic agents against microorganisms. At the root of each of these cases is the concept of a biological receptor understood in terms of its affinity for a chemical substance. In a sense, it was "applied" chemistry, for his chief purpose was to apply *chemical* principles to biological phenomena. In the process, he provided the vexing solution for quantifying toxins and their respective antibodies. Ehrlich explained that toxins bind to cellular side chains (now called antibody receptors) by chemical avidity. These side chains, when shed, would become the humoral (i.e., cell free) antitoxins, or what we now call free antibody. There are several highly evocative and prescient concepts inherent in Ehrlich's theory. First, he is credited with the earliest modern formulation of the "receptor," a notion which has proved to be extraordinarily rich in twentieth-century cell biology, particularly with respect to communicative functions. More specifically with respect to the history of immunology, Ehrlich rejected outright the notion of an adaptive response, which, by necessitating production of "new groups of atoms as required, would involve a return to the concepts current in the days of [an obsolete] natural philosophy." Instead he chose an approach involving what he called "normal enhancement of a normal cell function," that is, preformed antibody that would be secreted upon appropriate stimulation. This conception, the antigenic selective aspect of the theory, was successfully revived in a modern version during the 1950s, with the clonal selection theory expounded by Niels Jerne, Frank Macfarlane Burnet, David Talmage, and Joshua Lederberg. Finally, his basic understanding of immunological specificity, conferred by the unique three-dimensional structure of antibody, was later confirmed. The fate of Ehrlich's research had the most profound influence on what Svante Arrhenius defined as immunochemistry (Arrhenius 1907). As we will see, each of the subsequent theories of antibody generation had to contend with Ehrlich's formulation.

It is not surprising that Ehrlich and Metchnikoff held strikingly different views of immunity (Tauber and Chernyak 1991; Tauber 1991c). Suffice it to note here that the specific issue of defining the relative role of humoral factors versus the phagocyte was not Ehrlich's preoccupation; he was more concerned with defining the chemical nature of the humoral factor, a problem that drew him into vigorous debate with Jules Bordet, Metchnikoff's protégé (Bordet 1909; Silverstein 1989). But Ehrlich's intellectual position did lead him into speculations that illustrate a very different orientation from that held by Metchnikoff. Where Metchnikoff always saw immune defense as an active response, Ehrlich allowed for passive encounters; for instance, pathogens and tumors might be starved in competition with the host as an alternative mechanism of sustaining self-integrity. Ehrlich viewed tumors and certain bacteria as competitive for crucial nutritive or growth factors; because of higher avidity, they might grow at the expense of normal tissue or conversely be starved if tumor (or bacterial) receptors could not competitively obtain those factors (Ehrlich 1908, 1960; Tauber 1991c). Metchnikoff, in contrast to this passive, nonimmunological theory, viewed tumors and microbial pathogens as intruders that were relatively autonomous and were subject to phagocyte (i.e., immune) surveillance. Like any struggle between host and pathogen, outcome was determined by the ability of host defense mechanisms to prevail; in his terms these were always active immune responses (Tauber and Chernyak, 1991). In this regard Metchnikoff's vision of immunity was unique, if not radically different, from that of other founders of the field.

Another example concerns *horror autotoxicus,* the improbable situation of the immune system turning upon itself. Ehrlich did not disallow the formation of autoantibodies; rather, he saw them as being prevented from acting. However, the general community of immunologists misinterpreted his view of *horror autotoxicus* as prohibiting autoantibodies altogether. That observation in itself is most revealing. From Metchnikoff's point of view, the immune reaction was based on warring disharmonious centers: The immune response arose from the disharmonic state. Thus, phagocyte surveillence of diseased or damaged cells was integral to physiological inflammation, and immune mediation was but a special case. If antibodies were in fact only facilitators of these processes, then autoantibodies were expected. In fact, the persistent search for such factors in Metchnikoff's laboratory shows that they were of great interest (Tauber and Chernyak 1991). The experimental background of this issue in the late 1890s was based on the first demonstration of an antiantibody. Alexander Besredka, Metchnikoff's adoring protégé, pursued the issue. He viewed the existence of autoantibodies as the logical extension and natural consequence of Metchnikoff's theory. These self-destructive factors were then postulated as being held in check by an antiantibody minimizing self-attack, a theory revived eighty years later.

C. AN APPARENT SYNTHESIS

Whereas immune reactivity to pathogens and foreign insult was generally accepted by the immunological community by 1900, the philosophical infrastructure of that understanding was certainly diverse. Beyond the different scientific traditions of immunochemists and Metchnikoff there resided a different orientation toward the basic teleological vector of the organism. How, then, were the humoral and cellular perspectives to be integrated? There were essentially four positions: (1) Metchnikoff's view that the phagocytes were fundamentally the exclusive agents of immunity, and even cases of the efficacy of humoral factors arose from stimulated phagocytes; (2) followers of Richard Pfeiffer who conversely viewed the phagocyte as secondary, with complement and antibody as the critical agents of defense; (3) compromise positions first held by Hans Buchner and later advanced by Alfred Pettersson that under some conditions cells were primary, whereas with other infections the humoral factors were the crucial element (this eclectic theory was able to account for the viability of Metchnikoff's interpretation in the case of immune reaction against anthrax and suppurative microbes, whereas the response to cholera and typhoid would be explainable by Pfeiffer's approach); and (4) the theory of the English physician Almroth Wright, which attempted to establish the humoral factors not only as primary bactericidal compounds under certain circumstances, but as universal facilitators of phagocytosis (Tauber 1992a). In this scheme the phagocyte might function autonomously, but under typical circumstances it became subordinate to the opsonin, whose "strength" would become the measure of immunity.

Opsonins are proteins that coat the microorganism through a pathogen recognition unit; a separate phagocyte recognition unit of the protein allows the cell to attach more readily to the bacteria or virus and thus enhance its engulfment. (George Bernard Shaw described opsonins as a kind of butter that made the pathogen more savory and therefore more aggressively attached – see "Doctor's Dilemma" [p. 338 and p. 351, 1906].) Opsonins may be the specific antibody generated in the acquired immune response; there are also classes of opsonins that are less specific but do recognize classes of bacteria or viruses by certain generic carbohydrate characteristics. (These proteins are only one member of a complex group of an ancient phylogenetic class of humoral defensive proteins that recently are coming under close scrutiny [Sastry and Ezekowitz 1993].) Thus, opsonins may be either part of a pre-immune repertoire of proteins or specific antibody. The distinction between natural and acquired immunity refers to the resistance to infection in animals that have never been exposed (natural) or have already been exposed (acquired) to the invading pathogen. Generally, natural immune responses utilize mechanisms that are less specific, because the antibody generated in a primary infection and available (through a memory capacity) to combat the microorganism upon a second invasion is not part of the initial, or natural, armamentarium. Much of the debate concerning the importance of phagocytes depended on the role assigned natural defensive reactions. Metchnikoff regarded the phagocyte as crucial to both (1901), but the humoralists emphasized the issue of specificity as central to immune function. The phagocyte is an effective agent in certain instances without specific antibody and is the beneficiary of the antibody generated in the acquired setting, where the antibody opsonizes the pathogen for augmented phagocytosis. But the humoralists were correct in surmising that the crucial event in acquired immunity was the production of specific antibody, not augmented phagocyte function. In other words, it was the antibody – then called antitoxin – that distinguished protective acquired immunity. And note, it was acquired immunity that had caught the imagination of the early immunologists and excited the expectations of the public.

"It is still a matter of uncertainty whether the blood fluids perform any role in connexion with phagocytosis." So begins Wright's famous 1903 opsonin paper (Wright and Douglas 1903), which conspicuously minimizes the contributions of preceding research (not cited until 1906 [Wright and Reid 1905–6]). Acknowledging that Metchnikoff had shown that "bacteria may be ingested in the living condition," testing phagocytes apart from blood fluids and then judiciously mixing the two elements seemed to Wright and his co-worker Stewart Douglas crucial in deciding whether there was cooperation between cellular and humoral bactericidal mechanisms. In fact, there was already an interesting literature on the matter: Metchnikoff himself a decade earlier had shown that serum from immunized animals (to swine bacillis pneumoenteritis), although neither "antitoxic" nor "antibactericidal," augmented the phagocytosis response and was protective (Metchnikoff 1892). Reports between 1895 and 1902 similarly showed enhanced phagocytosis with immune serum and correlation of agglutinin titres with survival had been documented (Tauber 1992a). Wright interpreted these studies, as well as his own, as further support for an exclusive defensive function of humoral factors against certain bacteria (e.g., typhoid and cholera) but not others (e.g., staphylococcus) (Wright 1910, pp. 36–72). His general position, however, was not sympathetic to Metchnikoff's phagocytosis theory, for Wright rejected the central Metchnikovian tenet that the serum factors were leukocyte products. These studies at least showed complementarity between phagocytes and immune humoral factors. In the final analysis, Wright's opsonization studies championed an important conceptual bridge between Metchnikoff and the humoralists, which most histories regard as decisive in reaching an apparent synthesis. Wright's opsonic index was used in the assessment of tuberculosis and taught on medical wards (e.g., Davenport 1987). He was knighted in 1906 and nominated for the Nobel Prize, but how did the experts of the time regard his theory as resolving the debate between humoralists and cellularists that had been waged since George Nuttall's first studies in 1888?

There is evidence that by the time Metchnikoff wrote *Immunity*, in 1901, he was friendly with von Behring, Ehrlich, and Nuttall. The old animosities and rivalries had died, and an appreciation of complementarity between cellular and humoral immune elements was mutually apparent (Ernst 1903; see Bibel 1988; Silverstein 1989). Certainly there is such balanced assessment in the literature (Ricketts 1908; Hiss 1908–9; Vaughan 1915), but the debate was far from over even when Metchnikoff and Ehrlich shared the Nobel Prize in 1908. The tentativeness of the resolution between Metchnikoff and the humoralists at the turn of the century is richly documented in the Nobel Archives (Tauber 1992a).[7] None of the letters written to the Nobel Committee explicitly addresses the genesis of the immune concept as the first theory of active host defense. Consistent, then, with the later versions of how Metchnikoff developed his theory, the embryological studies are only mentioned in passing by a few critics. Similarly Metchnikoff's general theory of inflammation is recognized as part of his critical intellectual contribution by only a handful of admirers. The 1902 Secret Report (a summary of the candidate's scientific virtues)[8] explicitly recommends *not* awarding the Prize to Metchnikoff because the relative importance of the cellular and humoral immune theories was still unresolved. In fact, no one embroiled in the controversy was considered suitable. (It is of interest that Behring received the Prize in 1901 for his studies of antitoxin, reflecting the optimism for serum therapies.) The report carefully reviewed Metchnikoff's *Immunity in Infective Diseases* (1901), a work no doubt written to lobby for the phagocytosis theory and to ensure its author the recognition he believed it deserved; the reviewers, however, were not convinced by Metchnikoff's assessment of the humoral theory, nor were they convinced about the exclusive role of the phagocyte in either natural or acquired immunity. Accepting that phagocytes could kill live bacteria, the report remained skeptical that other immune mechanisms could be so easily dismissed but concluded with a rather remarkable statement:

> Regarding the rewarding of this year's Nobel Prize to Metchnikoff, you cannot assert that he has solved an important question. On the other hand it must be acknowledged that Metchnikoff did a considerable ser-

vice since he was the first to consciously and seriously present the immunity problem to be solved. It was however here that opposition to his theory brought the problem forward. The compromise between the main directions which Metchnikoff's report, *Immunity* [1901], presents with the main stress on the cellular theory has a lot to thank this opposition for. Any award of the Prize to Metchnikoff would certainly not lack for criticism. We therefore consider his candidacy for the Prize should remain until the position of the phagocytosis theory in regard to the humoral theories becomes more clear than it is now. (Nobel Archives, 1902)

Acknowledging that Metchnikoff "was the first to consciously and seriously present the immunity problem to be solved," the Nobel Committee denied him the Prize because the relative importance of the phagocyte was unresolved. This might be viewed as equivocating over the issue of immune mechanisms.

Metchnikoff had no less than authored the modern concept of immunity, that is, that the host mounts an *active* response to a pathogen. This was a totally novel notion. As late as 1880 Pasteur and all others concerned with the germ theory regarded the host's immune status as due to passive factors; in Pasteur's view it was due to exhausted nutrients consumed in the first infection.[9] Similar ideas were expressed in what Sauerbeck (1909) called passive theories. When immunity had been shown to be conferred by dead bacteria (Salmon and Smith 1884), a major effort was expended to establish the mechanism. Metchnikoff had formulated the question of the immune response and furthermore established the parameters of inquiry. The debate between the mid-1880s and the early twentieth century was over the mechanism of the immune response, rather than over the fundamental concept itself. The 1902 Secret Report is another confirmation of this interpretation.

But another issue is raised by this report, which credits the controversy over Metchnikoff's theory for propelling the field "forward." On this reading it is the humoralists' accomplishments that brought the descriptive science (the cellular theory) to a chemically sophisticated level. The primacy of the *immune idea* of organismic definition has been subsumed under the more superficial (not to say unimportant!) issue of host defense. It is on this basis that the critique is offered; it is reiterated in a different way in 1907:

It was so much easier for the humoral view point of the bacteriocide to penetrate as the fundamental working hypothesis which the antitoxic immunity, through the work of Behring and Kitasato, gave humoral pathology the strongest support. (Nobel Archives, Secret Report 1907)

The phagocytosis theory

In fairness, those of the humoralist persuasion were struck by the clarity of these antitoxin experiments, in contrast to the well-documented cases of phagocyte failure to kill certain pathogens. But von Behring's studies were obviously contrived to stimulate an immune response, and a theory of first encounter was not presented by the humoralist school. Thus, very early on the difference between natural, that is, preimmunized (presensitized), immunity and acquired immunity was recognized. Even so, in the bewildering confusion of immune theory of the 1890s, with the inability to discern the mechanism responsible for either form of immunity, the chemically definable humoral response was a seductive choice as the foundation to further erect the discipline. But soon a compromise position was offered.

The 1907 report was more sympathetic to Metchnikoff because of findings relating to the possible complementarity of the humoral and cellular immune theories, but the report was still dissatisfied concerning the true role of the phagocyte and the enigma of opsonins:

> Wright's endeavors initiated a number of important issues: are the opsonins the universal cause for the absorption of bacteria into leukocytes? Are the substances homogenous or multiple? How do the opsonins relate to immunization in immune sera? Are the opsonins unknown substances or identical to already known antibodies? Does the incorporation into phagocyte cells, which has come about with the aid of the opsonins actually bring about the destruction of the bacteria through the leukocytes and so on. (Nobel Archives, Secret Report 1907)

Opsonins were either simply new characteristics of the known humoral bactericidal substances – complement and antibody – or genuinely new factors, relegating the previously described factors to secondary significance. The report remained undecided as to the relative importance of the humoral and cellular immune theories but noted in closing that K. A. H. Mörner (author of another report on Ehrlich) did not regard Ehrlich's views as undermining Metchnikoff's phagocytosis doctrine. In the next year they in fact shared the Prize, so the reports of 1908 are of particular interest.

Carl Sundberg and Alfred Pettersson wrote separate reports in 1908. Sundberg remained skeptical toward the claim of the centrality of the phagocyte at the expense of the humoral factors. No further mention is made of an integrated view, with opsonins serving as the bridge between the two viewpoints, and no recommendations for a resolution is given. Pettersson, a humoralist ally of Pfeiffer, had published studies that were highly critical of Metchnikoff (Pettersson 1905–6). According

37

to Pettersson, (1) leukocytosis was a secondary phenomenon to the effect of increased blood flow; (2) leukocytes lacked bactericidal substances; (3) when ingested, bacteria were destroyed, the effector mechanism still being due to serum bacteriolysis; (4) phagocytosis was useful by sparing more sensitive tissues from the effects of bacterial toxins; (5) the main bactericidal mechanisms were humoral, and the work of Metchnikoff and Wright should be viewed cautiously. With such a published record one might expect a highly critical report. He reviewed the genesis of the phagocytosis theory as Metchnikoff described it in *Lectures* (1892, 1893) and *Immunity* (1901, 1905). Like Sundberg, he thus cites the early studies showing cellular bacterial killing, the humoralist reaction, and Metchnikoff's inconclusive defense, but he ends his report with a list of substantial achievements and an enthusiastic endorsement:

> He has to a large extent [identified] the various phagocytes and in which cases they are active. He has discovered the immune serum's ability to promote phagocytosis. He has shown the ability of phagocytes to neutralize substances of bacterial origin, toxic to the organism. He has actively contributed to the discovery of the complex structure of the hemo- and bacterial lysins. He has in a new way shown how to influence the blood-forming organs in their ability to form blood corpuscles. It seems to me in his discoveries regarding immunity to be fully deserving of the Nobel Prize Award. (Nobel Archives, Secret Report, Pettersson 1908)

Pettersson's opinion obviously prevailed.

Metchnikoff was awarded the Nobel Prize in Physiology or Medicine for being what Count Mörner, rector of the Royal Caroline Institute, described in the presentation speech as

> . . . the first to take up consciously and purposefully, by means of experiments, the study of the question so fundamental to the question of immunity; by what means does the organism vanquish the disease-bearing microbes. . . . Certain kinds of cells . . . *are supposed to have, in addition to the other functions,* the task of catching and destroying disease-producing microbes. (Mörner 1908, 1967, p. 277–8; emphasis added; no further mention is made of Metchnikoff's comprehensive theory)

Then the apparent rationale for the Prize is offered, with due regard to the lingering controversy as to the relative importance of the phagocytosis theory:

> One can safely predict that even if other features are of more immediate importance in this doctrine, nevertheless the abundant actual observations which have been made with regard to the importance of the cells to the problem of immunity will always remain of great and permanent

value. . . . The *research of recent years into the question of immunity has thrown the importance of Metchnikoff's work into strong relief.* As a recognition of Metchnikoff's accomplishment in initiating modern research into the question of immunity, the direction and development of which, *particularly in its early stages,* he profoundly influenced, the Caroline Institute wishes to honour him with this year's Nobel Prize. (ibid., p. 278; emphasis added)

Mörner acknowledges the importance of Metchnikoff's research in proposing active defense as *the* immune reaction, but then he draws a major caveat, already well delimited in the confidential reports, namely, that Metchnikoff's contribution becomes important as a result of the humoralist findings. Mörner then describes the importance of antibodies to immune phenomena, simply presents Ehrlich as "organizer and leader in this field," and closes with the bestowal of the award jointly to Metchnikoff and Ehrlich for "their work on the theory of immunity" (ibid., p. 280). The genesis of Mörner's presentation speech may be found in the minutes of the Nobel Committee for September 19, 1908. This document clearly reveals how a compromise decision was made. There was apparently strong concern that the Prize not be "diluted," but inasmuch as the judges viewed both Ehrlich and Metchnikoff as deserving recipients, there was disagreement as to who should first receive the Prize. Since there was no assurance of either candidate prevailing in the following year's deliberations, the award was made jointly. Metchnikoff was acknowledged as the "founder" of immune research (Almquist and Sundberg) and Ehrlich as the better experimentalist (Sundberg), but ultimately the Committee judged both as critical contributors to immune theory and coleaders of immunology.

The textbooks of the period reflected these deliberations. Some, like Victor Vaughan, gave unabashed accolades to Metchnikoff: "We owe our knowledge of the most important facts in the new science of immunology to the labor and genius of Metchnikoff" (Vaughan 1915, p. 183). There was a general unstated consensus that the humoralists and the cellularists advocated complementary positions (despite Metchnikoff's persistent claims of the priority – and primacy – of the phagocyte). Some took an ironic view. For instance, H. T. Ricketts noted that, comparing the theories of Ehrlich and Metchnikoff, "one finds little more in common than the general purpose of explaining the phenomena of immunity. Yet it is remarkable that where there is so little in common there are so few contradictions of an essential nature" (Ricketts 1908, p. 212). Ricketts contrasted Metchnikoff and Ehrlich regarding antibody formation (biological versus chemical orientation), the chemistry and structure of antitoxins, the nature and origin of complement, and the effect of antitoxin (that is, stimulatory effect on phagocytes versus chemical neutralizing). He concluded that "the two theories do not stand to each other in relation of antitheses, and in the light of present knowledge it would seem unwarranted to cling to one view to the absolute exclusion of the other" (ibid., p. 214). Like most immunologists of the period, he viewed phenomena related to opsonization as the likeliest bridge between the two schools.

D'Este Emery wrote a more critical assessment (Emery 1909). Although sympathetic to Metchnikoff's theory with respect to natural (presensitized) immunity and proclaiming that *Lectures* (Metchnikoff, 1892) "must ever remain a great medical classic," he viewed the phagocytosis hypothesis as inadequate to explain acquired immunity: "The theory has now but a historic interest. It has served its purpose: it has been the means of suggesting many researches which have helped greatly in the elucidation of a most difficult subject" (ibid., p. 243). In detailing the literature on opsonization Emery correctly concluded that complement and antibody coordinately act to enhance cellular killing and assigned the phagocyte a delimited role (ibid., pp. 257–97). But more important than these testimonials and textbook assessments are the development of the major immunologic themes in the twentieth century. In short, what was the historical fate of the cellular and humoral immune theories?

At the Pasteur Institute Metchnikoff's followers – Besredka, especially – endeavored to support the master's theses, but in their resolute defense (in the case of local immune mechanisms and immunity bereft of antibody), they were awash in the tidal wave of immunochemistry (Moulin 1991b). Clearly, those interested in following Metchnikoff's research by focusing on the phagocyte per se will not in fact pursue the fundamental thrust of his theory – immunity as the process of self-definition, that is, as the purveyor of identity. That problem arose from a totally different sector of concern, and the phagocyte became circumscribed as an important object of study in divergent areas of investigation, unrestricted to the basic tenets of immunity. As an agent of inflammation and host defense, the phagocyte was studied in relation to effector mechanisms, and thus a body of research independent of immune definition could claim the phagocyte as its central phenomenon, utilizing it as a model for various problems of cell biology (e.g., movement and secretion) and as a suitable system to study cell growth and differentiation. Only through the study of lymphocyte biology was the phagocyte seen, in affinity with Metchnikoff's approach, as a central element in immune recognition and regulation (Unanue 1981). Thus, not surprisingly, the phagocyte's experimental course was soon divided between various research agendas, and the history of its twentieth-century fate lacks a unified coherence.

Phagocyte research largely resided outside the horizon of the principal problems of twentieth-century immunology, although its physiology and biochemistry have attracted wide interest. The physiology of phagocyte numbers was not carefully studied in Metchnikoff's time. Ironically, the first description of agranulocytosis was in 1902 (P. K. Brown 1902)! The correlation of absent phagocytes with infection (surely strong ammunition for Metchnikoff) was not widely recognized until 1922 (Schultz 1922), and careful study of this correlation was made possible only after its cause from studies of drug-related toxicity was described in the mid-1930s (Plum 1935). Systematic studies of phagocyte production, maturation, and kinetics were only initiated after World War II, and the discovery that humoral factors regulated their growth was appreciated between 1965 and 1975 (Craddock 1980; Metcalf 1988). These issues became highly relevant with the advent of aggressive chemotherapy and bone marrow transplantation, as the success of these procedures was limited in large measure by bone marrow toxicity, prolonged neutropenia, and resultant life-endangering infections. Concomitant to studies of phagocyte numbers were those concerned with phagocyte metabolism.

Metchnikoff had already discerned the successive stages of the phagocyte–pathogen encounter: diapedesis through the vascular wall, chemotaxis to the inflammatory locus, attachment to pathogens and their engulfment, and two of the final killing mechanisms – degradative enzymes and acid formation (Metchnikoff 1901, 1905; Tauber 1990). A third microbicidal mechanism, generation of reactive oxidative species, is now recognized as crucial defense against a wide variety of organisms, including staphylococcus and pyogenic bacteria.[10] Mechanisms utilized in the destruction of pathogens have been discerned likewise to inflict damage on host tissue, and the variety and distribution of phagocytes in what Ludwig Aschoff first called the reticuloendothelial system in 1913 (Aschoff 1924) have been linked as sharing common effector functions but specializing in the particular microenvironment of their residence.[11] Each of these various effector functions has been subject to sophisticated molecular investigation, but the general understanding of these questions has not been fundamentally altered in their conceptual orientation over the past century.

Metchnikoff's original observation concerning the role of phagocytes in tissue remodeling, first in embryological development and then in wound repair in adult animals, spawned research that remains active today. Tissue regression, remodeling, and accompanying cell death are essential to growth and repair, but relatively little is known about the cellular and molecular interactions that regulate such processes – although renewed interest is evident (for example, Lang and Bishop 1993). Such processes as bone resorption, wound repair, postischemic injury in adults, and the development of a wide variety of organs in diverse animals have been shown to be dependent on phagocytosis of either effete or degenerative cells (Tauber and Chernyak 1991, pp. 194–6). However, this area of developmental biology and pathology is not generally regarded as the domain of immunology proper. For many years the phagocyte found no secure home in the immunology of infectious diseases. This is not to say that phagocytes were not accorded their due role as first-line defenders against pathogens; however, the more sophisticated secondary immune processes caught the scientific imagination of most immunologists as more fundamental and central to their discipline. The phagocyte was to remain a largely neglected stepchild in the immunological family until the macrophage was recognized as possessing central roles in regulating hematopoiesis, the inflammatory reaction, and immune recognition. That discussion we defer; suffice it to note here that our understanding of phagocyte function per se has closely followed the outline Metchnikoff proposed. However, phagocytic functions alone do not qualify as the central themes of immunology in the latter half of the twentieth century. Ironically Ehrlich's immediate heirs did not directly address these questions either.

Before exploring those issues, I must offer a cautionary note. Metchnikoff, in this interpretation, would not argue the relative importance of the phagocyte and the lymphocyte, natural and highly specific acquired immunity, respectively; clearly lymphocyte biology has come to define immunology in the past forty years. But irrespective of that division, the underlying concern is how immune recognition *defines* the identity of the organism. Obviously natural and acquired immunity each play their respective roles in that overarching problem. Metchnikoff, I believe, resisted the humoralists because the metaphysical basis of their research program was at odds with his own (see Chapter 2), and thus the way in which the facts of immunity were placed in their theoretical edifice took on a meaning quite different from his own. Quite simply he was asking a different question, and the polemics were less confrontational than tangential, although they *appeared* to be framed within the same structure.

The phagocytosis theory

Considering the discovery of an augmented antibody reaction, autoimmunity, and diverse immunopathology by 1910, one might well ponder why the biology of these problems was not more actively pursued after the first decade of the twentieth century. If the issue was solely a matter of recognizing Metchnikoff's observation that the host mounts an active response to pathogens, then the debate would have continued over the question of mechanism, that is, of the relative roles of cellular and humoral components and the basis of humoral elaboration from immunocytes. But by 1920 those issues were dormant and the question of immune specificity became paramount. The chemistry of the immune reaction was of central concern. Metchnikoff's highly abstract and often nebulous question of how the organism was defined by evolutionary dynamics was not of immediate *experimental* concern. In fact, few understood his philosophical framework. The humoralists tackled the simpler question of the mechanics of recognition.

Immunology, in the physicochemical reductionist approach of the twentieth century, chose for the next fifty years to follow the humoralists' angle. Only in the 1950s did developments in medicine and biology call for the reopening of the Metchnikovian problem, and we see then a turn to such questions as the nature of immunological tolerance and transplantation. Such issues are the modern expression of Metchnikoff's original developmental questions. His central concerns of organismic identity and the primacy of form were not problems of interest to the early immunological community. To put it simply, immune processes were generally regarded not as those that establish (i.e., ontologically define) individuality, but as the processes that "merely" protect the organism. In this view the immune reaction does not establish what was later referred to as the self, but at best discriminates the host from the foreign. For Metchnikoff, however, immunological processes are, above all, those activities that constitute organismic identity, and only as a consequence of secondary phenomena do they also *protect*. Correspondingly these processes cannot be simply reduced to functions of recognition of other and/or protection of host. Rather, in Metchnikoff's more far-reaching view, self-identity arises from the dynamics of these immunological activities. The heuristic value of this formulation has yielded extraordinary insight. Before we examine how little this conceptualization was understood, let us begin to place Metchnikoff's concepts in their broader biological and philosophical contexts.

2

The triumph of immunology

A. A BIOLOGY BASED ON THE WHOLE

The conflict between Metchnikoff and the humoralists must be regarded as a particular case of a more general struggle surrounding the
ascendency of reductionism, defining both the research strategies and
the metaphysical foundations of biology and medicine in general. Reductionism was declared as the program of German physiology in the
1840s to combat vitalism and establish the exclusive domain of physics
and chemistry as the sciences to describe the living realm (Galaty 1974;
Gregory 1977; Lenoir 1982; Kremer 1990). The application of physicochemical theory and methods to biology has a clear history dating from
the end of the eighteenth century in the maturation of combution theory (Lavoisier), the application of atomic proportions (Jeremias Rich-
ter and Dalton), and the extension of inorganic chemistry to organic
matter. More broadly the reductionist program in physiology arose
from the growing appreciation that energy was interconvertible from
one form to another and was conserved. As fundamental laws governing physical forces were defined for electrical, chemical, and mechanical
energies, biologists would seek to place organic functions within the
same scientific construct. Thus, the initial efforts to formalize the research agenda of physiology to chemistry focused on the problem of
accounting for animal heat. In the process teleology and holism (the
latter we would describe as organismic biology) were assaulted as allied,
if not intimately linked, with the vitalistic doctrine.

In the twenty-five-year period of 1883–1908 (the years of Metchnikoff's active immunology career), the emergence of new standards and
expectations can be discerned in almost all biological disciplines,
molded by what might well be regarded as a modern-age scientific
ethos. It is not my intent to delineate carefully the intellectual or social

44

forces that gave rise to this conflict, but I will offer certain parameters to place the early source of immunology's disciplinary borders in perspective. I have already alluded to the reductionism–holism debate, which in the case of immunology was a struggle between descriptive biology and a new chemically-based research program. We quite readily discern an argument of scientific legitimacy, where reductionism was proposed as an attempt to better "objectify" the life sciences. In philosophical terms it must be regarded as a metaphysical concept, where the nature of reality, in this case the biological realm, is sought in defining the most basic units or identifying the most fundamental laws that govern complex behavior. This research program would objectify the organism to its barest constituent elements, namely, its physico-chemistry.

The self-righteous zeal of the new ethos of scientific objectivity of reducing living processes to chemistry and physics blocked the development of other scientific method or conjecture. However, we would be amiss to suggest that reductionism had subsumed and completely overpowered the descriptive methodologies in biology. Note how John Merz, in his influential (and authoritative) review of nineteenth-century thought, summarizes the situation at the end of the century:

> Clearly, besides the abstract sciences, which profess to introduce us to the general relations or laws which govern everything that is or can be real, there must be those sciences which study the actually existing forms as distinguished from the possible ones, the "here" and "there," the "where" and "how," of things and processes, which look upon real things not as examples of the general and universal, but as alone possessed of that mysterious something which distinguishes the real and actual from the possible and artificial. These sciences are the truly descriptive sciences, in opposition to the abstract ones. They are indeed older than the abstract sciences, and they have, in the course of the period under review in this work, made quite as much progress as the purely abstract sciences. In a manner, though perhaps hardly as powerful in their influence on practical pursuits, they are more popular; they occupy a larger number of students; and inasmuch as they also comprise the study of man himself, they have a very profound influence on our latest opinions, interests, and beliefs – i.e., on our inner life. (Merz 1896, 1965, vol. 2, pp. 203–4)

This testament reflects at fin-de-siècle a persistent balance between the reductionist and descriptive sciences, but Merz underestimates how the next century would yield a more damning verdict. Metchnikoff was an early victim of the ascending reductionist standards. Descriptive sciences were indiscriminately associated with a former age, disparaged as riddled with romantic subjectivism and susceptibility to observer bias. If a scientific question was not definable in quantitative terms, ipso facto, its credibility immediately fell into question under the auspices of a reductionist strategy that claimed exclusive intellectual access to objectivist criteria. Reductionism was buoyed as a prevailing scientific approach because it claimed a more rigorous and "scientific" method of explanation than its competitors. This struggle of how to define the legitimacy of competing approaches to biology dates to the early nineteenth century. The term *biology* was first coined independently by Gottfried Treviranus and Jean Baptiste de Lamarck (Coleman 1977) at the beginning of the nineteenth century, when a new consensus was emerging among "natural philosophers" that animate nature required a different level of study than simple description taxonomy, or classification. The responses to the challenge of defining *functional* processes and of creating an explanatory science were disparate and passionately defended. The interest here is to address biology's scientific origins, that is, how biology was defined as a science, and why during its formative period the definition of biology's intellectual strategy generated such intense debate. So, when historians ask, "When did biology begin?" there is predictable disagreement (Caron 1988), since the answer clearly depends upon one's particular vision of biology.

Biology grew out of earnest and explicit philosophical argument, and discord dating from the early nineteenth century remains unresolved, if disguised, exerting a powerful influence. The modern presentation of bioscience was articulated in the 1840s, almost twenty years before the publication of *On the Origin of Species,* when a group of German physiologists made an explicit commitment to a new reductionism. The achievement – or better, the ascendancy – of this scientific attitude was on the most superficial level an attack against vitalism, but the casualties were wider. Other theoretical programs fell. Immanuel Kant commanded a profound influence over the biological sciences in this period. At the risk of oversimplification, it could be argued that during the first half of the nineteenth century, Kant served as the prominent intellectual inspiration of the various movements of biology, which may be classified into three schools of thought: *naturphilosophie,* teleomechanism, and reductionism (Lenoir 1982). Ironically, although all three were indebted to particular readings of Kant, as rival schools of thought, they fought bitterly.

Naturphilosophie was highly speculative and abstract; its advocates invoked Kant's theory of mind and developed its rationale in close connection with German idealism and romanticism. *Naturphilosophie*

46

might be viewed as an extreme systematization of Romantic ideas, seeking to explain how all natural phenomena were organically related. Nature was seen as a world of active beings and dynamic productivity, labeled *natura naturans,* in opposition to *natura naturata,* which denoted nature as product – passive, blind, and atomistic. *Naturphilosophie* took on various forms, such as Hegelian metaphysics, Jakob Fries's transcendentalism, the romanticism of Friedrich Schelling and Samuel Coleridge, and Goethe's distinctive subjectivism (Gode–von Aesch 1941; Levere 1981; Amrine et al 1987; Cunningham and Jardine 1990). At the same time, these approaches shared certain general principles: All had a central belief in nature's unity and interrelatedness, as opposed to atomization; the material independence of parts was countered by the view of connectedness, and nature was ultimately manifest as the action of polar forces.

Proponents of *naturphilosophie* used Kant's *Metaphysical Foundations of Natural Science* (1786) to build their program. The naturphilosophen derived support for a dynamic nature deploying the concept of polar forces, which they derived from Kant, as nature's foundation.[1] Also influenced by evolutionary progressionism, the naturphilosophen viewed nature's flux as unidirectional, from undifferentiated chaos to sentient Man, and this single direction of all inorganic and organic development accounted for their view of all processes being governed by the same laws. But such logic led to forces and objects becoming inextricably interwined, and because the organic and the inorganic were on a continuum, physiological models were freely applied to physical phenomena, the exact reverse of reductionism, which sought to formulate biological processes solely in terms of the laws of physics and chemistry. The search by the naturphilosophen for harmonies, symmetries, and parallelisms in nature resulted in both fanciful speculation and productive discoveries in comparative anatomy and embryology. However, the dominance this school of thought gave to mind (after the influence of Kantian categories) in comprehending nature was regarded as confluent with vitalistic tendencies and eventually served as a vulnerable target of the reductionist's attack.

Another group of early nineteenth-century biologists, the teleomechanists, also claimed Kant as the intellectual source in formulating their program, which, although rationalistic, was also committed to mechanistic explanation of nature (Lenoir 1982). In many respects this school expressed an intermediate perspective, firmly wedged between the romantic and reductionist philosophies. The teleomechanists opposed *naturphilosophie* because of its lack of scientific rigor and also resisted the reductionists' efforts to strip life down to its physicochemical rudiments. The teleomechanists differed from their reductionist critics in their unremitting focus on establishing what distinguished the organic from inorganic phenomena. This school, like *naturphilosophie,* was composed of heterogeneous points of view, ranging from frankly vitalistic to firmly mechanistic in the later phases (1830–60). But generally the teleomechanistic program was indebted to Kant's *Critique of Judgement* (1790, 1987), in which he expounded a teleology where cause and effect were self-referential (i.e., the effect would inevitably influence the initiating cause: $A{\to}B{\to}C{\to}A'$) (McFarland 1970). The key issue was that organism not only had purpose, but was structured by all components incorporated under the auspices of an organizing principle, the integrity of the organism. *What* conferred that integrity was not specified, and teleomechanism, in its uncontaminated form, was only used as an intellectual structure to appropriate the data in a meaningful construct.

According to the teleomechanists, living organisms are distinct from the inorganic precisely because of their purpose and a resultant ordering of their structure and function. Biologists working with such principles in conjunction with strong descriptive techniques (for instance, Johannes Mueller and Karl von Baer) were instrumental in establishing physiology and embryology on a firm experimental basis. Their basic conceptual attitude resided in an assumption that the parts of the organism were secondary to the whole; some organizational principle – or more directly, *telos* – conferred a distinctive property to animate being, distinguishing it from inert matter. Romantic sentiment extended this argument to the central dogma of "Nature's unity," and a second theme sought a special vitalist principle that conferred life to the organic domain. Thus, beyond vitalism, there were two intellectually closely linked yet at the same time distinct conceptual schemes, namely, holism and teleology. But why, then, were vitalism, holism, and teleology conflated in the later reductionist assault?

The holistic conception was shaped in the course of seventeenth-century debates over the metaphysical structure of nature. In response to the dualistic construction of mind and body proposed by Descartes, Baruch Spinoza endeavored to unify the schism by transcending the alternative primacy of either mind or body with a new concept, "substance": absolute, infinite, and unknowable. The finite expression of substance was the "mode," known only by our cognizing abilities – thought and extension. Thus, body and mind are the only conceivable manifestations of substance as mediated through modes, and although

48

each appears distinct, in fact, according to Spinoza, both are derivative of primal substance and thus complementary aspects of one and the same reality. How is the organic, then, to be defined? Following Spinoza's lead, under the rubric of Nature's unity, Romantic biology invoked a dual response: vitalism, on the one hand, as a physiological explanation, and on the other, teleology as a philosophical account. Logically independent, they were and continue to be viewed as interrelated. The escape from mechanistic speculation in the organic sciences with a vitalistic visa might well be assigned to George Stahl, whose *True Theory of Medicine* (1708) argued for the complete separation of living beings from inorganic matter on the basis of an imbred "anima sensitiva," which could not be detected by physical means. Thus, the early eighteenth-century solution to the mind–body dualism was to infuse matter with soul. In a sense this was a revival of the Greek Stoic concept of *pneuma,* which endowed all organic matter with life, and it was also similar to the later Enlightenment solution, of choosing activity and change (force and motion) over structure and permanence as nature's foundation (Hankins 1985). By 1740 the antimechanistic sentiment was in full swing, and it was essentially complete by 1802 when William Heberden declared: "[To] living bodies belong many additional powers, the operations of which can never be accounted for by the laws of lifeless matter" (Schofield 1970, p. 191). Newton's atomism had been replaced by a nebulous dynamicism: Corpuscularity succumbed to vital force, vital energy, or simply "life." However, in the process of imbuing the organic with a special and mysterious property, the holistic construct so crucial to Romantic science now became (unnecessarily) entangled with the confounding metaphysics of the vitalist perspective.

Vitalism was to have a pervasive influence even after the reductionist revolt (Benton 1974), but certainly prior to 1840 it obfuscated teleological arguments that, at least in certain cases, were scrupulously devoid of vitalistic tendencies. Certainly there was no logical compulsion to link vitalism and teleology. But certain teleomechanists did invoke an incipient vitalism, and as the nature of living processes became the focus of scientific inquiry, teleology, as a defining principle, was consistently rejected along with vitalism. This trend became especially powerful after the concept of spontaneous generation was discarded; the demonstration of the conservation of energy further discredited vitalism, since the distinctive "animateness" of the organic could be accounted for in the combustion of metabolites. This was the signal contribution of the reductionists, who sought to measure conclusively the conversion of matter to energy.

The reductionists, led by Hermann Helmholtz, ignored the *Critique of Judgement,* where Kant argued that biology as a science must have a completely different character from physics, and (like the naturphilosophen) fastened onto Kant's *Metaphysical Foundations of Natural Science* (1786, 1985) as the philosophical basis of their program (Galaty 1974). Their assault on Romantic biology, and particularly on vitalism, was initiated on philosophical rather than on empirical grounds. They did not dispute the argument concerning the uniqueness of organic phenomena, but only maintained that all causes, including the biologic, must have certain elements in common. They connected biology and physics by proclaiming the identity of the ultimate basis of their respective explanations in physical laws. Like the naturphilosophen, they sought unifying principles of all phenomena, but they expunged the notions of the mind and vitalism from their lexicon. The earliest declaration of the reductionist position is found in a letter of Emil du Bois-Reymond, who wrote in 1841: "Brueche and I have sworn to establish the truth that only common physical-chemical forces are at work in the organism" (ibid., p. 299). The reductionists focused on the problem of *Kraft* (force), which in the form *Lebenskraft* (vital force) had been used by some teleologists as an explanation rather than a problem of research. The reductionists proceeded to analyze organic processes within the framework of attractive and repulsive forces in order to link the physical sciences to the biological. This Cartesian strategy ignored the dominant concern of their Romantic predecessors.

During the Romantic period there are rich examples of scientists who sought to integrate all knowledge, whether in science or letters, within a holistically conceived natural order, where science and poetry were "complementary ways of seeing" (Abrams 1953, pp. 308–9). One of the major architects of this vision was Johann Wolfgang von Goethe, who, as poet and scientist, typified a vision still operative in the Victorian period but largely eclipsed by century's end. While Goethe did not embrace vitalism, he was strongly sympathetic to the teleological conception of natural processes. As a theorist of science he regarded with skepticism the ascendency of a reliance solely on reductionism; instead, he sought a more comprehensive holistic strategy that presumed a continuity of nature through its inorganic and organic dimensions. The notion of an integrated experience of nature was typically Romantic in that its inquiry was not limited to establishing the workings of nature but had a more grandiose mission: to counter a mechanistically conceived science.

Goethe attempted to unify varied experience by an idealized and comprehensive appreciation of nature. He certainly did not object to abstraction per se, but he rejected a science that would not encompass a personalized relation to the natural phenomenon (Tauber 1993). The characterization of subject–object relations in Goethe's understanding is problematic. On the one hand, he espoused a rigorously detached view from the observed phenomena. The scientist is supposed to observe and survey objects of inquiry with an impersonal and objective attitude, with a "quiet gaze"; he "must find the measure for what he learns, and the data for his judgement, not in himself, but in the sphere of what he observes" (Goethe 1792, 1988, p. 11). Yet there are numerous cases where Goethe is guilty of succumbing to a "subjective" view, relating phenomena as pleasant or unpleasant, useful or useless, and so on. There is, then, a complex tension between Goethe the detached observer, supposedly divorced from theoretical presuppositions, and Goethe the creative scientist (Tauber 1993). The mingling of the subject and object is contrary to the ideal of the scientist as independent from the part of the world under study – the austere observer and collector of data that are uncontaminated by projected personal prejudice. How, then, to account for those crucial elements that allow creative intuition, deduction, and assemblage of disparate information to create "objective" reality?

Shortly before Goethe's death (in 1832) a formal reply to this question was offered in positivism. Although the term *positivism* was coined by Auguste Comte in the 1820s, its complex history may be traced from Francis Bacon and most directly to the seventeenth-century scientific revolution (Simon 1973). The application of the scientific method not only to the sciences, but also to human affairs, served as the basis of Comte's philosophy, which had an enormous influence on the social sciences in particular. Unlike contemporary positivism, nineteenth-century positivism exalted science without concern about the limits of its validity. Philosophy, religion, and politics were exhorted to become scientific disciplines. From our perspective the issue was that an ever more rigorous format of science was espoused: Science was confined to the observable and the manipulable.

Perhaps the best example of a positivist in biology just preceding the birth of immunology was Claude Bernard, who espoused what he believed to be the neutral position of the dispassionate observer. He professed a physiology based on controlled conditions that allowed for results that were repeatably and rigorously consistent. The more varied and carefully designed the experimental manipulations, the more precise would be the characterization of the phenomena and the all-important interrelations between the object of study and the instruments (i.e., the observing scientist) would be minimized (Coleman 1977, pp. 154–8). He believed physiology and medicine might aspire to the standard of objectification already achieved in physics and chemistry. Although Bernard's relationship to vitalism and reductionism was complex, he was in closer affinity with the latter philosophical perspective: "In a word, physiologists and physicians must seek to reduce vital properties to physico-chemical properties, and not physico-chemical properties to vital properties" (Bernard 1865, p. 202).[2] As a positivist, Bernard embraced the objectivity of science: "The experimental method is the scientific method which proclaims the freedom of the mind and of thought" (ibid., p. 43). Rather than be caught in the webs of competing ideologies, Bernard argued for an idealized strict neutrality for science:

> The theories which embody our scientific ideas as a whole are, of course, indispensible as representations of science. . . . But as these theories and ideas are by no means immutable truth, one must always be ready to abandon them. . . . In a word, we must alter theory to adapt it to nature, but not nature to adapt it to theory. (Ibid., p. 39)

The strict positivist confines himself to phenomena and their ascertainable relationships, and it was through the strict standards of Bernard's experimentational methodology that his positivist ideal of knowledge was to be attained:

> Experiment only shows us the form of phenomena; but the relation of a phenomenon to a definite cause is necessary and independent of experiment; it is necessarily mathematical and absolute. Thus we see that the principle of the criterion in experimental sciences is fundamentally identical with that of the mathematical sciences, since in each case the principle is expressed by a necessary and absolute relation between things. Only in the experimental sciences these relations are surrounded by numerous, complex and infinitely varied phenomena which hide them from our sight. With the help of experiment, we analyze, we dissociate these phenomena, in order to reduce them to more and more simple relations and conditions. (Ibid., p. 54)

Positivism joined reductionism and eventually displaced a descriptive biology with one based on physics and chemistry. When Goethe wrote, "[In] organic being, first the form as a whole strikes us, then its parts and their shape and combination" (1790, 1952, p. 11), he was arguing that the phenomenon must be appreciated and understood first as a whole and then in its subordinated parts. He sought a comprehensive science, which was also shared by the *naturphilosophie* and teleomechanistic program: Each sought integration as a principle of their science, with the organism as the central basis. The reductionists prevailed over their competitors, although their battles have continued into our own century under various guises (Tauber 1991b). The history of these competing points of view is a prominent theme of this essay: The unresolved tension between an organismically based biology and one dominated by a reductionist paradigm frames the conceptual development of immunology.

B. REDUCTIONISM PREVAILS

The power of reductionism in twentieth-century biology is not to be denied, although its goals are coming under increasingly critical scrutiny (e.g., Schaffner 1974; Wimsatt 1976; Lewontin, Rose, and Kamin 1984; Sarkar 1991b; Tauber and Sarkar 1992; Sarkar 1992a); its success as a research program has dominated biology in virtually all its venues of investigation and explanation.[3] Reductionism may be divided into three broad categories: constitutive, explanatory, and theoretical (Mayr 1982; Sarkar 1992a). Constitutive reductionism asserts that the material composition of animate and inanimate matter are the same. Since the defeat of vitalism following the final attempts for reassertion before World War I in the arguments of Hans Driesch and Henri Bergson (discussed later in this chapter), this position is noncontroversial. Explanatory reductionism argues that to comprehend the whole, one must understand its components. But this approach does not account for the interaction of parts, and it is this interdependence that ultimately characterizes the various components in relation to the whole. Thus, in Michael Polanyi's terms, the lower levels of a hierarchy offer only limited information on the characteristics of higher levels (1969).

It is well recognized that new characteristics of a system emerge from its connected constituents. The issue of emergence of new properties at higher levels of integration was also debated in the nineteenth century, accompanying the drive for the sufficiency of reductive explanation (Tauber 1991b). Finally, with theoretical reductionism the theories and laws of one field of science are shown to be a special case of another branch of science, or one set of scientific proposals are claimed to be logical consequences of another set. This perspective allows for emergent properties by explicitly invoking "dictionaries" or "correspondence" or "translation" rules to go from one level of organization to the next (Schaffner 1967). As Mayr (1982) illustrates, although meiosis may be explained as a biophysical process, it is only meaningful as a biological concept, and a courtship ritual can only be understood in the context of behavioral or reproductive biology and loses its intelligibility when viewed simply as a musculoskeletal gyration.

Perhaps the most formal issue is to maintain the epistemological and ontological questions as distinctive. For example, is a reductionist argument given to explain some theory, law, or fact, or is it employed to demonstrate the more fundamental composition of a given entity (Shimony 1987)? Whereas the former is an epistemological question, the latter is an ontological one, and they obviously address different levels, though there may be deep connections between their underlying assumptions (Sarkar 1992a). During the course of scientific investigation, what entities are sought might well influence the type of explanation provided, and conversely, the explanatory success of a theory might well increase confidence in its ontology (i.e., in the reality of the entities explained or described in its propositions). Other than invoking a pluralistic appreciation of biological processes and warning that these cannot simply be analyzed into their simple elements without transformation rules or systematics to explain the complexity of the organism, biologists have yet to erect the scaffold of an alternative "new biology," where laws governing complex hierarchical systems cannot rely solely on characterizing isolated phenomena. It is from the historical and philosophical perspectives that the restrictions of reductionism are best appreciated. The experimental power of the reductionist program has driven the study of biological phenomena to the genetic and molecular levels of investigation, but in the process of establishing its hegemony, the holistic basis of an earlier biology was lost. The limitations of these discarded scientific programs left us bereft of certain strengths of their metaphysical structure. We carry a vague intuition that the organism is more than the sum of its measured functions; the contrived definition of particularities without substantive grasp of their interrelationship reduces the phenomena to isolated *essentia* (Polanyi 1969; Koestler and Smythies 1971; Salthe 1985; Yates 1987; Tauber 1991b; Sarkar 1992a).

In the twentieth century the reductionism debate has continued. For instance, Alfred North Whitehead (1925) argued for an "organic mechanism," a *process* biology in which understanding the *interactions* of parts was of more central interest than analyzing the components themselves. Studying more complex levels of organization (e.g., organs, tissues, systems), holistic materialists (Allen 1978) like Walter Cannon and Lawrence Henderson accepted the mechanical interpretation to a point, but believed that certain properties belonging to a component could only be described as it interacted with other parts in the context of the intact organism. These properties derived from the interaction itself. Examples include nervous system function, the buffering capacity of the blood, and even the complex array of stimulating and inhibiting factors on muscle construction:

> [T]here can be no doubt that one of the important tasks of general physiology, which differentiates it from the physical sciences, is that it has to take account of simultaneous activities of widely different kinds harmoniously interacting.
>
> Such interactions become more conspicuous as structures become more differentiated and finally take the extreme forms of hormonic and nervous integrative actions. But it is evident that *in the organism every activity is more or less integrative.* Or, more precisely stated, every physiological phenomenon must be studied, not only in isolation, but also in relation to other phenomena with which it will in general be found to be related in a manner usually called adaptive. (Henderson 1928, pp. 14–15; [emphasis added])

For the holistic materialist "new characteristics emerge from the interaction of parts," which "are not merely quantitatively more complex," but also qualitatively different (Allen 1978, p. 106). Although Cannon and Henderson were concerned with integration of function, this in itself did not distinguish them from orthodox mechanists like Jacques Loeb (1912). Loeb argued that chemical interactions were not only the most fundamental aspect of processes of life, but they also *determined* all higher levels of organization and thereby were ultimately predictive. In short, the whole was the sum of its parts; thus, the components, once defined, could be placed together into a coherent whole, and the properties of the whole were derivable without residue from the properties of the individual constituents.

Today the reductionist–holistic debate continues, and we will have occasion to witness this debate concerning immune function. In the simplest sense the reductionist argues that organismic identity is the genetic complement of the organism. In doing so, he may be conflating two forms of reductionism: physical and genetic. The former has already been described as the attempt to explain all properties of a complex entity entirely in terms of its constituent parts and their physical and chemical interactions. As such explanation proceeds, it goes to lower and lower levels of organization. On the other hand, genetic reductionism is the claim that all biological properties of an organism can be explained, and are determined, solely by its genes (Tauber and Sarkar 1993). Genetic reductionism does not remain at the DNA level alone, for the coding and replicative properties of DNA are connected with its structure, and the interactions of the DNA molecule are physically based, whether with reference to the double helix model or to transcription functions. In this sense the reductionism is not committed to understanding DNA as an isolated constituent; interaction with other cellular constituents (RNA, proteins, lipids, and so on) determine its function as an interactive "species," that is, as one of several components determining development and growth (ibid., 1993). The success of deciphering the mechanisms of Mendelian inheritance has ultimately led some enthusiastic proponents to argue that all traits are genetic. In this extreme position an organism is regarded as merely a vehicle for genes to make more genes (Dawkins 1976).

In genetic reductionism DNA *qua* DNA has claimed dominion – everything of biological interest is deemed genetically determined. So, beyond the obvious physical characteristics under immediate control of specific genes, complex behavior involving mentality has also been viewed as amenable to such reductive analysis (Lewontin, Rose, and Kamin 1984; Tauber and Sarkar 1993). A remarkable telescoping effect is usually used by genetic reductionists in discussing complex mental behavior as based on genetic determinism. A smooth transition is made first from biological to the inherited and then from the "inherited" to the "genetic." Such thinking has become particularly visible in debates concerning the utility of the human genome project (see Koshland 1990). In this view the environment is the almost passive, if necessary, background, as genes are allowed to play out their biological mission, albeit in a social setting. The DNA code embeds the potential of development, interaction with the environment, and reproduction. But to leave identity at this level, claiming only genetic rudiments (i.e., a particular allotment of DNA) is to deny the entire spectrum of the life process as just so much fuss to preserve and pass on that encapsulated genetic morsel. Advancing a hegemonic claim for the essential primacy of DNA is to adhere to a restricted vision of biology:

> It is raining DNA outside. . . . The cotton wool is made mostly of cellulose, and it dwarfs the capsule that contains DNA, the genetic information. The DNA content must be a small proportion of the total so why did I say that it was raining DNA rather than raining cellulose? The answer is that it is the DNA that matters. . . . The whole performance – cotton wool, catkins, tree, and all – is in aid of one thing and one thing only, the spreading of DNA around the countryside. . . . It is raining in-

structions out there; it's raining programs; it's raining tree-growing fluff spreading algorithms. This is not metaphor, it is plain truth. It coudn't be plainer if it were raining floppy discs. (Dawkins 1986, p. 111)

We intuit, however, that this is a distortion. Genetics offers *a* definition of the organism, one that fails to encompass interactive selves. The genes establish certain boundaries of identity (the potential for behavior), but the *actualization* of the organism only *begins* at the gene. There are several organismic-based models that compete with the one posited by genetic reductionists as fundamental.

The most radical contemporary view of biological holism is the organism as a *cooperative*. The Gaia hypothesis (proposed by James Lovelock and Lynn Margulis in 1970s), where the entire biosphere is viewed as a single ecosystem, an all-encompassing organismal-type entity constantly influencing and adjusting to the vicissitudes of global climatic and geological change, is but the latest vision of the organism in balance, both forming and accountable to its generalized environment, even far removed from its immediate concern (Schneider and Boston 1991). The ecologic perspective subordinates the individual to the collective's well-being and struggle. Individual identity is thus conferred by relation to the inclusive whole, and in this sense the subject is defined by its greater context – its other. This might be labeled the cubist vision of the subject, although it has deep roots in Romanticism (Thomas 1983). I designate this view as O_s. O_s must be distinguished from another form of a collective, one that is defined as multiple units inexorably linked together as a functioning entity we might call the *collective organism*. Examples include coral reefs or bacterial collectives, where differentiated morphology and function might distinguish individual elements, but the aggregate is viewed as the entity constituting the subject (Sonea and Panisset 1983; Shapiro 1988). I define such a subject as $S_1:S_2:S_3 \ldots S_n$; each such collective competes with other like collectives, as well as engaging in interspecies competition and struggle with environmental forces. The collective entity extending to seemingly individual, self-sustaining units, originates from the holistic notion of the organism, whose parts function within an individual as integral to a total design that confers character – and specific function – on each constituent. This was a widely appreciated view of early nineteenth-century biologists. For instance, Goethe (1790, 1952) eloquently expounded in his various biological writings how "each living creature is a complex, not a unit; even when it appears to be an individual, it nevertheless remains an aggregation of living and independent parts" (p. 24).

This collective orientation was further refined as science continued to divide the organism into smaller parts and more basic constituents. By mid–nineteenth century, the cell theory was so applied toward understanding structure/function relationships in physiology, anatomy, and pathology, and there was a general awareness (even by philosophers!) of how integrated wholes must maintain their integrity vis à vis each part: "Life itself recognizes no solidarity, no 'equal rights,' between the healthy and the degenerate parts of an organism; one must excise the latter – or the whole will perish" (Nietzsche 1904, p. 389). The configuration of the collective organism is the overall view of an ecosystem where there is a balance of predatory animals, mutually beneficial species, and in the last instance a symbiotic relationship, where a potentially lethal or parasitic encounter can be converted into a mutually dependent relationship that can be designated $S{:}s$. The symbiotic individual is distinguishable in its history as comprising distinct entities that have merged into a single, interdependent organism (Margulis 1981). Although such an $S{:}s$ organism must be viewed differently from other collective models, such as O_s and $S_1 : S_2 : S_3 \ldots S_n$, each shares an essentially cooperative or balanced vision of the biosphere.

Finally, there is the *evolving* model in which the responsive, challenged organism instantiates a different version of biological individuality from those previously described; it can be designated $S{:}S_O$. $S{:}S_O$ is an abbreviation of $S{\rightarrow}S_O{}^1{\rightarrow}S_O{}^2{\rightarrow}S_O{}^3 \ldots S_O{}^n$, where each encounter of the organism alters its previous identity. This organismal truism is relevant whether the context is developmental (turning a particular gene on or off), phylogenetic, cognitive-emotional, or, as discussed below, immunological. The characteristic of $S{:}S_O$ is that the organism is constantly engaging its other (whether in the external environment or simply in the course of homeostatic/developmental processes), incorporating the engulfing of that object, and then assimilating or responding (equally applicable to information, food, infection, and so on). It is a formulation emphasizing the experiencing nonstatic, dynamic, and dialectical organism. As opposed to a given entity, that is, S (probably best exemplified by the conception of the fixed genome), $S{:}S_O$ identity is unstable, and its figurative boundaries or characteristics are always changing. It is the subject in time, and as a living entity, $S{:}S_O$ is always different in its temporal flux.

The organism, as defined by the immune system, most closely approximates this last model. The novel insight of Elie Metchnikoff (discussed in Chapter 1) was to apply such active dialectical behavior to the inner workings that govern the organism and maintain its integrity (Tauber and Chernyak 1991, pp. 109–56). The foundation of his thinking was the insight into the indeterminateness of organismal identity, which was attained only by an ongoing process of self-aggrandizement and self-definition. The organism was no longer so clearly delineated as a given entity. Its "boundaries" were constantly remolded and reestablished under the assault of temporal change and environmental challenge. From this perspective the organism is ceaselessly reconstructed, changing, and adapting. It requires special faculties to deal with the entire biological spectrum of interaction with its environment and within itself. As the object of inquiry, the individual organism becomes the focus of a particular question of biology: What confers self-identity to an interactive, dynamic, challenged entity? If we seek a science that can address the nature of this *process*, then immunology offers a more comprehensive approach, with its conceptual array of perceptual mechanisms, cognitive processing, and effector mechanisms. Such an organismal biology takes on a very different flavor from that recognized by those who view our science with an exclusively reductive bent (Levins and Lewontin 1985). The DNA offers the blueprint of action; the organism does the living.

Metchnikoff's underlying scientific ethos was formed by the problematics of an organismal biology. This philosophical issue is profoundly relevant to the history of immunology, which as a modern discipline can fairly be dated from his phagocytosis theory. As already described, Metchnikoff's hypothesis evoked the deepest hostility. He was aggressively attacked by the German microbiologists, who challenged every element of his thesis (Tauber and Chernyak 1989; 1991). The debate with the early detractors hinged on issues of the phagocyte's effectiveness, its bactericidal capacities, and its universal application to infectious models. Debates also centered on the definition of the natural course of disease, as well as on philosophical questions about the nature of teleology; in fact, the accusation of teleology was a disguised attack on what was perceived as a vitalistic hypothesis. Metchnikoff dealt with the twin accusations of teleology and vitalism in the famous Paris lectures of 1891. In his defense vitalism was simply subsumed under the autonomous behavior of organisms. Volition of primitive animals became the precursor of the mind. Thus, in defending the apparent self-willed behavior of the phagocyte, he did not deny a vitalistic innuendo, but assigned its function to an early form of psychic behavior. Although it was hardly a defense against the vitalist charge, since he did not adequately explore the mind–body relationship, he nevertheless was content with construing an evolutionary connection between the "freedom" of the phagocyte and psychic free will:[4]

This biological theory has often been considered too vitalistic in its tendency. I need only quote Frankel's outspoken criticism of my theory from this point of view. "The phagocyte theory presupposes extra-ordinary powers on the part of the protoplasm of leucocytes, to which are attributed sensations, thoughts and actions, in fact a kind of psychical activity." The sensibility of the phagocytes is not an hypothesis which can be admitted or rejected at will, but an established fact, which cannot be ignored, as it is by Frankel. Whether they possess powers of thought and volition, as this author accuses me of assuming, is quite beside the question, though we are justified in considering that they possess a germ of these qualities and that their sensibility, like that of various vegetable and animal unicellular organisms, represents the lowest stage in the long series of phenomena which culminate in the psychical activities of man. . . . The accusation of vitalism and animism, which is unjustly cast at the phagocyte theory, might really be more appropriately applied to my opponents, who maintain that the psychical acts of the higher animals are fundamentally different in their nature from the more simple phenomena peculiar to the lower organisms. (Metchnikoff 1892, 1893, pp. 192–3)

He then immediately turns to the question of phagocyte autonomy as teleological, and argues that this property of life – purposeful activity – is blindly selected for in the course of evolution:

It is equally erroneous to attribute a teleological character to the theory that inflammation is a reaction to the organism against injurious agencies. This theory is based on the law of evolution according to which the properties that are useful to the organism survive while those which are harmful are eliminated by natural selection. Those of the lower animals which were possessed of mobile cells to englobe and destroy the enemy, survived, whereas others whose phagocytes did not exercise their function were necessarily destined to perish. In consequence of this natural selection the useful characteristics, including those required for inflammatory reaction, have been established and transmitted, and we need not invoke the assistance of a designed adaptation to a predestined end, as we should from the teleological point of view. (Ibid., p. 193)

Beyond the antimechanistic attitude exhibited by Metchnikoff, there was an alarming indictment against a struggling reductionist science that would ignore the teleological character of biological function. It is interesting to note that Jacques Monod (1972), a key pioneer of molecular biology, would embrace teleology in the "new" notion of "teleonomy" – "as if" purposeful, but evolved blindly. Eighty years after Metchnikoff's spirited defense, modern biology recognized, as had Aristotle, the fundamental characteristic "common to all living beings without exception: that of being objects endowed with a purpose or

60

project, which at the same time they exhibit in their structure and carry out through their performance" (Monod 1972, p. 9). But in addition to the issue of teleology, Metchnikoff, in his great enthusiasm, at times presented the phagocyte as almost autonomous; that is, phagocytes determined organismal integrity simply by striving for their aggrandizement. Phagocyte activity as competitive with other cell lines was Darwin's struggle for survival turned inward into a struggle within the organism. Metchnikoff's argument, its rationale and how it arose from embryological studies, has already been documented (Tauber and Chernyak 1991), and suffice it to note here that the hypothesis was doomed: Neither was it based on a reductionist research program, nor could it fulfill the new positivist standards of the period.

Metchnikoff was vilified as an atavist, contaminating the attempts of microbiologists firmly to establish new experimental positivist criteria. It is of particular interest that the humoralists and the Metchnikovian cellularists used the same methods until Ehrlich quantified the antibody response in 1897. At issue, of course, was not whether Metchnikoff could repeat analogous experiments (e.g., modify the experimental conditions by changing the organism or the nature of infection) and explain his results consistently with his position; the very aims of research were discrepantly defined by Metchnikoff and his detractors, and this difference was soon expressed in terms of the increasingly dominant role of immunochemistry. The humoralists had always aspired toward a biochemical framework as providing more reliable method and explanations more consistent with mechanical models. The irony is that the alternative humoralist challenge was over *mechanisms* of active host defense, and in this sense Metchnikoff had established the first two tenets of his theory. First, there is the issue of "active mechanisms" (that in the Metchnikovian construct of phagocyte surveillance was initially viewed as vitalistic), and second, the organism is defended by an adaptive system of complex humoral and cellular factors that are *evoked* by immune challenge. Despite the origins of this immune theory in an organismic-based biology, the first twenty-five years of modern immunology (ca. 1883–1908) were marked by a debate apparently concerned with the mechanisms mediating the immune reaction and not directly dealing with the organismic issues raised by Metchnikoff. These problems were simply ignored. Today we might appreciate the novelty of Metchnikoff's formulation and perhaps understand that as an organismically oriented biology it employed conceptual ideals quite different from those of the humoralists to orient its scientific posture.

If Metchnikoff was guided by the broadest concerns of how the *or-*

ganism was defined in its *totality,* then we perceive how that program was at variance with a biology that esteemed a different research goal: The early humoralists, later called immunochemists, were committed to establishing the biochemical basis of organic function. Their concern was reductionist, and any endeavor to utilize a holism reminiscent of Romantic influence, with all of its discounted methods and discredited theories, was to be purged from the new biology.

But the crucial issues of dispute ramified beyond arguments over particular mechanisms of immune defense. The debate arose from diverging research traditions that connected with different philosophical assumptions and from the different fundamental concepts of research goals of the respective camps. Beyond the nature of the mechanism of immunity, Metchnikoff's theory encompassed a response to an even more fundamental biological question: What constitutes organismal identity? Herein lies the distinction that Metchnikoff made with respect to immune function: immune identity, established in the context of developmental principles, and secondarily immune integrity, mechanisms whereby the organism is defended. *Identity* and *integrity* represent distinct functions. The issue of integrity pertains to those activities that defend the host from foreign offense, repair damaged parts, remove effete cells, and destroy malignancies. These integrity-preserving activities depend on immune effector functions. But central to this behavior is the issue of identity: What is to be protected, preserved, repaired? The mechanisms employed by the immune system to preserve identity must logically follow the establishment of that identity. Not only the logic of his thinking, but the historical development of these notions is explicitly found in Metchnikoff's scientific biography (ibid.).

Identity functions represent the afferent arm of the immune reaction, that is, the recognition phase, where matters of immune perception are mediated. Integrity is what is protected by the efferent limb of immunity. It is at this juncture that immunity becomes definitive. Clearly the afferent and efferent arms of the immune process are locked together in a highly integrated fashion, analogously to the nervous system's sensory perceptive modalities, which are inexorably connected to the cognitive and motor responses. In both cases the anatomic structures are separate but functionally efficient only as linked entities. Metchnikoff never sought to separate these functions explicitly, nor did he employ the terminology of this modern interpretation in his writings, but the underlying logic of his formulation reveals in fact two repositories of immune activity: those that arose from his embryological concerns over the emergence of organismic identity and, in the particular context of infectious disease, a subordinate concern with the protection of the host by integrity-preserving immune effector functions.

It is only in hindsight – after discerning the course of modern immunology – that we can appreciate the heuristic value of this theoretical construction. It is of some interest to note that a similar fate would have befallen Metchnikoff had he remained in developmental biology, rather than engaging the immunochemists in his newly adopted field of pathology. For we may discern the same conflict between a descriptive tradition and a newly emerging reductionism in the embryology of the period. Metchnikoff was no longer an active embryologist when, in 1894, Wilhelm Roux proposed a novel scientific agenda: *Entwicklungsmechanik* (developmental mechanics). Roux provided developmental biology with a *new* manifesto that could just as easily have served as a proclamation of German immunologists: "[T]he task of developmental mechanics would be the reduction of the formative processes of development to the natural laws which underlie them. . . . *We must reduce the processes of organic formation to the fewest and simplest modi operandi*" (Roux 1894, p. 109; emphasis added). These simple modi operandi were to be sought as in a mechanical system, that is, under the assumption that they were "identical with those which underlie inorganic or physico-chemical processes" (ibid., p. 111). Until the 1880s embryology was essentially a descriptive science, where concern with tracing embryonic layers and the fate of primordial cells was actively stimulated by the problematics of recapitulation. The genealogy of species was pursued by attempting to define the correspondence of morphological traits of different animals during development. Thus, soon after *Origin of Species* was published, developmental biology became a major vehicle for establishing evolutionary relationships, and it was from this *descriptive* research tradition that Metchnikoff practiced his science. Embryology suddenly became an experimentally based discipline following Roux's 1888 experiments on frog embryos.

Roux's studies emerged from his mosaic theory of development, which postulated that hereditary particles of the cell were divided unequally during the division that formed the multicellular embryo. As growth and differentiation continued, multipotentiality would correspondingly diminish because of restricted genetic information available to daughter cells missing the full, that is, original, hereditary library. Roux tested the hypothesis by destroying one of the blastomeres of a two-cell-stage frog embryo, allowing the other one to develop. He believed the mosaic theory was vindicated because the growing embryo was abnormal and in fact could not attain maturity. Hans Driesch performed an analogous experiment (in 1891) with another species, the sea urchin, and instead of destroying one of the cells, he separated them and allowed each cell to produce a normal, although somewhat smaller, larva. These studies were in direct contradiction to Roux's hypothesis and were actually of more general interest, for Driesch correctly surmised that differentiation might depend on cellular responses to both internal *and* external factors. The progeny cells of each blastomere give rise to different parts of the embryo, and thus differentiation arises both from inherent internal characteristics and from the immediate morphological environment. Later it was shown that if the injured cell in Roux's experiment was removed, a fully viable embryo would develop from the remaining cell. But beyond the particulars of this experimental argument, the resulting philosophical debate is of most interest here.

Driesch despaired of finding a causal-mechanical explanation. He saw development not as a programmed, mechanical process, but as the response of a living organism to life experiences. He thus *began* with an organismic approach to his embryological studies: "The individual organism is of the type . . . that it represents a factual *wholeness,* if we may express its most essential character in a single technical word" (Driesch 1914, p. 3). He further argued that wholeness depends on teleological or purposeful processes (embryological, restitutive, and adaptive), characteristics that a mechanistic modeler abhors. Primarily, however, Driesch used his experimental data to dismantle the machine analogy for an organism:

> *Every* cell of the original system can play *every* single role in morphogensis; *which* role it will play is merely "a function of its position." In face of these facts the machine theory as an embryological theory becomes an absurdity. These facts contradict the *concept* of a machine; for a machine is a specific arrangement of parts, and it does not remain what it was if you remove from it any portion you like. (Ibid., p. 18)

In a strong sense, Driesch's biology was holistic: The embryo interacting with its environment was effected by that experience, incorporated it (i.e., "knew" or "perceived"), and responded accordingly. He could not envision a physicochemical explanation for such an activity, and as a consequence he viewed the process as somehow dependent on a vital force that could not be analyzed. By "vitalism" Driesch meant some type of explanation that could not be captured by a mechanical model: "[T]o establish the doctrine of the *autonomy of life,* i.e., the doctrine of so-called vitalism, at least in a limited field: there is some agent at work in morphogenesis which is not of the type of physico-chemical agents" (ibid., p. 19).

Driesch, along with Henri Bergson, led an enthusiastic and powerful European antireductionism until the beginning of World War I. Each viewed life not as a contingent by-product of physical or chemical laws proceeding by blind mechanical Darwinian rules, but as a manifestation of creative energy. Determinism was denied and "life was opposed not only to inert matter, but to the supremacy of calculating reason and to the monopoly of the analytical spirit" (Kolakowski 1985, p. 9). Each sought to discuss "the world as permeated by quasi-subjective energies, and . . . to stress the unity of this all-pervading spirit in which the human personality seemed to dissolve" (ibid.). Driesch carefully avoided substituting psychical for mechanical forces, although the relationship is rather complex (see, e.g., Driesch 1908, vol. 2, pp. 277–86). He used the Aristotelian word *entelechy* to name the autonomous agent at work in his vitalism that, being neither substance nor energy, allows – in a most nebulous and undefined fashion – for life to order itself (Driesch 1914, pp. 33–40).[5] But by the turn of the century Driesch was essentially ostracized from the scientific community and was left to address those of the general public who were sympathetic to what they perceived as neo-Romanticism (also reflected in the wide appeal of Nietzsche, Bergson and other antirationalists).

In developmental biology Roux had effectively shifted the experimental standards from a descriptive science to one based on a program totally opposed to Driesch's philosophical perspective, namely, an embryology committed to mechanistic principles. Evolutionary questions were replaced by the concrete concern for developmental processes: What were the mechanisms by which embryos grew and differentiated? Roux's "experimental morphology" was not totally novel, for earlier embryologists had manipulated the embryo to produce specific structural changes in the adult. Roux, however, effectively shifted the theoretical concerns of developmental biology. As a student he was influenced by the concerns of other biologists in physiology and cytology, who were already committed to a reductionist approach in their research (Allen 1978, pp. 25–8), and he became especially indebted to the mechanistic interpretation and the physicochemical methods espoused by Haeckel. The view of the organism as a mechanical and chemical system arose from Haeckel's rather vague and confused theories of causation and his misguided notion of recapitulation. The irony is that Haeckel in fact was not an experimentalist but rather a somewhat biased theoretician (Tauber and Chernyak 1991, pp. 53–66).

Although Roux ultimately rejected Haeckel's specific theories, his original sympathetic leanings prepared him for the mechanistic attitude of his teacher Wilhelm Preyer and for the work of Wilhelm His, who also heavily influenced him. Preyer, a student of the Helmholtz school, applied physiological methods to embryology, showing that functional as well as anatomical changes occurred with development. He was also a physiologist who was interested in the physical and chemical forces that caused movement of embryonic cells, and devised mechanical models to show how various stresses and strains could cause folding, bending, and specific cell movements associated with growth (ibid., p. 28). Roux then applied this general reductionist approach to the problem of differentiation by employing experimentalism as the crucial method of research. As quoted above from the introductory essay of his new journal, by manipulating physical and chemical conditions Roux sought to establish an explanation for developmental changes. In the opening salvo the program is announced:

> Developmental mechanics or causal morphology of organism, to the service of which these "Archives" are devoted, *is the doctrine of the causes of organic forms, and hence the doctrine of the causes of the origin, maintenance, and involution (Rückbildung) of these forms.* (Roux 1894, 1986, p. 107)

"To Roux, cause meant material cause – that is, molecular and mechanical interactions," and research of this new agenda would "seek the same kind of explanation as the physical sciences" (Allen, 1978, p. 34):

> [T]he causal doctrine of the movements of masses has been extended to coincide with the philosophical concept of mechanism, comprising as it does all causally conditioned phenomena, so that the words "developmental mechanics" agree with the more recent concepts of physics and chemistry, and may be taken to designate the doctrine of all formative phenomena (Roux 1894, 1986, p. 108)

Roux's attitude was widely shared in the various biomedical sciences, and his method gained almost immediate support amongst embryologists. *Entwicklungsmechanik* had far-reaching influence that revolutionized the fields of embryology in particular and biology in general. Although one may attribute the authorship of experiment embryology to Roux, clearly he was most innovative in *applying* a research agenda well established in physiology and biochemistry over the preceding fifty years. Friedrich Wöhler accomplished the laboratory synthesis of urea in 1828, showing the shared chemistry of the organic and inorganic realms, although this insight has a long history and assumed a wide credence well before the actual experiment (Coleman 1977, pp. 146–7).

As already discussed, the deliberate, polemical reductionist program in physiology of Helmholtz and other German investigators of the 1840s was devised as a response to vitalism. By the end of the century vitalist doctrines had been essentially demolished by Pasteur's demonstration against the spontaneous generation of life (1864) and by the increasingly influential positivism of Claude Bernard, who led the rising ascendency of a chemically based physiology. When Roux was adopting the research philosophy of physiology, that science was already firmly committed to an agenda of biochemistry. For instance, Max Rubner, in a series of classic experiments between 1889 and 1894, completed the initial studies of the earliest reductionists on heat production by metabolically active tissue. He showed that animals placed in a respiration calorimeter generated heat entirely from the energy source of their nutrients, concluding with Helmholtz that the animal is only a heat machine (ibid., p. 142). This general notion of a machine model had enormous influence on perpetuating and legitimating the emergence of biochemistry and organic chemistry as the foundations of understanding life processes. With his discovery in 1897 of zymase, a yeast sugar-fermenting enzyme, Edward Buchner and his followers argued that biochemistry was born as a distinct discipline.

The attempt to interpret all life processes as the direct function of intracellular enzymes subordinated the organizational structure of the cell's cytoplasm and placed the descriptive sciences of cytology and histology in a position of secondary importance. Thus, by the early twentieth century cell function could be discussed in chemical terms and studied by chemical means. The test tube was to replace the intact cell or integrated organism as the object of study. Only by focusing on particular reactions would a more complete physiology emerge, and to that end a heady optimism was ushered in by the early descriptions of enzyme kinetics. A. J. Brown (1902) and Victor Henri (1903) recognized that enzymatic reactions differed from ordinary chemical reactions, where by the law of mass action, the reaction rate was proportional to each of the reactants. They independently and correctly hypothesized that the enzyme forms a compound with the substrate that subsequently breaks down into the native enzyme and the substrate's reaction products; as long as the enzyme is saturated, increasing the substrate concentration does not increase the reaction rate. A decade later Leonor Michaelis and Maud Menten (1913) generalized the theory and gave it more systematic and detailed treatment. (For a historical review of early biochemical kinetic theory see Sarkar [1992b].) In any case, by

1910 several tissue extracts were shown to yield only a small proportion of in vivo respiratory activity, and thus an opposition to the new dogmatic reductionism generated lively debate, paralleling the controversies in physiology (Henderson 1928), immunology, and embryology of the same period.

In the same year that the Nobel Committee decided to award Metchnikoff the Prize, Svante Arrhenius was attempting to apply biochemical reaction kinetics and biophysical analyses to the reaction of antibody and antigen (1907). This development was obviously fueled by the general interest in biochemical reactions, but it also gained impetus from the growing desire of clinical medicine for serological diagnostic tests. Ludwig Fleck explored the changing definition of syphilis during this period of emerging immunodiagnostics and clearly demonstrated how beyond altering the "thought collective" concerning the prevailing definition of this particular disease, the introduction of the Wassermann reaction (published in 1906) was a signal triumph of new criteria for a medical nosography based on immunochemical methods (1935, 1979).[6] Thus, reductionism as a governing ethos is easily discernible beyond the specific research programs of particular sciences, as it also infiltrated into the very fabric of the applied clinical sciences.

A strong case might even be made that the impact of reductionism in medicine was its most significant manifestation, for the patient as a holistic construct was to become a dissection of laboratory values and penetrated images. Immunology was to play a central role in twentieth-century medicine, as it strived to conform to the prevailing ideal of a view of pathology reduced to aberrant physicochemical functions (Canguilhem 1966, 1989). The new construct of the normative as defined by reductionist tenets propelled diagnostic serology, which was generally viewed as the focus of immune research until the 1940s (Landsteiner 1933, 1945; Silverstein 1989). The immune chapter was but one example of "a new scientific medicine," which was exemplified by the revolution in medical education that largely occurred between 1890 and 1910. Twentieth-century standards for medical education and practice were established in the first decade of this century as encapsulated in the Flexner Report (1910). The modern physician was to be an active and skeptical medical scientist, trained according to German reductionist standards. The conflict between this new ideal and a patient-based clinical medicine provides an important case study of the conflict between holistic and reductionist programs during this period (Tauber 1992c).

C. THE PROBLEM OF SPECIFICITY

Returning to the immunology of the 1890s, we find the reasons for the ascendency of immunochemistry at the expense of Metchnikoff's vision of the discipline reside in the domain of the dynamic biology he advocated. The fundamental animosity between Metchnikoff and his detractors was not simply over methodological differences, but stemmed from a deeper divide over the very nature of immunity. The view of the phagocyte as possessor of its own destiny and mediator of the organism's *selfhood* was simply viewed as vitalistic; the specific chemical reactions were of central concern. Metchnikoff's highly abstract and nebulous question of how to define the organism was not grasped as an immediate *experimental* problem. In fact, few understood his program. The humoralists asked a simpler question: What are the mechanics of recognition? In fact, as already noted in Chapter 1, Metchnikoff's well-known philosophical orientation, corollary to his views of biology, were scarely mentioned in the files of correspondence endorsing his nomination for a Nobel Prize or even in the secret critiques offered by the Committee's reviewers (Nobel Archives 1900–8). These documents exhibit an apparent ignorance of the underlying intellectual basis of Metchnikoff's theory, even though he had published several books on his musings. In order for Metchnikoff to share the Nobel Prize in 1908 with Paul Ehrlich, an emerging synthesis was apparent: Complementary roles for humoral and cellular (phagocytic) mechanisms were recognized; moreover, humoral augmentation of phagocytosis by oponization of targeted bacteria was viewed as having therapeutic promise – Almroth Wright (1910) proclaimed, "The Physician of the Future will be an Immunisator" – and an important contribution to understanding the respective roles of these two limbs of the immune response to pathogens. But this view of Metchnikoff's role in the establishment of immunology obfuscates an understanding the very intellectual foundation of immune theory.

The immunochemical approach does not address the same question posed by Metchnikoff, who wanted to know not only how the organism is defended (that is, how its integrity is preserved), but also how identity, self-definition, is established. The immunochemists did not address this biological question until midcentury, when modern biologists, spurred by medical challenges, revived Metchnikoff's question in terms of immune tolerance, organ transplantation, and autoimmunity (Landsteiner 1933, 1945, Mazumdar 1976; 1989; Silverstein 1989; Moulin 1991a). The question posed by the humoralists (what constitutes immune specificity, and how is it mediated?) only admitted a chemical answer. This problem has received a comprehensive solution that began with the theoretical explanation of antibody formation (Buchner's notion that antitoxin might be formed directly from toxin (1893), Ehrlich's alternative view of the side chain theory (1897), and the early serological characterizations of the immune response [Landsteiner 1933, 1945]), and concluded in the 1970s with the discovery of how gene components were shuffled to make antibody (Leder 1982; Tonegawa 1985).

To outline the theories of antibody formation is at the same time to trace the apparent congruent conceptual history of immunological specificity (Mazumdar 1974, 1976; Silverstein 1989; Moulin 1991a; Kay 1993). Neils Jerne dissected that history by focusing on two competing scientific communities: The biologists, whom he called "cis-immunologists," were concerned with characterizing the biological response to antigen, whereas the chemists, or "trans-immunologists," were preoccupied with the nature of the antibody molecule in itself, that is, with the structural basis of specificity, the size of the antibody repertoire, and quantitative relationships (Jerne 1967a). (He viewed the dividing point between the "cis" and "trans" positions at the receptor on the antigen-sensitive lymphocyte, an issue discussed below.) This history, which bespeaks a highly complex interplay of various competing agendas and purposes, does not in fact easily fit into such neat compartmentalization, but Jerne's scheme *is* useful in broadly outlining how successfully the molecularization of immunology fulfilled the biological part of its assignment. This is a complex matter to which we will return in Chapter 4, after the historical developments in the two conceptual camps have been summarized. Suffice it to note here that each field of immunological investigation today is oriented, if not committed, to formulating molecular definitions and solutions.

But the immunologist must be concerned with matters beyond the chemistry and molecular biology of the immune encounter: Characterizing how the immune system is regulated is the obvious next step. This characterization must include the deciphering of the *source* of immune activity – its "inner control." It has always been immunology's preoccupation to account for the discrimination between the foreign and the endogenous and to establish the theoretical framework that supports such identification. The issue underlying this discussion is to integrate the broadest theoretical concerns of organismic identity with the molecular particulars of immune recognition. I believe the history of immunology in our own period has largely followed the original divisions between Metchnikoff and the humoralists, where their respective theoretical concerns reflected divergent conceptual interests. We will now review the history of the immunochemical program.

With Behring's demonstration that neutralizing antitoxins (antibody) to diphtheria and tetanus toxins might be elaborated and passively used to protect against deleterious effects (1890), there emerged the kernel of what was to become the molecularization of immunology. Behring's studies served as the foundation of immunochemistry, but its first comprehensive theory was offered by Ehrlich.[7] As already discussed, to account for the antibody response, Ehrlich postulated that cells projected so-called side chains from their surface that served as the receptors for what we now call antigens, and that upon challenge the host reacts by shedding excess side chains to combine with the offending substance by chemical affinity. There are several highly evocative and prescient concepts inherent in this theory. More specifically for the history of immunology, Ehrlich (1897, 1957) rejected outright the notion of an adaptive response, which, by necessitating production of "new groups of atoms as required, would involve a return to the concepts current in the days of [an obsolete] natural philosophy." Instead he chose "normal enhancement of a normal cell function," that is, preformed antibody that would be secreted upon appropriate stimulation. Each of the subsequent theories of antibody generation were juxtaposed against this hypothesis.

Ehrlich had originally proposed the side chain theory as an alternative to Hans Buchner's proposal (in 1893) that antitoxin might be formed directly from toxin. This antigen incorporation theory was soon dismissed (although revived in various forms as late as 1930) when antibody was shown to be continually formed in the absence of antigen or in augmented amounts upon second stimulation with minute quantities of antigen challenge (Silverstein 1989, pp. 61–4). A different version for antigen participation in forming antibody were the "instruction" theories, which appeared in many guises and constituted the major countervailing alternative until the revival of Ehrlich's basic concept by the clonal selection theory (discussed in Chapter 3). The rationale was based on the obvious need for a highly varied antibody repertoire, one whose information content was so enormous that the basis for antibody generation arising de novo to each new challenge appeared a daunting biological task. A seemingly simpler mechanism utilized antigen to instruct antibody synthesis. In 1905 Karl Landsteiner explicitly stated that following the stimulus of immunization, differently constituted

71

products would be formed (Landsteiner and Reich 1905). Antibody in this view was not a given, preformed cell product (i.e., side chain), but a complete new substance.

The genesis of the instruction theories – and the basis of rejecting Ehrlich's side chain hypothesis – was the application that chemical treatment of proteins would similarly alter them, and more saliently for immunologists, an antibody raised against one chemically modified protein would react to another protein treated in the same fashion (Pick 1912). Landsteiner then adopted the strategy of examining the specificity of cross-reacting antibodies against acyl-substituted proteins (Landsteiner 1917). Aside from important observations concerning the library of antibody cross-reactions with antigens subtly altered, which was summarized in *The Specificity of Serological Reactions* (Landsteiner 1933, 1945), Landsteiner demonstrated that immunity against a universe of artificial antigens was possible. Given the number and variety of such challenges, prior knowledge and presence of a responding native antibody library was viewed as highly unlikely. His work thus corresponded theoretically with the basic concept of antigen serving as a template, as proposed by Oscar Bail and co-workers (Bail and Tsuda 1909). They believed that persistent antigen impresses its specificity on host factors, is released, and continues its mating to enhance the titre of specific antibody. (Not surprisingly, such a process was thought potentially capable of being mimicked in vitro, and efforts in that regard were made [Manwaring 1930a; 1930b; Pauling and Campbell 1942a, 1942b]). The chemical nature of such a process was left unspecified until the globular protein nature of antibody was determined. In the 1930s various investigators (William Topley, Friedrich Breinl, Felix Haurowitz, Stuart Mudd, Jerome Alexander), recognizing the protein nature of immunoglobulin, suggested that antigen, carried to the site of protein synthesis, served as a direct template for the synthesis of the antibody in a unique amino acid sequence (Topley 1930; Breinl and Haurowitz 1930; Mudd 1932; Alexander 1931). As Silverstein observed, this theory fulfilled several criteria: (1) The host was not required to contain explicit information to code for the multitudinous antigen library of the environment; (2) the high ratio of antibody molecules to antigen was explained by this synthetic process; (3) it accounted for immunological specificity (Silverstein 1989, p. 69).

Upon this stage Linus Pauling introduced modern chemical theory to immunology in an attempt to rationalize the template theory. There was a degree of congruence between the interests of molecular biology and those of immunochemistry in the 1930s and 1940s (Kay 1993). The study of the biological specificities of "giant protein molecules" in this period eventuated into one of the major schools of molecular biology. In this pre-DNA era proteins were generally considered the primary agents of heredity (as well as of cellular regulation) and thus were thought to be copied from preexisting protein prototypes. This template concept was in turn based on complementarity (having nothing to do with Niels Bohr's complementarity principle) and weak intermolecular interactions. Viruses, enzymes, and antibodies were each subject to such study, and thus immunochemistry attained a central role in the research program of early molecular biology. Pauling's major innovation was the introduction of stereochemical complementarity as the mechanism of biological specificity, including that of the putative antigen template and the antibody formed on it. It is interesting to note that he convinced Max Delbrück, who came to molecular biology from a radically different background, of the validity of this position (Pauling and Delbrück 1940; Fischer and Lipson 1988).

They were reacting to Pascual Jordon's hypothesis that macromolecules had a special quantum mechanical resonance attraction for each other, so that replication was accomplished by homologous "synapse." Synthesis by coupling complementary structures, rather than identical replication, was in fact first suggested by J. B. S. Haldane (1937). Pauling's interest in immunology dated from 1936, when he gave a seminar at the Rockefeller Institute and was queried by Landsteiner on possible mechanisms of antibody formation (Pauling 1970). It was Landsteiner who, twenty years earlier, had conjugated artificial antigen (hapten) to protein carriers and elicited immune reactions in animals (Landsteiner 1933, 1945). Because under normal circumstances, such an encounter would not occur, Landsteiner discounted preexisting antibody, à la Ehrlich, and assumed a de novo chemical synthesis. Based on the earlier template theorists (Mazumdar 1989, pp. 13–32), Pauling offered his own sophisticated chemical version: Assuming identical polypeptide chains, antibodies might differ only in configuration of the antigen-combining portion at the end of the molecule; antigen, serving as a template, generated a complementary configuration (again, analogous to Fischer's lock-and-key model, originally employed by Ehrlich), allowing the synthesis of antibody around any antigen – natural or artificial (Pauling 1940). This instructional theory propelled research in serological genetics at Caltech, supported by the Rockefeller Foundation (Kay 1993), and boosted attempts to generate antibody in vitro (Pauling and Campbell 1942a; 1942b). (Pauling's venture is a superb case study of early venture capital speculation in biotechnology.)

As discussed in detail in the next chapter, Frank Macfarlane Burnet (1941) raised fundamental biological concerns that were not addressed by the template model: (1) the general kinetics of antibody production, both in the case of initial antigen challenge and most especially with augmented secondary responses in terms of quantity and type of antibody produced, and (2) the persistence of antibody production as a function not only of the cells originally stimulated, but of their descendants as well. Jerne's distinction of cis- and trans-immunologists was heralded. By the late 1940s, with the appreciation of protein synthesis as being under the control of an information-laden genome, the spectre of the antigen interacting with that genome was raised (Silverstein 1989, pp. 72–5). The incorporation of genetic processing into the picture allowed for maintenance of information in the cell and transfer during proliferation, which would explain both the persistence of antibody formation and the accentuated booster response seen upon repeated antigen challenge. The second focus of biologic concern in this period was the resurgence of interest in natural antibody, that is, in that preexisting pool of immunoglobulin that was formed independent of overt antigen challenge and thus unexplained by instructional models. Finally, the template theories could not address the problem of tolerance, the failure to launch an immune reaction against a given antigen and the fact that tolerance can be acquired for foreign antigens administered before or at birth (Ada and Nossal 1987).

In 1955 Jerne restated and revised Ehrlich's original hypothesis that the antibody repertoire was preformed and that small amounts of antibody were synthesized under normal conditions, with an augmented response being initiated upon antigen challenge (Jerne 1955). Under "natural selection" conditions antigen served as a "selective carrier" of antibody and signaled the production of the appropriately reactive antibodies by an unspecified mechanism. Despite ignoring the chemical basis for antibody specificity, Jerne's model offered the basis for a biological alternative to instruction theories that was soon elaborated into a full-fledged cellular-based theory by Burnet, David Talmage, and Joshua Lederberg, who developed the clonal selection theory (CST). Both Burnet and Talmage had suggested that cells are selected for multiplication when the antibody they synthesize matches the invading antigen (Burnet 1957; Talmage 1957). Like Ehrlich, Burnet placed on the surface of the cell a phenotypically restricted "natural antibody" that upon interaction with antigen, triggered (by an unknown mechanism) cellular differentiation and unique antibody synthesis (Nossal and Lederberg 1958) through clonal multiplication. Other aspects of the theory

The triumph of immunology

suggested mechanisms for tolerance (deletion of self-reacting clones during ontogeny) and autoimmunity (reappearance of sequestered antigen), but a mechanism to explain antibody diversity was not yet forthcoming. The history leading to CST will be postponed until Chapter 3; I will now summarize the main currents leading to our present notions concerning the mechanisms of immune recognition.

Fittingly the dialogue between cis- and trans-immunologists allows the story to be completed by the molecular biologists (Leder 1982; Tonegawa 1985). The (William) Dreyer–(J. Claude) Bennett hypothesis of 1965, deduced from known immunoglobulin structure (light and heavy chains each contained one invariable and one variable region), postulated that the protein was encoded by a single constant (C) gene and the variable (V) region was encoded by a discontinuous segment composed of several hundred or thousand separately encoded variable region genes (Dreyer and Bennett 1965). The hypothesis proposed that as the cell matures, it randomly selects one of the V genes and combines it with the C gene to create a single strand of DNA that encodes the complete immunoglobulin chain. A mechanism was then required to bring the genetic information together by joining two sequences of DNA to form a single sequence in an antibody-producing cell by so-called somatic recombination. This hypothesis was formulated to account for the power to generate extraordinary diversity at the genomic level. With the discovery of restriction endonucleases, experimental proof of the hypothesis was advanced by Susumu Tonegawa, who demonstrated that the arrangement of light-chain genes was different in antibody-producing tumor cells from that in embryonic cells. The activation of these genes during development was accompanied by somatic recombination or gene shuffling (Hozumi and Tonegawa 1976). Using cloning techniques, Tonegawa showed that the constant and variable regions were indeed encoded far apart in nonproducing immunoglobulin cells, whereas in antibody-producing tumor cells the genes were close together, with an intervening sequence. Thus, after showing that the constant and variable region genes were separate in embryonic DNA, Tonegawa, like Philip Leder and Leroy Hood and their respective colleagues, was led to noting that the variable region polypeptide chain contains more amino acids than is encoded in the DNA V region alone; this extra genetic information was called the J (joining) segment.

Antibody diversity derives from heterogeneity in the V genes (ca. 150), J genes (ca. 5), and a hypervariable region, adding another order of complexity to the gene product; the latter actually forms the direct antigen binding site to yield 7500 (5 × 150 × 10) genetic combinations

75

forming various light chains. The heavy chain has a similar selective strategy among V, J, and a hypervariable D region that yields 2.4 million possible different heavy chains. Thus, from approximately three hundred separate genetic segments in embryonic DNA, the combinational possibilities yield 18 billion ($7.5 \times 10^3 \times 2.4 \times 10^6$) possible antibodies, which may further be supplemented by the extraordinarily high rate of somatic mutation in these genes (1×10^{-4}/generation) (Leder 1982; Litman et al 1990). A molecular basis for clonal selection is now available: Antigen, after processing (discussed below), "selects" from the available antibody repertoire and by clonal stimulation drives the proliferation of the appropriate antibody-producing lymphocytes. Our current theory thus appears to vindicate Ehrlich's original hypothesis offered almost one hundred years ago.

G.O.D. (generation of [antibody] diversity) is a mechanistic but, critically, not a fully deterministic explanation of a secondary *immune* phenomenon, namely, how immune specificity is achieved. What is not addressed is how the host and the foreign agent are first differentiated.[8] Addressing that issue, a biological model was offered by Burnet, to the effect that the purging of "self-reactive" cellular clones in ontogeny defines post-partum identity. Although the molecular understanding of that issue has achieved enormous sophistication, its biological meaning remains ambiguous; I will argue that the final integration of cis- and trans-immunology (what Jerne [1967a] appropriately called "waiting for the end" in reference to a system model of an immune network) remains incomplete. But first, what is the genetic definition of immunological identity? The issue of immune selfhood was implicit from the first descriptions of autoreactive antibody and Ehrlich's *horror autotoxicus* at the turn of the century. But the molecularization of self-identification as it is currently understood originated in studies of the 1930s and 1940s that ultimately defined the antigens determining immunological compatibility for organ transplantation, and which serves as the immuno-genetic signature of the organism. These proteins are coded in clustered loci termed the major histocompatibility complex (MHC). Obviously the MHC does not exist to frustrate our attempts to transplant, but it is thought to serve as a mechanism to introduce the foreign substance (viz. antigen) to a thymus-dependent pool of lymphocytes (T cells); the T cell receptor (generated by molecular mechanisms similar to those utilized to manufacture immunoglobulin) recognizes antigen complexed (presented) to MHC; nonself is thus recognized in the context of a unique signature of identification.

The triumph of immunology

The history elucidating the character and function of MHC has been chronicled as emanating from disparate fields of study, namely, work on tumor transplantation (dependent on the development of inbred mice strains) and blood group typing (Klein 1986; Shreffler 1988). With successfully inbred mice strains, the genetic factors affecting tumor biology were capable of being examined. These studies developed from the research of Carl O. Jensen, which found that tumor originating in a mouse stock strain did not survive when transplanted into wild mice, from which he concluded that there were "race" restrictions on transplantation (Jensen 1903). (Leo Loeb, who performed the first studies of transplantable tumors, in 1901, made similar observations, not knowing Jensen's work; however, he did not understand the genetic implications [Loeb 1901, 1908].) Ernest Tyzzer pursued Loeb's findings and explicitly concluded that although tumor susceptibility was heritable, it did not follow simple Mendelian segregation; after crossing the two mice races, he showed that all the F1 hybrids grew tumor, but none of the F2 generation (Tyzzer 1909).[9]

After the demonstration that tumor susceptibility was controlled by several genes, any one of which could cause rejection, the immunological basis for transplantation was linked to the second tributary of blood group research in the late 1930s. Blood incompatibility between species had been suggested in the mid-1870s, when hemolysis of red cells was noted with heterologous blood transfusions, for example, of lamb's blood given to humans (Diamond 1980). Bordet, working in Metchnikoff's laboratory, showed that hemolysins were produced when he injected red cells of one species into a different one (Bordet 1909; Tauber and Chernyak 1991, pp. 161–2). Ehrlich and Julius Morgenroth confirmed this finding but also discovered that there were hemolysins present normally in sera for erythrocytes of other species. They defined four blood groups by immunizing four goats and recovering antibodies, which they called isolysins, that reacted differently against unimmunized animals (Ehrlich and Morgenroth 1900). At the same time Landsteiner observed that certain human sera clumped the red cells of other individuals, a finding he pursued by systematically performing a hemagglutination matrix study of various human sera and erythrocytes. These studies resulted in the definition of the first three human blood groups (Landsteiner 1900; 1901). (Note: Landsteiner used preimmunized antibodies, and his detection system was agglutination, not lysis.) The genetic aspect was investigated by Emil von Dungern and Ludwik Hirshfeld, who showed, first in dogs and then in humans, that blood group antigens followed strict Mendelian laws (von Dungern and Hirshfeld 1910). Felix Bernstein later proposed that the antigens were controlled by allelic genes at a single locus (Bernstein 1924).

The integration of the genetics of tumor transplantation, on the one hand, and of blood groups, on the other, is credited to that most eclectic and profound biologist, J. B. S. Haldane, who in 1933 postulated that tumor rejection was due to an immune reaction directed against alloantigens. On this view, tumor resistance factors were analogous to blood group antigens, and tumor rejection was comparable to destruction of incompatible blood cells. Haldane did not perform the experiments himself, but successfully encouraged Peter A. Gorer, who had just completed his medical studies, to pursue the project; Gorer successfully showed that one of the blood group genes codes for antigen shared by malignant and normal tissues (Gorer 1936a, 1936b, 1937a, 1937b, 1938). The major issue in deciphering individual resistance genes was to identify individual loci, which were both numerous and had the same effect. George Snell, a geneticist, undertook the laborious task of backcrossing mice so that strains would differ at only one resistance locus. By selecting tumor-resistant animals, he could define the histocompatibility genes that control the immunological fate of transplanted tissue (Snell 1948; Snell, Smith, and Gabrielson 1953).[10]

Gorer and Snell utilized two complementary approaches in defining the MHC (Klein 1986). Gorer used serologic methods to identify MHC antigens, demonstrating antigenic complexity encoded by several genes. Snell's histogenetic approach, based on cellular immunology, also demonstrated genetic complexity. However, this latter approach also measured in vivo compatibility of transplanted tissue and eventually fused with Peter Medawar's reexamination of transplant biology during World War II, when surgical skin grafting had significant medical urgency. In his careful studies Medawar established three characteristics of rejection: (1) accelerated response to second grafts, (2) specificity, as seen by no acceleration if the donor was different, and (3) the systemic character of sensitization, such that no site was protected. He concluded that allograft reactions were immunologically mediated (Medawar 1944). But only in 1953, when Avrion Mitchison, Rupert Billingham, Leslie Brent, and Medawar demonstrated that transplantation immunity could be successfully adopted by transference of cells but not serum from sensitized donors, were mechanisms of transplantation rejection appreciated as analogous to phenomena dependent on lymphocyte reactivity (Billingham, Brent, and Medawar 1953; Mitchison 1954). Shortly thereafter, the graft-versus-host reaction was shown to be caused by immunocompetent cells, and the entire problem concerning the regulation of the immune response was conceptualized in terms of an analysis of lymphocyte reactivity (Billingham, Brent, and Medawar 1955; Trentin 1956; Billingham and Brent 1957; Grebe and Streilein 1976).[11]

The science of immunogenetics is now highly sophisticated, and a confident molecular definition of the immune self, based on the MHC, has emerged. More is known about the MHC, which contains 4 million base pairs (0.1 percent of the genome), than about any other genetic region of similar size (Trowsdale 1993). It is functionally divided into two major classes (I and II), further subdivided into two subclasses (alpha and beta), and finally portioned into families and subfamilies. In each individual only two or three gene pairs are functional (the rest of the DNA is believed to serve the purposes of evolutionary flexibility), with each of the gene classes, I and II, occurring as multiple alleles and thus acting as alloantigens on transplanted tissue to be recognized by recipient T and B lymphocytes. What function MHC has had in phylogenetic history is still unfathomed, and the debate concerning its primitive biological function remains unresolved (Kelso and Schulze 1987).

The function of MHC is not to obstruct transplantation per se, but rather to provide the context for the recognition of antigen by T lymphocytes.[12] When T cells encounter a molecule on the surface of an antigen presenting cell (APC), they do so by "seeing" the antigen complexed to MHC, a process called MHC-restriction (Kourilsky and Claverie 1989; Saper, Bjorkman, and Wiley 1991; Fremont et al. 1992; Guo et al. 1992; Matsumura et al. 1992; Silver et al. 1992). However, how T cells recognize nonself in the context of self (unlike B lymphocytes) remains an enigma (Kelsoe and Schulze 1987). The elucidation of MHC function is an exciting chapter in modern immunology, and we may expect important insight into the process of immune perception as the molecular details of how lymphocytes "see" antigen complexed in the MHC are more fully explored. But there is a crucial distinction between perception and cognition, the latter referring to higher-order functions. These pertain to what I have previously referred to as identity, and it is in the definition of immune identity that some hierarchical structure beyond MHC is required. This is where the biological, as opposed to the reductive chemical, domain becomes operative.

In the elucidation of the genetic-immune signature in the MHC, molecular biology has offered crucial insight. But we may ask whether the problems to which molecular biology has been applied so successfully address the deepest issue of immunology, namely, what the source of *immune* identification is and what controls evolving immune identity. I will argue that such a question in fact can hardly be answered by a reductionist approach alone and must be fitted into a biological construct at a higher level of complexity. To develop this critique, we must carefully examine how the theoretical immune construct of organismic identity developed after World War II, and to do so a short scientific biography of its major architect, Sir Frank Macfarlane Burnet, follows.

3

The immune self declared

The concern with arriving at a science of immunology that addresses both the pathogen challenge *and* the surveillance of normal/abnormal body economy expanded immunology into the science of self/not-self discrimination. In this narrative I have so far scrupulously avoided using the term *self* as a noun, but now we are prepared to evaluate critically how it entered the language of immunology and consider what its conceptual utility has been. The concept of the self was formally inaugurated into immunology by Sir Frank Macfarlane Burnet (Figure 2). In the second edition of *The Production of Antibodies* (1949), Burnet and Frank Fenner introduced the "self-marker" hypothesis, which eventually evolved into the theory that tolerance is a result of the elimination of lymphocytes reactive to autologous constituents:

It is an obvious physiological necessity and a fact fully established by experiment that the body's own cells should not provoke antibody formation. Minor exceptions to this rule concern only tissues which are "unexpendable" parts of the central nervous system and the eye. An animal's own red cells are non-antigenic. This is not due to any intrinsic absence of antigenic components; the same cells injected into a different species or even into another unrelated animal of the same species may give rise to active antibody production. The failure of antibody production against autologous cells demands the postulation of an active ability of the reticulo-endothelial cells to recognize "self" pattern from "not-self" pattern in organic material taken into their substance. The first requirement of an adequate theory of antibody production is to account for this differentiation of function by which the natural entry of foreign micro-organisms or the artificial injection of foreign red cells provokes an im-

81

Figure 2. Sir Frank Macfarlane Burnet. (Courtesy of The Walter and
Eliza Hall Institute of Medical Research, Melbourne.)

munological reaction while the physically similar autologous material is
inert. (Ibid., pp. 85–6)

Simply to assign the birth of the self/not-self concept to this statement
without exploring its intellectual origins is to ignore the rich sources
from which Burnet developed the idea. This chapter is largely devoted
to tracing how he was led to the articulation of the self as a scientific
problem, after embarking on research questions seemingly quite far removed from such issues. Since Burnet was a virologist, the problems
concerning the evolutionary significance of the virus–host relationship
dominated his formative research career. Having wide biological interests and an ambitious personal agenda (Sexton 1991), Burnet attempted to synthesize apparently disparate areas into a comprehensive
theory that became dominated by the question of organismal identity.
Diverse conceptual threads from virology, ecology, genetics, develop-

mental biology, and, of course, immunology were to be woven into a fecund theory that remains the crucial conceptual foundation of the discipline today.

Burnet noted in his autobiography that he had actually introduced the distinction between self and not-self in a semipopular book, *Biological Aspects of Infectious Diseases* (1940), which was written "from a consistently ecological and evolutionary angle" (Burnet 1969c, p. 189). Referring to an amoeba digesting another microorganism, he wrote:

> The fact that the one is digested, the other not, demands that in some way or other the living substance of the amoeba can distinguish between the chemical structure characteristics of "self" and any sufficiently different chemical structure which is recognized as "not-self." (Burnet 1940, p. 29)

Burnet extended the primitive property of self-identification to dispersed cells of two species of sponges that reaggregate to form new organisms according to a "dim recognition of 'self' structure as opposed to 'not-self'" (ibid., p. 30). As he recalled in his autobiography, "[H]aving introduced the concept of difference between self and not-self, the important thing in 1937 was the nature of antibody" (Burnet 1969c, p. 190). (The *Infectious Diseases* manuscript was finished in 1938 but not published until 1940. That is why Burnet refers to self/not-self as first introduced in 1937, well before the first edition of *Antibodies* was published in 1941.) But as will be clear from the discussion, the conceptual richness of self did not come to full fruition until the second edition of *Antibodies* was published in 1949. It was his frustration in publishing his findings concerning the secondary logarithmic antibody response that stimulated the first edition of *Antibodies*. This work appears unconcerned with the experimental issues that generated Burnet's theory of tolerance and that underlie his conception of the self. "Self" emerges to solve the problem of *tolerance* explicitly in the second edition of *Antibodies*. In this reading, then, the significance of *self* remained hidden for another decade. But in fact, and not surprisingly, Burnet's interest in what he called tolerance dates to this same earlier period – between 1936 and 1941:

> A particularly striking incidence of the tolerance of immature tissues for foreign material is seen in the chick embryo on which fragments of a mammalian tumour, the Jensen rat sarcoma, have been grafted. The foreign tissue becomes vascularized from the chorioallantoic vessels and grows freely without any of the leucocytic reaction that it would provoke in an adult alien host [Murphy, J. B. *J. Exp. Med.* 1913, 17:482]. In

some way the embryonic cells seem to be unable to recognize and resent contact with foreign material in the way adult cells do. It is therefore not unexpected that no antibody response takes place. (Burnet 1941, p. 45)

With a succinct and prescient comment Burnet concluded by noting that "immunological reactions of every sort are the result of a process of 'training'" (ibid., p. 46). The notion of training was the springboard for the elaboration of the self/not-self distinction in the second edition of *Antibodies* (1949), and any comprehensive history of this distinction must ponder what Burnet implied with these terms in 1941.

Prior to 1936 Burnet's primary scientific interests were in bacteriophage research; from this corpus of work the seeds of what would later emerge in the self concept can be discerned.[1] The evidence I cite is exclusively from his published work – his private papers remain to be critically examined. From the mid-1920s, soon after he graduated from medical school, until the writing of the *Biological Aspects of Infectious Diseases,* in 1936–8, Burnet was concerned with two major questions: (1) to discover how the virus enters the bacterium and specifically to ascertain whether the site of entry is the same as or similar to that through which the antibody is absorbed; (2) to quantify toxin–antitoxin dynamic ratios. As mentioned, Burnet's chief research interests were in virology during this early period of his career. His work in mainstream immunology stemmed from his first antibody–antitoxin studies, which arose from staphylococcus experiments initiated in the wake of a staphylococcal epidemic, the so-called Bundaberg fatalities (Sexton 1991, pp. 65–6). Most of these papers are experimental and explicitly nontheoretical, but there is an underlying evolving framework that becomes clearly articulated in *Infectious Diseases.* (The comprehensive paper, "The Immunological Reactions of the Filtrable Viruses" [Burnet, Keogh, and Lush 1937], written in the same year, is essentially a summary of these experimental findings.) Burnet's research may be divided into several stages, each of which reflects his relationship to the dominant theories of bacteriophage research of the period. Situating Burnet in this way allows us to trace the scientific development that eventuated in the publication of *Antibodies* in 1941. So, first, a brief summary concerning bacteriophage research in the 1920s and 1930s will serve to orient the discussion.

Bacteriophages were the last of the major classes of viruses to be identified. (Plant viruses were discovered in 1892 by Ivanowski, and animal viruses by Loeffler and Froesch in 1902 [Duckworth 1987].) They are characterized by their lytic properties; that is, as they multiply, the host bacteria are killed, releasing the virions that had usurped the bacterium's reproductive apparatus for viral replication. But bacteriophage may also lie "dormant" in bacteria by a phenomenon called lysogeny, which is recognized when the virus is cultured with different strains of bacteria that are lysed. The lysogenic phage are thus present in a noninfectious form in their natural host and are capable of persistent, nonlysing association with the bacterium. These so-called temperate phages are reproduced during bacterial multiplication and thus passed on to progeny (a phenomenon originally described by Bordet). The quiescent "prophage" may shift to a vegetative form and then reproduce just like a virulent phage, with resultant lysis of the bacterial host. (The analysis of the phage "life cycle" has been an important chapter in the deciphering of molecular genetic control mechanisms, exquisitely detailed in Ptashne [1992].)

The discovery of bacteriophage was made by Frederick W. Twort, who, in growing vaccinia viruses on artificial medium, "observed that colonies of a contaminating micrococcus often appeared to be afflicted with some disease" (Twort 1949). He found that the transforming principle could be filtered out and concluded that it was an agent that killed bacteria and multiplied itself in the process. After considering the possibility that he had discovered a bacterial virus, he thought it equally possible that he had found an "enzyme with the power of growth" (Twort 1915). The paper was ignored for five years, but in the meantime Felix d'Herelle described the filtrable agent that killed a dysentery bacillus found in the intestines of recovering patients but not present in the early phase of the disease (d'Herelle 1917). Because this lytic agent could protect rabbits from a lethal dose of the bacillus, d'Herelle hypothesized that the virus might play a part in immunity; his discovery was thus widely celebrated as a newly discovered mechanism of host defense (Duckworth 1987). However, Bordet contested d'Herelle's claim of priority in favor of Twort for the discovery of bacteriophage, and more important, a rigorous polemic developed between them concerning the role of the virus in immune reactions.

In d'Herelle's *The Bacteriophage: Its Role in Immunity* (1921) an important assault was made against the humoralists. The position Bordet had developed since the 1890s was that humoral immunity was composed of two kinds of soluble factors: antitoxins (or antibody), which specifically identified targets for immune destruction, and alexine (or complement), which served as a common means of lysing red blood cells or killing bacteria. In other words, antibody would attach to the cell or microorganism, and as complementary action, after the antibody binding, complement would destroy the target (Bordet 1909). The assay for antibody was agglutination of bacteria or red blood cells, and for complement, lysis or cell destruction. (Incidentally, this essential theoretical framework is still in place today.) Bordet's agenda was to link the humoral and cellular components of immune theory. By proposing that leukocytes could stimulate what would be a humoral bacteriolytic activity, Bordet was endorsing Twort's original hypothesis, namely, that the bacteriolytic principle was in fact a self-replicative enzyme (Bordet and Ciuca 1921). We now view such a conception of a protein gene as wrong, but in the 1920s it was reasonable to speculate that the genetic information molecule was a protein.

Thus, a lytic enzyme (with mechanistic characteristics possibly similar to complement) was Bordet's candidate, in contrast to d'Herelle's candidate, a bacterial virus. D'Herelle disputed Bordet's theory on the basis that the targets (bacteria and red blood cells) were so different that only analogy linked the two experimental systems, and in support of his claim he cited instances in which sera aids, rather than inhibits, bacterial activity. "These antibodies, like agglutinins, can only be considered as indices of infection" (d'Herelle 1921, p. 168). He then discounted the attribution of antibacterial activity to any known antibodies and concluded that "all organic immunity is reduced to antibacterial immunity, assured by the antitoxins [antibody]" (ibid.). Bacteriophage becomes an internal bacterial phagocyte, so to speak, and it appears then as no accident that *phage* ("to eat") appears both in *phagocyte* ("eating cell") and *bacteriophage* ("eater of bacteria"), despite d'Herelle's disavowal of any affinity between the two phenomena (ibid., p. 21). As he wrote, six years after Metchnikoff's death:

> ... the role which the partisans of the theory of "bactericidal humoral immunity" have desired that these simple indices of infection – the antibodies – play, is based upon a non-existent phenomenon of bacteriolysis by immune sera. Could it not in reality be played by the bacteriophage, that principle endowed with a powerful bacteriolytic action, operating upon the most varied bacteria? ... Can not the bacteriophage play, in addition to its direct action, an important role in phagocytosis itself, in bringing about what might be called a phagocytic education? (Ibid., p. 169)

Next d'Herelle articulated a theme to be reiterated in many different contexts for the next seventy years, that is, that serology is not necessarily a reflection of immunity:

> *Immunity refers to actual resistance to infection, and serology refers to antigen-antibody reactions. They should never be brought together until*

*they have been thoroughly inspected separately, and rarely then....
[K]nowing that he can control serology, man speculated that antibodies
neutralize everything right and left, and that hardly anything else is needed
to explain the whole of immunity. This simply is not so.* (Marshall 1959;
emphasis in the original)

From d'Herelle's point of view, antibodies may "mark" microbes, but
they do not necessarily destroy them. The assault on the humoralists
was no less than an attempt to replace the humoral arm of immunity
with what d'Herelle called heterogeneous immunity, mediated by bacteriophages:

One cannot avoid the conclusion that it is impossible to attribute any
active role in the production of antibacterial immunity to any actually
known antibodies. All *organic* immunity is reduced to antibacterial immunity, assured by phagocytosis, and to antitoxic immunity, assured by
the antitoxins. (D'Herelle 1921, p. 168)

Bordet resisted this assault on his theory, and regarded the evidence
from lysogeny as supporting the appearance of phage as the result of
an effect of the body defenses on the infecting bacterium. Irrespective
of the details, the debate concerning the nature of bacteriophage was
conducted between competing theories of immunity, and from our perspective, d'Herelle is one of the few scientists of the period to have
explicitly framed his ideas within the Metchnikovian worldview. The
notion of the bacteriophage–metazoan relationship as symbiotic (derived phylogenetically) is at the core of his entire theory:

A being which evolves is necessarily a being which lives, which adapts
itself, and which conquers. From the instant that it ceases to adapt itself –
to evolve – it dies. Evolution is always conducted according to the law of
least effort. The multicellular organisms have profited by securing for
their defense the parasitism of the bacteriophage for the bacteria, which
is only a chapter in the universal struggle. If, among all living beings, the
bacteria alone escaped parasitism, where would we arrive? It is very simple. One of two things would take place. Either evolution would not extend beyond the stage of the unicellular being, or evolution would be
accomplished in another manner and immunity would be assured by
other means. The bacteriophage does not exist for the defense of the
superior organism against the bacteria, it exists simply because in the
course of evolution certain germs have parasitized others. (Ibid., pp.
171–2)

Thus, immunity arises directly from evolutionary relationships, where
the processes of natural selection and competition between species
serve to frame the phenomena we observe as immunity.

The lysis of bacteria suggested that bacteriophage fulfilled an important role in normal symbiotic relationships between bacteria and their more complex hosts (Sapp 1994). Thus, early bacteriophage research was preoccupied with two major themes: immunity and symbiosis. We might well search for Metchnikoff's legacy in such rich intellectual soil. The d'Herellian perspective is in fact strikingly Metchnikovian in both its affinity with evolution as its guiding framework *and* in its implicit rejection of an immunochemical construct essentially defined according to a mechanistic blueprint:

> It is certain that a theory of immunity based on the bacteriophage, that is, on an autonomous organism, is so far outside of all present opinion that it will stir up at first incredulity and will be called a "finalistic theory" – a synonym of "anti-scientific." I affirm that from my point of view this theory can not be "finalistic." "To be is to struggle, to live is to conquer," a very just statement by LeDantec. It is all contained in a single word – evolution. (D'Herelle 1921, p. 171)

This almost defiant, if not self-conscious, proclamation clearly set d'Herelle in the margins of the dominant immunochemical community. On this stage Burnet would have to choose sides. There are two general influences to examine carefully: First, what *was* Burnet's position with respect to d'Herelle's views, and second, was there any direct linkage to Metchnikoff's biological vision? In fact, there is an unexpected scientific geneology among Metchnikoff, d'Herelle, and Burnet that clearly reveals Burnet's intellectual indebtedness to these predecessors. In summary, by 1933–7 Burnet, after some wavering, accepted the d'Herellian argument that bacteriophage originates from a diverse lineage of autonomous organisms that have come into symbiotic relations with bacteria. Burnet thus adopts this ecological/symbiotic orientation championed by d'Herelle, who in turn explicitly formulated his own theory in the context of Metchnikoff's concepts of immunity. Beyond specifically choosing between Bordet's and d'Herelle's perspectives in the dispute over the nature of bacteriophage, Burnet borrowed directly from Metchnikoff when writing his work *Infectious Diseases,* in 1936–8, to describe the general biological principles that guided his views toward what was eventually to develop into his own comprehensive theory of immune identity. With this short orientation, we will now look in more detail at the record.

When Burnet actually began his bacteriophage research in the mid-1920s, he was closely aligned with Bordet's orientation. Burnet viewed bacteriophage as an enzyme normally required as part of bacterial fusion and the d'Herelle phenomenon (bacterial lysis or lysogeny) as a pathological manifestation of the normal process:

> The normal fission process therefore seems to be closely related to the sensitivity or resistance to different bacteriophages, and the simplest hypothesis is that the portals of access of the various lytic agents are the different forms of enzyme in the bacterial cell that are available for the lytic activities of normal fission. There are as many fundamental types of phage as there are distinct bacterial enzymes available. (Burnet 1925a, p. 417)

However, Burnet viewed bacteriophage as corpuscular and put some effort into reconciling this understanding in relation to how he situated himself in the Bordet–d'Herelle debate (1925b). These early papers were soon eclipsed by a more sophisticated understanding of the evolutionary context of biological phenomena. For example, in 1925 Burnet wrote on the evolution of antiperoxide mechanisms in bacteria, conceiving of such systems as inseparable from adaptation to aerobic conditions, and thereby recognizing the teleological significance of this finding (1925c).

After reading *The Bacteriophage and Its Behavior* (d'Herelle 1926), which he cited in most of his post-1926 bacteriophage papers, Burnet championed the biological independence of bacteriophage particles. Although agreeing with d'Herelle as to the general biological nature of the phage, he disagreed on particular crucial points. First, he viewed bacterial resistance not as an instructive process, but as a strictly passive, selective one: "It will probably be admitted by all that the capacity to develop resistance must depend primarily on some degree of variation pre-existent in the bacterial population. . . . The action of the phage is purely selective" (Burnet 1929a). Hence, his thinking would foreshadow the type of selective process at work with clonal selection theory (CST); this way of thinking arose through the ecological perspective he was now starting to develop. A second difference from d'Herelle's views, important at this stage, was that Burnet dissociated himself from making any type of explicit analogy between bacterial resistance and metazoan resistance. Eventually he would in fact set d'Herelle's perspective in antithesis with his own in this regard. Other differences with d'Herelle included Burnet's view that bacteriophage was similar to a gene; moreover, he would slowly dissociate himself from being an advocate of bacteriophage therapy against dysentery. This is the seed of his sensitivity to the ecological perspective. Specifically Burnet viewed the test tube as an artificial environment, and the body as a complex ecological community to which findings of in vitro experiments should not be extrapolated (Burnet 1929b, p. 409).

By the late 1920s the emergence of Burnet's ecological concerns is quite evident. However, before fully developing such an ecological perspective, Burnet would go through a brief period in which his investigations of lysogeny would, on the one hand, force him to somewhat distance himself from d'Herelle and reach a ground closer to Bordet and, on the other hand, compel him to ponder the nature of the organism at the bacterial level. This intermediate stage, lasting between 1929 and 1932 (Burnet and McKie 1929; Burnet 1932), would again prepare the grounds and adumbrate several important ideas he would later develop. With respect to lysogeny, Burnet modified his views to what he considered a compromise between Bordet and d'Herelle. He regarded Bordet as correct in that a lysogenic phage is self-produced. However, once the virus leaves the first bacterium and attacks others (especially bacteria of another strain), it behaves like an autonomous living agent; on this he regarded d'Herelle as correct. Thus, Burnet thought he had struck an intermediate position, but as he later admitted (Burnet 1932, p. 859), it is evident how close he was to Bordet's camp at this time. The principle issue driving the argument was to reconcile the possibilities that bacteriophage could induce bacterial lysis or rest dormant during several generations of bacterial replication:

> The difficulty of reconciling these two aspects of bacteriophage phenomena has been responsible for all the current controversy on the intimate nature of phage, whether it is an independent parasite or a pathologically altered constituent of normal bacteria. In our view both these contentions have been completely proved, and the current attitude on both sides of regarding them as irreconcilable alternatives is quite unjustified. According to the particular type of bacterium that is reacting with the phage concerned, it may be useful and convenient to regard the phage as an independent parasite or as a unit liberated from the hereditary constitution of some bacterium, the usage being determined wholly by its functional activity at the time. (Burnet and McKie 1929, p. 284)

Burnet recognized that bacteriophage might possess a dual nature, each aspect expressed under different conditions. He might well have begun to ponder the nature of "selfhood" of the bacterium: Was it lysing itself or being lysed by another, and what was the symbiotic relation of this other to the host? In an uncharacteristic extrapolation from protozoan biology to animals, he drew an interesting analogy:

> Transmissible mammalian cancer affords an analogy which is perhaps more than superficial. The original spontaneous tumour is composed of

cells which are genetically derived from the fertilized ovum that gave rise to the host organism. When a few of these cells are inoculated into a young animal of the same species they act as independent parasites reproducing their kind. By successive transfer an unlimited number of animals may serve as hosts for cells which are in their biological relationship true parasites yet are ultimately derived by genetic descent from the fertilized ovum of an animal similar to the hosts. At a higher level this analogy almost exactly parallels the only other hypothesis [the one from Burnet and McKie 1929] which seems to interpret all the facts of bacteriophage phenomena as adequately as that of d'Herelle. (Burnet 1930, pp. 649–50)

We see an important seed of his later theory. By 1930 the bacteriophage research forced him not only to consider phage in the context of what later would be referred to as selfhood, but also to consider phage–bacteria ecological relations in relation to a "higher" immunological phenomenon. Thus, it was a brief, but perhaps important, interlude for Burnet.

The path from the bacteriophage to the self expressed in *Infectious Diseases* is clearly marked by Burnet's 1932 and 1933 papers. Utilizing lysogenic evidence taken from studies of *Salmonella* strains, he realized that unrelated strains can derive lysogenically from the same bacterium. With this finding he finally shifted irrevocably to d'Herelle's position:

D'Herelle regards the strains showing no gross evidence of lysis, i.e., the great majority of Salmonella cultures, as representing stable symbioses between phage and bacterium. This hypothesis is not at variance with general biological experience and at least offers fertile suggestions as to the significance of the phage types obtained in the present work. (Burnet 1932, p. 859)

In his next paper physical and serological evidence is invoked in support of d'Herelle's views (although the latter differ in that Burnet regarded bacteriophages as a diverse lineage of independent species, not a wide-ranging variety of a single species):

It is clear that with these species at least lysogenicity is not an essential characteristic of the species. It is hard to conceive why if the weak phages detected by the technique I used, represent a normal by-product of metabolism those obtainable from cultures of a single, well-defined species should be so heterogeneous. On the other hand, a table of almost precisely similar form could be used to express, say, the results of a survey of the intestinal protozoa of a number of related mammalian species. Perhaps a closer parallel still would be a table showing the various species of alga which could form stable symbioses with a series of lichen-producing fungi. A relationship of this numerical type between two sets of biological entities can, I think, only exist when the evolution and differentiation of the two sets has been essentially independent. In other words, the phenomena of lysogenicity indicate that bacteriophages form a group of independently evolved living organisms capable of entering into parasitic or symbiotic relationships with bacteria. (Burnet 1933a, p. 413) [2]

Here we have strong evidence of Burnet's shift to the full-scale ecological perspective of the phage–bacterium relationship, a relationship he would reiterate in many of his later bacteriophage papers. Hoping to classify the phages as a phylogenetic lineage, Burnet explicitly notes "the conception of phage–bacterial interaction in nature as an extremely complex oecological problem, many varieties both of phage and bacteria being concerned and every degree of parasitic and symbiotic relationship being established" (Burnet 1933b, pp. 316–17). He also wrote an entire section entitled "Some Ecological Considerations," wherein he observed of phages and bacteria: "As in all host–parasite relationships a form of dynamic equilibrium must be set up by which a fairly constant proportion of both species manage to survive" (Burnet 1934, p. 345). Thus, the ecological groundwork that would inform Burnet's conception of the self was now fully in place, based upon his own experimental work concerning bacteriophages.

Burnet's ecological viewpoint would culminate in his 1940 publication (written between 1936 and 1938), *Biological Aspects of Infectious Disease;* in this book he introduced the terms *tolerance* and *self.* This was no accident. The second chapter, "The Evolution of Infection and Defense," wherein *self* is introduced, cannot be read apart from the first chapter, "The Ecological Point of View," where *tolerance* first appears. Both chapters were intended as the cornerstone of a biological approach to immunology. Part I of *Infectious Diseases,* wherein both chapters are contained, is entitled "Biological Considerations," an orientation that Burnet would continue to invoke even in his preface to the second edition of *Production of Antibodies* (Burnet and Fenner 1949). They may even be seen as the foundation of his entire later theoretical formulation. The first chapter recapitulates the general ecological principles current at the time, expounding especially on the concept of the "climax community."

The idea of a climax in the development of vegetation was first suggested by Hult in 1885 and was applied to designate a more or less permanent and final stage of vegetative succession in a restricted area (Clements 1936). Animals are, of course, considered members of the climax, and the word *biome* was proposed to stress the mutual roles of plants and animals. *Biome* and *climax* are synonyms when the entire biotic community is indicated, but *climax* is employed for the matrix when plants alone are considered (ibid.). Frederic Clements was a key architect of climax theory and thus an important founder of ecology. Arthur Tansley disagreed with this organismic metaphor and after using the transition term *quasi-organism,* coined the word *ecosystem* in 1935 as a more accurate characterization of the vegetational unit. (This shift from a biological to a physical model for ecology opened the way to mathematical analysis of the system and set the conceptual orientation of current analyses [Kingsland 1991, pp. 5–6].)

Burnet regarded the climax community as the culminating dynamic equilibrium created in any given environment:

> The process by which an approximately constant population of living organisms is developed has been studied in more detail by botanists than by animal ecologists, but the broad principles are the same for both great classes of living organisms. The final constant condition of the vegetation in any particular region is called by plant ecologists the "climax" state. (Burnet 1940, pp. 9–10)

Of fundamental importance, Burnet then proceeded to conclude this first chapter and lay the foundation for his entire subsequent thinking by extending the concept of a "climax community" to the human body itself. In other words, he conceptualized the body as a derived ecosystem evolved over time into a dynamic equilibrium. Moreover, the ability of the system as a whole to allow the existence of particular component members is termed *tolerance:*

> It is a conflict between man and his parasites which, in a constant environment, would tend to result in a virtual equilibrium, a climax state, in which both species would survive indefinitely. Man, however, lives in an environment constantly being changed by his own activities, and few of his diseases have attained such an equilibrium. The practical problems of prevention and treatment demand principally an understanding of the results of new types of infection on the individual and the community, and of the stages by which a condition of tolerance is reached. When such knowledge has been gained it is a simple matter to apply it to the interpretation of those diseases which have reached their climax state. (Ibid., pp. 23–4)

Here, Burnet first uses *tolerance,* which will serve as a cornerstone of his later immune theory. The etymology notably derives from observations of an ecology conceived as a "social" community.

The remarkable genealogy of the concepts of self and tolerance is their derivation from ecological concern. Each is telescoped into the individual organism. I would call this extrapolation from one level of complexity to another, a "collapse of organizational boundaries," where relationships of various components of an organism are structured analogously, whether the "organism" is an ecological community, an individual animal, or a bacterium. The ecological view of the self had been implied in Chapter 1 of *Infectious Diseases;* it would take Chapter 2 to call the ecosystem a self. However, between Chapters 1 and 2 there is a missing link. How does Burnet get from the dynamic view of the organism to the term *self?* The answer lies in Burnet's intellectual debt to *The Science of Life,* by H. G. Wells, Julian Huxley, and George Wells (1929, 1934). In his autobiography, he acknowledges the influence of this work:

> My adult education in animal behaviour and ecology probably started with Julian Huxley's *Essays of a Biologist.* Not so long afterwards, Wells and Huxley's *The Science of Life* appeared and had a major impact on my thinking. It made me particularly interested in the ecological aspects of microbiology and epidemic disease, the field that I was then just entering as a laboratory worker. . . . In 1936, I started to write a semi-popular book on infectious disease and immunity in which the ecological approach should be dominant. (Burnet, 1969c, p. 23)

Moreover, in the preface to *Infectious Diseases* Burnet wrote: "In its own much more limited field I should like to think that this book expresses the same general point of view that runs through Wells, Huxley and Wells' *Science of Life.*" Indeed, Burnet espoused the views on ecology propounded in *The Science of Life.*

First, he extracted the ecological principles that would inform his own first chapter; in fact, his first chapter is essentially (except for its crucial application to the human body) a summary of Book 6, Section V, "The Science of Ecology" (Wells, Huxley, and Wells, 1929, pp. 961–1011). In Subsection 4 of that section (pp. 973–82, essential for noting parallels between this book and Burnet's first chapter), the development of the climax community, or what would later be called an ecosystem, is described as analogous to an organism:

> There is a progression of inhabitants, one set of animals and plants succeeding another in sequence, until finally a stable state is reached. In a state of nature, the animal and plant life of this stable phase is the same as the original life of the area. The life-community has reproduced itself. (Ibid., 1929, p. 973)

> Just as in an individual, reproduction must be by simple structures like egg or spore or gemmule, whose development into the adult state is through steps of increasing complication, so the pioneer group of organ-

isms by which alone the life-community can reproduce itself is much simpler than the climax phase. (Ibid., p. 977)

The analogy to developmental processes explicitly links such ecological concerns to the individual: "What we have been talking about hitherto are for the most part only the final stages of development, called *climax* stages by ecologists, and corresponding closely enough with the adult phase of an individual's life" (ibid.). Thus, this chapter served as a fertile source of information for Burnet, although he was obviously sympathetic to its point of view as a result of his own research. Moreover, many of his examples are directly lifted from *The Science of Life*, for example, the Krakatoa instance or the mouse plague of Australia (ibid., pp. 973–5, 997). It is also important that this chapter regards an ecological community in dynamic terms: "It is not enough to think of life-communities in this fixed and static way, for they are continually changing. For one thing, they themselves bring about that community-development we call succession, up to the climax which is in equilibrium with itself" (ibid., p. 987).

We may simply note those characteristics of the climax that might be broadly applicable to Burnet's thinking. The most noteworthy is the organismic quality of the concept. Clements and other proponents of the climax community explicitly invoked this quality by drawing the parallel between the climax and the "twin concept" of organism, with the identity and boundaries of both being founded on "emergent evolution and holism" (Clements 1936 [citing Phillips 1935]). Like an organism, a climax community has distinct developmental stages, and once it reaches maturity, there is a high degree of fundamental stability despite the superficial, constant, and universal changes at work. Like an organism, the climax is responsive to its environment (i.e., to the climate and other physicochemical conditions), and it is also able to reproduce itself. Thus, the climax has a life history "comparable in its chief features with the life-history of an individual plant. The climax formation is the adult organism . . . the final stage of vegetational development in a climactic unit" (ibid. [quotes from Clements, *Plant Succession*, 1916]). The key characteristics, then, were the ecological succession of plant formations, and the treatment of the plant community as a complex organism undergoing a life-cycle and evolutionary history analogous to the individual organism. *The Science of Life* clearly adopts this orientation and uses it as a highly dynamic and unifying construct to characterize life from its simplest form to the most complex ecological system. This vision is then explicitly related to the human body. And behold, in so doing, Metchnikoff's general notions reappear! In the next section, Book 7, Section I, Subsection 1, "Is Man Particularly Unhealthy?," the notions of struggle, continual adaptation, and the striving toward an ideal frame the discussion of health:

> Few if any living species are completely happy in their habitat, for exterior circumstances are always changing and are generally ahead of adaptation. A consequent stress and malaise rests therefore upon the surviving minority of each generation even if such species are holding their own in the struggle. (Wells, Huxley, and Wells 1929, 1934, p. 1033)

> One hears a lot of nonsense about "natural health" – about savages and wild things being almost uniformly fit, while only civilized man is disease-ridden; the suggestion generally follows that by turning one's back on civilization and by taking to the woods one could achieve a savage healthiness. In truth what would be achieved would be a swifter death. . . . "Perfect health" is a fantasy. The unstable health of man is only a case of the universal instability of life. (Ibid., p. 1034)

Consonant with the views of Metchnikoff, an active, struggling organism is being described here. Now only the term *self* need be introduced, as is done in Book 8, Section VIII, Subsection 2, "What Is the Self We Conduct?" (ibid., pp. 1391–5), which is a pseudo-Jungian account of the psychological self.

That the self was introduced as a psychological concept is not surprising. After Freud, psychoanalysis had been preoccupied, if not fixated, on defining both normal and pathological selfhood. The term was widely employed in that discourse. On the basis of the linguistic turn, or perhaps as a result of psychological motivation, the collapse of *self, climax community,* and *human body* into parallel uses simply reflected the underlying structure and overarching themes of biological unity noted by Wells et al. and Burnet. Not only was the existence of a chain of being reaffirmed, but the essential structure of biological organization was seen as shared at the cellular, individual, and ecological levels. In a strong sense, "self" as presented in a psychological context, "climax community" as used in the ecological section, and "human body" as understood in the health section all reflect the same biological ethos: Beyond the consistency of the narrative's structure, the "self" shares the same dynamic qualities assigned to the preceding "organism" (and climax community). This reflects a highly dynamic, holistic vision of biology, with the "self" taking its rightful place as the unifying concept of actively developing, diverse selective systems. It proved to be a fertile conceptual ground for Burnet's later immunological theory of selfhood. When the following quotes are read as an immune system, it becomes readily apparent that Burnet had found the perfect source for his term. As Wells et al. begin: "Where precisely in this seething mass of mental activity does the self begin and end?" (ibid., p. 1391). Considering the context of a struggling, selecting self, they continue: "Often but not always these conflicts resolved themselves into a struggle between what he was informed was his better and his worse self. . . . What is this second more *selected* self in Mr. Everyman and all of us?" (ibid., p. 1392). Introducing the persona they continue: "It is not a fixed thing. It varies with our audience and our mood. . . . Now something is out, now something is in" (ibid.). "At their first onset in a human being's life, many things are found to be attractive or repulsive in this way or that. Ideas are grasped or refused. The mind, one must conclude, is ready to receive or repel them" (ibid., pp. 1393–4). "It becomes manifest under such a scrutiny that, in the inner world, just as in the outer, the individuality comes and goes, that it changes, is now more and now less, now fainter and now intenser, that it assimilates and again rejects" (ibid., p. 1395). Thus, if Burnet was looking here (and it appears that he was), he certainly found the self well annunciated and applicable for translocation from the psychological to the immunological sphere.

Selfhood appears in a rich biological context, namely, the ecological perspective governed by evolutionary principles. Burnet fully endorsed this general perspective, and beyond the tantalizing scientific lineage to d'Herelle and secondarily to Metchnikoff, we have explicit evidence that Burnet was sympathetic to, and knowledgeable of, Metchnikoff's vision. In tracing the linkage between Metchnikoff and Burnet, I may be accused of overstretching my thesis. But in *Infectious Diseases* Burnet draws heavily on Metchnikoff in the second chapter, "The Evolution of Infection and Defence." Therein, citing Metchnikoff repeatedly, Burnet traces the path from amoeba to sponge to human being, much as Metchnikoff (1892) had done. Burnet's notion of "self/not-self" relations between the amoeba and bacteria apparently derive directly from Metchnikoff.[3] Burnet recapitulates Metchnikoff's phylogenetic survey of immunity, declaring the phagocyte theory to be compatible with and complementary to the humoral school, with the qualification that "if there has been any victory for either side, it must be with Metchnikoff. The phagocytic cells of the blood and of the fixed tissue *are* the final defenders of the body" (Burnet 1940, p. 36). Burnet not only read Metchnikoff's developmental approach to immunology, but was inspired by that earlier work to ponder the same questions. He concludes this second chapter by stating that "there is very little comparative knowledge to allow any speculation on the evolution of the antibody mechanism" (ibid., p. 38), but he returns to the discussion anew in Chapter 9, "The Wider Significance of Immunity" (ibid., pp. 114–33), where many threads at last tie together. There, Burnet renders explicit parallels between psychological and immunological "learning." The adaptive character of organismal behavior, with all its parameters, is emphasized. For instance, "The acquired immunity which follows many infections is just as much a matter of 'learning by experience' as is the acquired ability to swim or to ride a bicycle" (ibid., p. 116). He calls antibodies "reminders" and draws explicit metaphors with persons:

> The "reminders" are like plain-clothes detectives with perfect memories for criminal faces; when they come into context with their "opposite number," this is recognized as a dangerous individual which has on some former occasion penetrated the defenses either of the body or of society. (Ibid., p. 125)

Thus, the self/not-self distinction takes on its powerful parallel to our very sense of selfhood. Framed in terms of immune function, the immune self is characterized in highly personalized terms. Burnet's indebtedness to a phylogenetic analysis reiterates the logic of Metchnikoff's original formulation, and both thinkers are committed to explicating immunity in terms of an evolutionary framework. Burnet clearly articulates his hope for the phylogenetic investigation into the origins of immunity in writing: "Any organism which lives by digesting the substance of other organisms must in some way be able to distinguish between 'self' and 'not-self', and . . . this primitive differentiation has in some way been elaborated and given rise to the mechanism of immunity" (ibid., p. 117). He was to reap the full rewards of his theorizing a decade later.

B. TOLERANCE AND A THEORY OF THE SELF

As noted, the 1941 edition of *Antibodies,* written shortly after *Infectious Diseases,* makes no mention of autoimmunity, and Burnet had yet to articulate our current understanding of immune tolerance. The stimulus to reexamine the problem of tolerance is generally credited to Ray Owen's seminal study of dizygotic cattle twins (Owen 1945). He demonstrated that animals that shared an in utero circulatory system were blood cell chimeras and unable to mount an immune response to one another's antigens. The ancient Greek chimera was a monster composed of different animals. In contemporary biology *chimerism* refers to a clinical status in which a graft from one individual is accepted and tolerated by the immune system of another. Thus, a chimera is an animal that is a recipient of an organ (e.g., kidney, heart, or liver) from another animal or, as in the case of dizygotic twins, two animals that share a common circulatory system, each fetus partially constituted with blood cells (including lymphocytes) from its twin. In the latter case hematopoietic chimerism persists into adulthood because of in utero transplantation of allogenic (genetically nonidentical individuals from the same species) stem cells that populate the bone marrow for lifelong hematopoiesis. Tolerance of the twin's tissue is thus achieved.[4] This observation signaled the fact that immune tolerance was not genetically programmed; coupled with the understanding of antibody formation, it led Burnet and Fenner to infer that the body must assume "self-markers" at some point in ontogeny. Although the "self pattern" is inherited, the recognition of its constituents is acquired during development. These markers were, then, the basis to discriminate self from nonself, allowing for such recognition to inhibit an immune response where inappropriate. In retrospect Burnet, then, recognized that almost a decade earlier Erich Traub (1936, 1938) had reported that fetal infection of mice with the lymphocyte choriomeningitis virus rendered the animal incapable of immune reaction to the viral antigen after birth (Burnet and Fenner 1949, p. 104). The first edition of *Antibodies* failed to cite Traub's observation. The distinction between an immature immune response and a truly abrogated immune reaction (Murphy 1926) was now realized as qualitatively different from tolerance.

Tolerance is defined as a specific depression of the immune response induced by previous exposure to the antigen. Burnet used it originally in 1941 as a generally depressed or hyporesponsive immune reaction. Tolerance, as credited by Siskind (1984), was first discovered by Harry Wells and Thomas Osborne in 1911, when they reported decreased anaphylactic sensitivity following the feeding of a large dose of antigen (Wells and Osborn 1911). There were sporadic extensions of the idea (Glenny and Hopkins 1923, 1924; Sulzberger 1930), but none in the area of transplantation until after World War II, when tolerance was posed as a central immunological problem. We need only note that the implications of the concept were quickly pursued both experimentally and theoretically, and that Burnet and Fenner correctly concluded that any antigen present during a crucial developmental phase would be exempt from autoantigenicity: "These phenomena are obviously complex but there is the development of a tolerance to the foreign microorganism during embryonic life. . . ." (Burnet and Fenner 1949, p. 104).

This is the rough outline of the widely accepted account of how Burnet arrived at the notion of tolerance and selfhood; not surprisingly, the history is more complex.

To examine how Burnet developed the phylogenetically based, ecological notion of the self in *Infectious Diseases* to a radical new conception of organismal identity as expounded in the second edition of *Antibodies* is to trace a complicated scientific genealogy. There are several strands that must be followed and ultimately tied together into a coherent synthesis. The first of these is to combine Burnet's original dynamic view of the self with the more static view of genetic determinism. Genetics was beginning to dominate transplantation biology in the late 1930s (see Chapter 2), and it is perhaps useful to contrast Burnet's 1937 position with that of Leo Loeb (the younger brother of Jacques Loeb), a transplantation researcher. In "The Biological Basis of Individuality" (1937) an important summary paper written before Gorer published his crucial studies, Leo Loeb refers to graft rejection (the cellular reaction against transplanted tissue) as expressing the "individuality differential," the common denominator that distinguished one individual from another. Such "differentials" were regarded as genetically determined. Loeb, in a wide overview, concluded that "it is not organ differentials which determine these injurious reactions of the host cells toward the grafts, but the individuality differentials" (ibid.), that is, the commonly shared determinants of an individual that defined identity. Based on serological and transplantation studies, Loeb clearly identified immune mechanisms as controlling individuality. This was certainly not a novel insight. Graft rejection as mediated by immune cells had long been recognized. The role of immune cells in tissue rejection has a history dating to James B. Murphy's pre–World War I studies of transplantation-induced leukocyte reactions and spleen enlargement (Murphy 1926 [a compilation of papers, 1912–24]; Ford 1980; Klein 1986, p. 369; Silverstein 1989, pp. 275–304). He assigned the lymphocyte the predominant role in tumor graft rejection. After discarding Ehrlich's athrepsia theory, anatomic limitations of vascular supply, or rejection by antibody, Murphy correctly interpreted the cellular reaction as the basis of graft rejection. His primary evidence was the elimination of lymphocytes by treating laboratory animals with the newly discovered x-rays: "In the absence of these cells the foreign tissue meets with no check in its growth and with them [lymphocytes] even non-resisting organisms or organs, such as the chick embryo and the adult brain, resistance is practically complete" (Murphy 1926 pp. 27–8).

Burnet was well aware of Murphy's studies and cited them in the first edition of *Antibodies,* but the issue at hand is to contrast Burnet's 1937 view of selfhood, essentially autonomous from genetic considerations, with that of transplantation, where encoded individuality was consciously thought of in genetic terms. We must note that at the same time Burnet is vaguely referring to the self, Gorer is defining the genetic basis of such immune identity reactions. And Loeb, too, shares this view of the genetic basis of individuality:

> Certain experiments show that the similarity or difference between two individuality differentials corresponds to the similarity or difference in the composition of the gene sets in the host and donor, and that the host cells respond, so to speak, to genes which are strange to them. In reality, however, it is not the genes as such to which the host cells react, but the organismal differentials which develop in accordance with the gene sets (Loeb 1937, p. 3)

Loeb also noted (and Burnet was to address the issue) the specific role of embryonic development in tolerance reactions. Loeb cited how embryos do not reject transplants as rigorously as mature animals, and Burnet also emphasized this phenomenon, which he called *training* (Burnet 1941). Loeb's paper alerts us to the broader milieu of immunology in which Burnet wrote the first edition of *The Production of Antibodies* in 1941. There is a vague intimation in this work that genetic factors are the controlling elements that initiate immune reactions and, further, that these so-called differentials have a complex interplay during embryonic life that somehow programs later immune reactivity – a phenomenon variously called training or tolerance. Later, Loeb's *The Biological Basis of the Individual* (1945) is cited in the 1948 paper that Burnet and Fenner wrote as a prelude to their 1949 *Antibodies.* There is no direct evidence that Burnet knew of Loeb's 1937 paper. Loeb's 1945 treatise is an extensive expansion of the earlier paper, where the guiding theme, individuality differential, is defined in the same fashion: the "particular characteristic distinguishing one individual from another . . . common to all the various tissues and organs of an individual" (p. 5). From our general perspective what is particularly striking is how Loeb uses *individual* in contrast to Burnet's employment of *self:* "We apply the term '*individual*' to a human being to emphasize the distinctive unique features which such a person possesses. . . . Individuality may be conceived of as the original physical and psychical state of an organism, which has developed in accordance with the genetic constitution of this organism with the co-operation of a sequence of more or less fixed physical-chemical environmental conditions" (ibid., p. 3). Burnet would use the more elusive, and suggestive term *self* to deal with the same entity, but in so doing, he chose to follow his own lead and harvest its unique connotations.

By 1941 Burnet's agenda was a relatively straightforward combination of theoretical and experimental considerations. At the theoretical level the ecological conception of the self dominated his thinking. The organism was viewed as a system of order evolving from selective cooperation among its constituent elements, much like a climax community. In fact, on the ecological view, in addition to the organism as composite whole, each cell of the body could be understood as a self (and this, of course, is consistent with his amoeba analogy), with antigens upsetting the dynamics of cooperation and calling forth the resultant antibody formation. In a later paper the 1941 approach was characterized as follows:

> Antibody production must be regarded as in many ways equivalent to the development of an adaptive enzyme process. On this view immunological phenomena derive essentially from the intrusion of a foreign pattern into that nest of uniquely patterned and mutually tolerant enzymes, the living substance of the cell. . . . Within the living functioning cell their mutual destructiveness is curbed by some means which we cannot particularise, but which must be related in some way to their respective molecular patterns and to their spatial relationship to one another. . . . We assume that when an alien molecule or particle finds itself in such an environment it is somehow recognized as 'not-self' and is subject to destruction by appropriate enzyme systems. (Burnet and Fenner 1948, pp. 314–15)

This must be regarded as a more interactional or "mutually adapted" approach than a strictly genetically encoded one. To what degree one might invoke order-through-adaptation versus order-through-code as a dichotomy is problematic, but it is the telenomic emphasis that characterizes Burnet's orientation.

Beyond Burnet's relation to a genetic orientation, a second consideration concerns his view of antibody production. The motivation for writing *The Production of Antibodies* (as it was accurately reflected in the title) hinged especially on attacking the template theory of antibody formation rather than on defining selfhood. Burnet felt that the instructive hypotheses of the time were experimentally unable to deal with two facts: the logarithmic secondary response of antibody production and the persistence of antibody in the absence of antigen. The antibody secondary response, as noted in his autobiography (1969c), was recognized many years earlier by Burnet, ignored by others, and addressed as a crucial concern in this very monograph. This fact could not be explained by the instructive hypotheses, for as Burnet elucidated, such an approach would predict that "the successive liberations of antibody molecules should occur at approximately equal intervals" (Burnet 1941, p. 23). Referring to the different kinetic and quantitative characteristic of the primary and secondary antibody responses (ibid., Figs. 5 and 6, p. 14), Burnet made a prescient observation that would be revived sixteen years later as he modified Jerne's natural selection model:

> The production of antibody is quite certainly not a multiplication of antibody in the blood plasma, but a cellular phenomenon, and it is not easy to see any simple explanation for the observed course of production. . . . The simplest assumption would be that as each antigen molecule makes contact with a cell "sensitized" by previous contact with antigen, it sets going a change in the cell so that after a suitable latent period the cell liberates a series of antibody molecules. . . . The normal reason for a biological process showing a logarithmic rise of this type is because the entities concerned are either themselves multiplying or are being produced by multiplying agents. . . . The production of antibody is quite certainly not a multiplication of antibody molecules in the blood plasma, but a cellular phenomenon, and it is not easy to see any simple explanation for the observed course of production. (Ibid., p. 23)

Burnet would maintain the central role of cell biology (in this case cellular proliferation as controlling production of antibody) throughout his life. At this point, however, his views on cellular biology seem to be founded on ecological analogues rather than on strict scientific data. In any case, at this stage of his intellectual career this orientation is what chiefly distinguished him from the proponents of instructive theories.

Of crucial importance was Burnet's recognition that the persistence of antibody in the absence of antigen represented a serious problem for the template theory. A simple yet powerful solution was offered in his cellular hypothesis: "It is equally conceivable that once the antigenic stimulus has been applied, the cells continue to produce antibody long after the antigen has been completely disintegrated" (ibid., p. 42). This would be his chief point of contention with instructive theories, although ironically his own theory was heavily influenced by the template model. To account for the continued production of antibody, Burnet invoked an adaptive enzyme approach (ibid., pp. 58–64). This strategy was further elaborated between 1949 and 1956. Basically his explanation drew upon the dominant opinion regarding proteinase activity and enzymatic protein synthesis. Within the cell, such autocatalytic – and hence, as he referred to them, "living" (ibid., p. 61) – enzymes could produce either themselves or inert (nonreplicating) copies of themselves when needed. Antibodies were thus "something intermediate between the simple protein secretions and the living protein of the cell" (ibid.). He reasoned that when an antigen contacted and invaded the cell, proteinases would "switch" (much as bacteria could switch from one type of sugar fermentation to another) to a new type of activity that would produce antibody. Such a switch would persist for some time (hence the production of antibody in the absence of antigen) and permanently alter the cell so as to be receptive to future instances of the particular antigen contact. Moreover, such activity could be passed onto the cell's descendants:

> These proteinases, in virtue of their enzymic function, come into contact with any foreign antigens taken into the cell, and are lastingly modified by this contact. From analogy with other aspects of enzyme activity, this modification takes place simultaneously with the destruction of the antigenic particle. (Ibid.)

Turning to bacterial analogues, he continued:

> The modification of the proteinase unit which is produced by contact with the antigen is not to be regarded as resulting from a synthesis of a new unit in spatial contact with the antigen, but rather as a process analogous to the production of adaptive enzymes by bacteria. (Ibid., p. 62)

He then referred to "training" induced in such bacteria upon repeated subculturing:

> A soil bacillus was "trained" by repeated subculture in a medium containing polysaccharide as the sole source of carbon. . . . It is unfortunate that so little is known of the process of protein synthesis in bacteria, since some of the most spectacular results of "training" the metabolic processes of bacteria are concerned with the supply of nitrogenous material. . . . There has clearly been some deep-seated change in the enzymes responsible for protein synthesis, in this case one which is inherited indefinitely. (Ibid., p. 63)

Aside from the impact on his conception of tolerance, this view is, in short, an instructive theory modified to account for persistence of antibody. Adaptive enzymes could also be "trained," and so could the cells in their response to antigens, an analogy drawn from nervous system–muscular activity:

> Immunological reactions of every sort are the process of "training." All biological activities tend to improve with practice: this holds not only for reactions in which a complex neurovascular apparatus is involved, but

also for the simple feeding and avoiding reactions of protozoa. To take the very simplest form of living matter, even a virus can be rendered more efficient (i.e. virulent) by repeated passages on a given tissue. From the general point of view the immunological reactions of the higher inverte-brates are a specialized and efficient development of this inherent quality of living matter to respond to environmental simuli by more effective reaction. (Burnet 1940, p. 46)

Tolerance at this time was merely an immature immune response, and training involved a proper assortment of the self–other relationship that would have to be learned; this formulation is clearly akin to how we understand the mind to function. "Tolerance" is thus initially linked to a context of immaturity: "In some way the embryonic cells seem to be unable to recognize and resent contact with foreign material in the way adult cells do" (ibid., p. 45). It is therefore not surprising that no anti-body response takes place. Recall that this was written before the Owen dyzygotic twin experiment (1945), and in the corresponding 1949 text Burnet would draw out the implications of this experiment: "The im-portant implication of this work is that cells 'foreign' to the host may be tolerated indefinitely provided they are implanted early in embryonic life" (Burnet and Fenner 1949, p. 76). It is this notion of "encoding" the self that would distinguish the 1941 edition from the 1949 edition of *The Production of Antibodies.*

Although the notion of *self* does not appear in the first edition of *Antibodies,* it is implicit. Using the differentiating ability of the primi-tive amoeba to exemplify the universal property of selfhood in *Infec-tious Diseases* (Burnet 1940, p. 29), he eschews the term when using the same example in *Antibodies:*

An amoeba ingests a smaller living organism, and is able to digest the major part of it without damage to its own protein, a leucocyte takes in bacteria and either kills or disintegrates them or is killed and supplies pabulum for their growth. Any conception which makes such interac-tions clearer is worthy of consideration until some better alternative ap-pears, for there can be little doubt that the whole subject matter of immu-nology is founded on this intolerance of living matter for foreign matter. (Burnet 1941, p. 60)

He was clearly uncomfortable with the use of the term *self* at this point, but its shadow remains as a view of cells enzymatically harmonized internally, whether viewed as an ecological community, or as a more abstract "pattern."

In the second edition of *Antibodies* there is a conceptual shift from what heretofore has been referred to as the "ecological" self to the "marked" self. When did this occur? At the end of the 1949 book Burnet pondered how to corroborate the self-marker hypothesis by widening the observations made by Owen. One approach was to discover substances that would turn host cells into antigenic stimuli. Then he recalled that "it has been shown in this laboratory that a new antigen is produced on red cells by the action of influenza viruses or a vibrio enzyme [(Burnet and Anderson 1947)] and this may indicate a possible general approach to the wider question of self-markers on expendable cells" (Burnet and Fenner 1949, p. 129). However, in that 1947 paper there was no mention of such an application; so even if Burnet was surmising about self-markers, he left his musings unpublished. In a concurrent paper read in August of 1947 he stated:

> I think you will now see how the argument has come back to the problem of self and not-self with which I started. When an alien protein molecule finds itself within a living cell, it is in the midst of a coordinated series of enzymes, largely proteinases, whose mutual destructiveness is curbed by some system of specific checks that we cannot particularize, but some of which with or without preliminary adaptation can destroy the alien molecule. (Burnet 1948, p. 31)

Notable not only for the first use of the term *self* without quotation marks, this passage also shows that Burnet was still entertaining an ecological view of the self. In fact, the first discussion of the self-marker concept occurs in a coauthored paper with Frank Fenner in 1948, entitled "Genetics and Immunology." This crucial article, the prelude to the theoretical sections of the 1949 text, was actually even more comprehensive than its successor. It provides a revealing glimpse of how Burnet's thinking regarding the nebulous concept of the self was jolted into a new orientation by a modified understanding of genetics. Based on the discussion above, Burnet's 1941 views of genetics left its potential contributions subordinate to the concerns of a committed cell biologist. But by 1947 what had remained quiescent in the background was to be explicitly asserted.

Burnet was extremely well informed about contemporary genetics, and he attempted to apply current theory to search for the genetic basis underlying antigenicity. He even wrote of a "one gene–one antigen" relation (Burnet and Fenner 1948, p. 301) and remarked:

> This has led to the suggestion, supported by Haldane (1942), Sturtevant (1944), and others, that the molecular pattern responsible for antigenic specificity might be traced back to the gene itself, and that in the last analysis the "code" by which phenotypic characters are represented in

the chromosomal mechanism is equivalent in some way to the antigenic determinants which confer specific serological differences. (Ibid., p. 316)

He was now fully concerned with foreign "patterns," not just alien activity. As he wrote:

> ... it is of greater interest, and of potentailly greater practical importance, to attempt to elucidate the process by which foreign molecular patterns give rise to characteristic immunological responses. . . . The present approach is . . . largely influenced by the rapidly developing new conceptions of gene and enzyme action. (Ibid., pp. 307–8)

Therefore, by the late 1940s, in Burnet's view, order and selfhood no longer simply arise from a climax community, but have become linked to the genetic code of the individual. How this code might program immunological phenomena and allow for individual experiences was highly problematic, but the tide had turned.

This new orientation reflected an ambitious program. Burnet was attempting nothing less than to position the science of immunology prominently alongside genetics and developmental biology. His wide range of reading in genetics and his fascination with Beadle's biochemical genetic approach are impressive. Basically Burnet was attempting to broaden the scientific purview of immunology beyond the earlier somewhat restricted scope. He would thus seek "a combined biochemical, genetic, and immunological study of some convenient bisexual micro-organism such as Neurospora" (ibid., p. 320). In *Antibodies* he articulated a clear agenda:

> Until the various levels of biological thought can be brought together in terms of a unified system of concepts each sub-science has the right to introduce its own generalizing concepts. Immunology has through most of its history been remote from the general stream of biological discovery and generalization. This attempt to bring it more directly into that stream is equally an attempt to show that immunological phenomena and interpretations must be given due weight in any future formulations of the nature of living process. (Burnet and Fenner 1949, p. 133)

The ambition to introduce genetic mechanisms into these theoretical aspects of immunology were in fact not to be realized for another decade (Lederberg 1959; Burnet 1967). In any case, turning to the specifics of antibody production, Burnet took his vague 1941 proteinase theory and fleshed it out with all the evidence of adaptive enzymes and cytoplasmic inheritance discussed in the mid-1940s (Sapp 1987). Cytoplasmic hereditary units (killer, plasmagens, etc.) were being discovered and discussed in the 1940s, and Burnet was aware of these find-

107

ings. He quoted from Tracy Sonneborn and Sol Spiegelman, and was especially impressed with Carl and Gertrude Lindegren's work on "adaptive enzymes" (see Burnet and Fenner 1949, p. 88). Based on cytoplasmic heredity, Burnet could strengthen his proteinase theory into a theory of immunology as "soft genetics," of heritable changes (at least from cell to cell, if not from organism to organism) amongst cells within the lifetime of an individual:

> The weakness of such a formulation centres on the postulated intracellular enzymic unit and its capacity to undergo adaptive modification. This goes far beyond what is known of classical enzyme chemistry. There is no well studied instance of an enzyme with a specificity of action of a type comparable to antigen–antibody specificity but equally there are not present modes of studying such enzymes did they exist. There is a growing impression that self-replicating systems of the type postulated are an essential feature of cytoplasmic structure. (Burnet and Fenner 1949, p. 127)

It is thus reasonable to assess that Burnet was proposing to make immunology the third leg of fundamental biology, alongside genetics proper and developmental biology.[5] At the very least, he was undergoing an expansive scientific transformation. The horizons of his theorizing had noticeably widened.

Returning to the main theme, tolerance, it is generally acknowledged that the Owen experiment had a major impact upon Burnet, as did related experiments. Burnet's earlier understanding of tolerance as hyporesponsiveness had clearly changed to the notion of tolerance as nonreactivity acquired during embryonic development. Thus, he now had to model a self defined *during* embryonic development, programmed to modify its identity, that is, preadapted to respond to any antigen present during the crucial stage of embryonic life as self, and finally to become "hardened" after birth. He now also would replace the vague ecology of the self with one grounded in an a priori genetic patterning. It is a fundamental shift, even if somewhat anticipated in 1941. To this end, Burnet developed his "self-marker" hypothesis and applied it ontogenetically. The paper, "Genetics and Immunology" (Burnet and Fenner 1948), served as an important prelude to the 1949 *Antibodies* (also coauthored with Frank Fenner). After arguing that self-markers (encoded by a small number of genes) must identify host constituents,[6] Burnet and Fenner propose a mechanism for antibody production:

> Within the potential antibody-producing cells there are enzymic groups adapted genetically to "fit" a sufficient number of marker constituents to

allow differentiation of "self" from "foreign" (i.e., normally invasive micro-organismal) organic material. When such a "fit" occurs, normal disintegration processes go on without stimulation to immunological activity. An antigen is such in virtue of the fact that it corresponds in its general chemical character to one or other of the self-markers, but will only fit the corresponding enzyme when this is deformed to an appropriate adaptive configuration. It is this deformation which provides the stimulus for replication and antibody liberation. (Burnet and Fenner 1948, p. 318)

In other words, enzymes are present which are charged with recognizing and removing effete cells. To do so, they acquire a fit to these cells' markers and upon contact with such a marker "peacefully" destroy the cell without liberating antibody. However, when they come across a foreign marker, the enzyme must be deformed to attach to the antigen; antibody proliferation would then occur. In *Antibodies,* published the next year, Burnet and Fenner reiterated their hypothesis:

> It is our contention that the primary units (adaptive enzymes) on which antibody production is based are modifications of enzyme systems primarily adapted to specific adsorption to one or other of the self-marker components of the body cells. When, and only when, specific adsorption occurs the marker is enzymatically destroyed without disturbance of intracellular equilibrium. Keeping the blood group polysaccharide as our typical marker we can consider what happens when a chemically related but structurally dissimilar polysaccharide, e.g. from a foreign red cell or an invading pneumococcus, enters the cell. The same general type of enzymic action is called for but the new substrate does not quite fit the specific adsorptive pattern of the enzyme. If the pattern of the enzyme can be adapted to fit the substrate configuration then the primary unit of our hypothesis has been established and the process leading to antibody production set in train. On this view there are as many potential types of antibody as there are enzyme units tuned to particular marker components in those cells capable of initiating the production of antibodies. (Burnet and Fenner 1949, p. 101)

The 1948–9 self thus arises from complex cellular "learning" and "communication" based on a marker system. The dominant and governing problem remains the crucial role of the marker system, as yet undefined. Note that the hypothesis regarding antibody synthesis remains closely aligned to a template model. As further elaborated in *Enzyme Antigen and Virus* (written in 1955), Burnet explained: "[I]n Haurowitz's view specificity is determined by a controlled folding of the polypeptide chain against a secondary template. . . . This point of view is in many ways not very different from our own" (Burnet 1956, p. 66). The revision pertains to the role of RNA, and Burnet even at this late date was valiantly trying to incorporate the earlier template theory in a dynamic cellular model, while recognizing the shortcoming of Haurowitz's view to account for (1) the failure of embryonic animals to produce antibody, (2) prenatal acquired tolerance, (3) the difference between primary and secondary antibody response, and (4) the persistance of antibody production after disappearance of antigen (ibid., p. 67). "We contend that it is of the essence of the problem of macromolecular pattern relationships that orthodox chemical principles are inadequate to deal with questions of pattern tranfer and replication" (ibid., p. 68). The gauntlet had been thrown seven years earlier, and it was again reiterated: A biological theory was required.

Tolerance is given fairly short shrift in "Genetics and Immunology." Although it is noted therein with respect to the Owen findings that "cells 'foreign' to the host may be tolerated indefinitely provided they are implanted early in embryonic life" (Burnet and Fenner 1948, p. 300), they are clearly writing for geneticists. The section on self-markers of the second edition of *Antibodies* (Burnet and Fenner 1949, pp. 100–2) is presented here, but the section on the development of the markers in embryonic life (ibid., pp. 102–5) is totally neglected. Presumably this strategy reflected the approach to two different audiences, namely, geneticists and immunologists. In any event, tolerance is the central issue of the 1949 book, and the self-marker hypothesis is this work's critical insight. The view presented in 1949 is a far different type of tolerance than that of 1941. Tolerance is no longer a lack, that is, inadequate training; it is now the active ability to recognize "self" in embryonic development and to freeze this recognition pattern thereafter. Burnet also provides a mechanism for such recognition, linking his adaptive enzyme approach with tolerance as seamlessly (and imaginatively) as possible:

> Taking the red cell as the typical expendable body cell we may assume that in embryonic life phagocytosis and disintegration of worn out cells is actively taking place. During this phase appropriate intracellular enzyme systems are being adapted to deal effectively with those components which need to be broken down for reintegration into the metabolic activities of the body. . . . An adaptive enzyme specifically uniting with and initiating the destruction of the component is brought into being in the cells that destroy erythrocytes. This newly patterned enzyme becomes stabilized as part of the inheritable structure of these cells and is trans-

mitted indefinitely to their descendants. With the development of the free-living state this lability of intracellular enzymes is lost, the patterns engraved during embryonic life harden as it were and become permanent possessions. (Ibid., pp. 102–3)

In other words, for all the genetic encoding of the self-markers, the self is actively defined by the enzymes during embryonic growth. They are flexible enough at this stage to adapt to any antigens present (of course, most of these antigens are from the host and marked genetically), but antigens deriving from a foreign genetic system are likewise just as well accepted. At a certain stage in embryonic growth, this lability is hardened and the recognition ability of the self is frozen. What then remains labile is the population of the antibody-producing cells, themselves modified "genetically" – in the cytoplasmic sense (Sapp 1987) – during the life of the organism. With respect to such populations:

> . . . if we take a lymph node subjected to an intense antigenic stimulus so that virtually all macrophages lining the sinuses take up the antigenic particles, our interpretation would picture the transfer of a potential antibody-producing stimulus from such macrophages to most or all of the stem cells in the node. Many of these would proliferate and develop toward plasma cells from which most of the acutely produced antibody would derive. Others would give rise to lymphocytes in which cytoplasmic activity was much less marked. Still others would remain quiescent or give rise to cells like themselves. We can conceive no way in which antibody-producing power can persist far beyond the individual life of the "end-cells" which produce it unless the capacity is in some way implanted in the mother cell type from which the expendable cells are descended. (Burnet and Fenner 1949, p. 99)

The simple template theory is discarded: "We would reiterate our contention that in the dynamic constantly changing population of spleen or lymph node there is no storage place for either intact antigen or any hypothetical 'template' derivative" (ibid., p. 128). And more fundamentally, the self is declared to be only loosely encoded genetically. In fact, in his follow-up book, *Enzyme Antigen and Virus,* Burnet would write that "the significance of 'self' and 'not self' in relation to immunity is the basis of the whole approach. . . . The recognition of self is something 'learnt' during embryonic life and not genetically ingrained" (Burnet 1956, p. 67).

In fact, there is a fair degree of equivocation around the idea of the genetically encoded nature of selfhood. Judging from the paper "Genetics and Immunology," we might believe Burnet to be quite committed to a genetic orientation, but in subsequent writings, he qualified this view. First, concerning cell types in the immune response, Burnet and Fenner state:

> We would regard the lymph node as a repository of highly reactive primitive mesenchymal (reticulum) cells which under appropriate stimulation can proliferate and take on one or more of several functions. The function adopted will determine the histological appearance of the cells involved. There is a subtle but probably real difference between this statement and the more orthodox one that proliferation of reticulum cells gives rise to genetically determined cell types each with specific functions. (Burnet and Fenner 1949, p. 97)

Thus, the dynamic population of the self arises through process and is not defined a priori. Second, and almost rhetorically, Burnet reverts to a view of the self on a cellular-ecological model of the pre-1948 period:

> Weiss (1947) has even gone so far as to speak of molecular ecology by which the cell is viewed as a complex population of molecules and molecular groups. The cellular organization then becomes the resultant of interaction, competition, selective increase (proliferation) and regrouping of the units. (Ibid., p. 93)

Likewise, concerning Traub's famous lymphocyte choriomeningitis experiments, he wrote: "At the close of his studies the infected community of mice appeared to have reached a virtual state of symbiosis with the virus" (ibid., p. 104). These are organismically oriented issues, and in the ensuing development of the self concept Burnet again designates genetics a subordinate role. He was much more comfortable dealing with a dynamic self, as opposed to a genetically determined one.[7] Perhaps a fair comparison between the self of the second edition of *Antibodies* with that of the first edition is that the 1941 self resembles a Skinnerian model of language acquisition, with "training," evolving adaptive systems under constant stimuli, etc., whereas the 1949 self is much like a Chomskian model, with a strong degree of structure built in (namely, genetics), but with flexibility for a limited time despite such hardwiring.

Retrospective analysis perhaps has a propensity to attribute extraordinary creative insight to particular discoveries or hypotheses that only later are recognized as serving as the germinal nidus for major advances. Its signal importance is not doubted, but how novel was the 1949 edition of *Production of Antibodies*? This is not a subject of my concern, but I am struck by Paul Cannon's general opinion of immunology in 1949 in juxtaposition to Burnet's work. Cannon was president of the American Association of Immunologists in the early 1940s, a member of the National Academy of Sciences, and a senior scientist among those who were examining the immune response. When a young doctor approached him regarding his opinion of embarking on a career in immunology, Dr. Cannon said, "No young man with aspirations for becoming a scientist should stake his future in immunology; all of the important questions in immunology have been answered" (Rowley 1991, p. 27). Whether this was an assessment of immunology's impact on public health (i.e., immunization) or was a rather shallow opinion concerning the theoretical and experimental sophistication of the science need not be decided here, but I suspect it was a generally accepted view and reflects the assessment of this 1910–50 period as the "Dark Ages" of immunology (Silverstein 1989). In any case, Cannon clearly was oblivious to those issues that would soon spark an exciting acceleration of progress in the field. A careful and critical analysis of Burnet's originality in the broader context of related theorizing has yet to be done, but I believe the scientific genealogy outlined above reveals Burnet's synthesis of disparate issues into a comprehensive theory is stamped by a unique intuition. We readily discern the antecedents of the self-marker hypothesis, but it is the weaving together of diverse data and themes, together with anticipating the theoretical implications of the theory that marks *Production of Antibodies* of 1949 as a seminal and novel work.

C. THE CLONAL SELECTION THEORY

Support for Burnet's theory of the self soon followed *Production of Antibodies* (1949). The self-marker hypothesis was initially tested by Medawar and his co-workers, demonstrating acquired immunological tolerance to skin grafting by exposing embryos (mice and chickens) to foreign but homologous cells (Billingham, Brent, and Medawar 1953; 1954). Others followed up this research (Silverstein 1989, pp. 177–8), but an elaborated theoretical formulation was pursued by Burnet, who proposed an explanation based on what he called the clonal selection theory (CST) (Burnet 1957; 1959). Immune recognition selects lymphocyte clones capable of producing antibody or fulfilling cell-mediated immunity. It was central to this hypothesis that deletion of self-reactive clones protected the body from what Ehrlich had called *horror autotoxicus* (Ehrlich and Morgenroth 1901, 1957, p. 253) Naturally occurring antibodies against host tissue were described at the turn of the century

The immune self

and generated theoretical discussions concerning autoimmunity (Silverstein 1989). Ehrlich viewed such immune activity as "dysteleological in the highest degree" (Ehrlich and Morgenroth 1901, 1957, p. 206) and devoted considerable attention to its implications. (Although he used the terms "organism" and "own body," Ehrlich did not employ *self* in his immunological writings. For that matter, neither did Metchnikoff.) Antigens not catalogued as self were designated nonself and would be capable of stimulating an active immune response. David Talmage independently proposed that the selectable element in the immune system was the immunocompetent cell: "As a working hypothesis it is tempting to consider that one of the multiplying units in the antibody response is the cell itself. According to this hypothesis, only those cells are selected for multiplication whose synthesized product has affinity for the antigen injected" (Talmage 1957, p. 247). Support for CST was soon forthcoming: (1) Individual antibody-producing cells in culture made only one antibody (Nossal and Lederberg 1958); (2) newborn and fetal animals have immunoglobulin determinants on circulating lymphocytes (Binns et al. 1972); (3) destruction of spleen cell response to one antigen had no effect on response to another (Dutton and Mishell 1967); (4) antigen aggregated all the surface immunoglobulin on antigen-binding cells, indicating single specificity (Raff, Feldmann, and dePetris 1973). Although the theory was initially controversial, it has now become immunological dogma not only for B cells (antibody-producing cells), but for T cells as well (Klein 1990, p. 8). As Klein put it in his textbook, CST "is no longer a theory but a fact" (ibid., p. 335).

CST arose from combining the concepts of tolerance enunciated in *The Production of Antibodies* (1949) with Jerne's natural selection theory of antibody formation proposed in 1955. Serious attention has been paid to the genesis and critical modification of CST, and the principal scientists involved have published their own recollections (Talmage 1986; Ada 1989; Nossal 1989; Silverstein 1989; Moulin 1991a; Schaffner 1992). Burnet, in a lengthy autobiography, called CST his greatest scientific achievement (Burnet 1969c). *Enzyme Antigen and Virus* is an interesting but neglected source regarding this matter (Burnet 1956). Basically this book, though published in 1956, had been completed in September 1955, two weeks before Delbrück submitted Jerne's manuscript on the natural selection theory of antibody selection to the *Proceedings of the National Academy of Sciences* (Jerne 1955). *Enzyme* tackles two issues: The first, most general theme was Burnet's attempt to integrate all of biology into a system dependent upon pattern recognition and pattern production. As suggested in the subtitle, *A Study of Macromolecular Pattern in Action,* he was ambitiously attempting to reduce much of biology to a study of such patterns. He wrote of antibodies:

> ... it was postulated [in *Antibodies* (1949)] that the self-recognition system was developed during embryonic life. Any antigenic pattern that reached what we called the scavenging cell system before a certain critical point around the time of birth of hatching, would be accepted as "self" and in subsequent life its re-entry into the body would not provoke antibody production. (Burnet 1956, p. 41)

Second, and most important, the theoretical grounding of the antibody chapter did not change in its fundamentals from the 1949 book. Burnet considered Chapter 2, "Antibody Production," his revision of the 1949 monograph (see the preface [ibid.]). Burnet clearly admitted that the theoretical formulation was essentially unchanged: "This formulation of antibody production differs from that of Burnet and Fenner's 1949 presentation only in the adoption of a theory of protein biosynthesis more in line with recent work on adaptive enzymes" (ibid., pp. 64–5). (*Enzyme* is again discussed in a different context in Chapter 5.)

Thus, Burnet was no nearer CST than seven years earlier. It may be conjectured, then, through a reading of *Enzyme,* that it is unlikely that Burnet would have proposed CST without Jerne. Burnet and Jerne shared many of the same scientific influences and interests: Population thinking, Ehrlich's protoselection theory, the problem of antibody avidity, and the problem of deciphering the significance of natural antibodies. Although these elements were there, Burnet was wedded to a model he had been developing for sixteen years, and he was in a certain sense "stuck." Why Jerne, and not Burnet, seized upon the natural selective basis for immune recognition is an intriguing question. (See Söderqvist [1994] for an analysis of Jerne's discovery.) First, as has been noted, Burnet always thought of antibodies in terms of populations. Even in 1941 he had written: "[I]t is now well recognized that the population of specifically reactive antibody molecules in a given immune serum is not homogeneous" (Burnet 1941, p. 25). In 1948 he reiterated this idea more fully:

> It has, however, been shown in several instances that the population of antibody molecules which is produced by a single antigen A is not uniform, but contains a proportion of antibody which can be differentiated by its behaviour on absorption with related antigens A', A", etc. (Burnet and Fenner 1948 p. 292)

I have already cited relevant statements from *Antibodies* of 1949. Such a population orientation, then, reflects a deep commitment to ecological thinking that would presumably predispose him to a theory allocating natural selection a crucial role. Second, Burnet was certainly aware of Ehrlich's work, having mentioned him in *Enzyme* on at least two occasions. Early in the text he wrote: "The basis of antibody production has been widely discussed since the days of the side-chain theory of Ehrlich" (Burnet 1956, p. 40). Far more important, he later placed Ehrlich as the progenitor of all "pattern thinking" in biology, from immunology to molecular biology: "The approach to biology by the use of the concept of specific complementary pattern has been more or less consciously developing since the time when Ehrlich first put forward his ideas on immunological specificity" (ibid., p. 171). Third, Burnet was certainly aware of the avidity problem, although he lacked the extensive acquaintance with it that Jerne had:

> References to variations in avidity, range of cross reaction and proportion of monovalent antibody obtained from the same antigen according to species, age of animal and course of immunization will be found in Burnet and Fenner (1949). This again is biologically inevitable where we are dealing with complex patterns carried by molecules which, though functionally equivalent, are by no means necessarily identical in detail. (Ibid., pp. 72–3)

Burnet fit this into his own scheme, but it is interesting to compare how Jerne developed the same data. Finally, Burnet was certainly aware of "natural antibodies," namely, isoagglutinins, and he extensively discussed them (Burnet and Fenner 1949, pp. 119–21). "The pseudo-immunological patterns carried by normal antibodies, including the isoagglutinins," he wrote, "can only be ascribed to the 'accidental' presence of a configuration complementary to the antigen concern" (ibid., p. 120). However, Burnet saw this as a strange quirk of the human immune system and not as a fundamental problem to be solved. In his 1957 response to Jerne, he refers to Jerne's ability to deal with such "normal antibodies" as the red cell agglutinins as a check in its favor over the Burnet and Fenner model.

Jerne's recollection of how he devised the natural selection theory of antibody formation emphasized its roots in the bacteriophage research of the early 1950s (Jerne 1956, 1966a). In the *festschrift* for Max Delbrück that chronicles the Phage Group, Jerne wrote the following:

> I do not know whether reverberations of Kierkegaard contributed to the idea of a selective mechanism of antibody formation that occurred to me

one evening in March 1954, as I was walking home in Copenhagen from the Danish State Serum Institute to Amaliegade. The train of thought went like this: the only property that all antigens share is that they can attach to the combining site of an appropriate antibody molecule; this attachment must, therefore, be a crucial step in the sequences of events by which the introduction of an antigen into an animal leads to antibody formation; a million structurally different antibody-combining sites would suffice to explain serological specificity; if all 10^{17} gammaglobulin molecules per ml of blood are antibodies, they must include a vast number of different combining sites, because otherwise normal serum would show a high titer against all usual antigens; three mechanisms must be assumed: (1) a random mechanism for ensuring the limited synthesis of antibody molecules possessing all possible combining sites, in the absence of antigen, (2) a purging mechanism for repressing the synthesis of such antibody molecules that happen to fit to auto-antigens, and (3) a selective mechanism for promoting the synthesis of those antibody molecules that made the best fit to any antigen entering the animal. The framework of the theory was complete before I had crossed Knippels-bridge. I decided to let it mature and to preserve it for a first discussion with Max Delbrück on our freighter trip to the U.S.A., planned for that summer. . . . Immunology was not then an "in" subject, and I had to apply antibodies to bacteriophage in order to hang on to the fringe. My avidity observations strengthened my faith in the truth of antibody selection. Antibodies produced by an animal against one antigen appeared to increase in "goodness of fit" during the course of immunization. This was true both for antitoxin and for anti-T4 antibodies. The phenomenon had Darwinian overtones. (Jerne 1966a)

Söderqvist (1994) has critically reviewed this *eureka* story and has found it an inadequate account. The elements are all there: the experimental background of the 1940s and early 1950s, Jerne's work in serology, his dissertation on the avidity phenomenon, his acquaintance with the phage group, and his subsequent adoption of the bacteriophage as a tool to examine antibody–antigen kinetics. Less evident are the other factors: Jerne's demonstration of an antibody in normal serum as a "natural antibody" (a critical precondition for a selection theory of antibody formation), his negative attitude to template theories, his knowledge of Darwinian selection theory (and Ehrlich's side-chain theory), and finally, his long-term interests in biostatistics, physical chemistry, and random molecular movements.

In a 1992 letter Jerne described this story more fully. He had just finished his doctoral dissertation on the neutralization of diphtheria toxin by antitoxin serum, where he showed that neutralization was re-

versible by simple dilution. He then switched his attention to the inactivation of bacteriophage. (James Watson and Gunther Stent had joined
Jerne in Ole Maaloe's laboratory at the Danish State Serum Institute in
1950–1, bringing with them the bacteriophages.) The inactivation of T4
by anti-T4 serum had been found by Delbrück and Alfred Hershey to
be a one-hit affair, like the decay of radioactivity. They believed one
antibody molecule was needed for inactivating a T4 particle, and that
it was a reversible process. Jerne doubted both claims. He first established that the rate of T4 inactivation was dependent on the solution's
saline content (1000-fold faster in 0.001 normal saline than in 1 normal), but then he discovered that phage inactivation did not occur in
distilled water. He went on to prove that inactivation was due to the
kind of antibody present, namely, IgG found in hyperimmunized sera
and IgM from early immunization, whose titre was much lower. With
these studies as background, in 1954 (before joining Delbrück at Cal
Tech), Jerne immunized a horse, called B, with only one injection of
1013 T4 phages and obtained preimmune (B0) and sequential sera (B1,
B2, B3, etc.), every day for a week. The early sera did not inactivate T4
but abolished the tryptophan requirement of T4.38, which he called P*.
He ascertained that the P*-inducing property of these early sera was
due to an anti-T4 antibody, probably IgM, and then showed that the
rate of P* formation by these sera required four molecules of this early
antibody per phage to transform native T4 to P*:

> Moreover, (and this deeply impressed me), even serum B0 and the sera
> of all species of non-immunized animals (mice, rats, rabbits, birds, etc.)
> contained the P* inducing antibody in their globulin. It turned out that
> the concentration of this P* forming antibody in all such normal sera
> was low: these sera could not act on more than 10^8 T4 phages per ml.
> Suggesting a "normal" P* inducing antibody concentrations of perhaps
> 5×10^8, or at least less than 10^9 antibody molecules per ml. As normal
> serum contains more than 10^{16} globulin molecules per ml., I came to the
> generalized idea that normal serum globulin is a mixture of antibody
> molecules of a million or ten million different specificities, and that anti-
> gen used for immunization selects the fitting specificities present in this
> mixture, and brings these fitting antibodies to a system of cells that can
> reproduce them to large amounts. This is the "Natural Selection Theory
> of Antibody Formation." My manuscript with this theory was sent to the
> PNAS by Delbrück in 1955. At the same time, in Pasadena, I wrote up
> my manuscript on the formation of P*, with the Copenhagen experimen-
> tal results from which my Natural Selection Theory resulted. But this
> paper went to the *Journal of Immunology,* published in 1956 [Jerne 1956],

though its contents actually preceded the 1955 PNAS paper that contributed to my Nobel prize. (Jerne 1992)

As Jerne noted in this paper, there were two possible interpretations of the P* antibody in normal serum: "Either that they were spontaneously produced by the animal, or that practically all normal animals have been exposed to and have responded to T4 antigen" (Jerne 1956). But Jerne preferred the presence (at a ratio of less than one to one million) of a spontaneously reactive (natural) antibody. As Söderqvist writes (1994), natural antibodies had already been noted in the serological literature, but the template theory of antibody formation, which argued that antibody was formed de novo to fit the stimulating antigen, regarded such antibodies as anomalous. Jerne intuitively rejected Pauling's template model and sought evidence to refute it. Jerne thus believed that the natural selection hypothesis was built on this disgruntled attitude concerning antibody selection and his extensive prior interests in statistics and the role of random selection in biometrical phenomena.

Remaining steadfast to the task of tracing the self concept, I offer only summary treatment to the history of CST and thus need only note Burnet's major critical modification of Jerne's insightful hypothesis. Jerne proposed (1) that antigen selects among the circulating "natural antibody" pool, consisting of random specificities, those with appropriate affinity, and they are then delivered as a complex to cells capable of producing immunoglobulin; (2) upon delivery the antigen is released and induces the synthesis of "specific antibodies," identical to those from the original selected pool; (3) the continued presence of antigen results in finer selection and increased titres of antibody, and when cleared, the selected immunglobulins slowly revert to an earlier equilibrium; (4) those antibodies that "attach themselves to structures in the body of the animal itself will be removed and will therefore not be available for reproduction" (Jerne 1955). The notion of self does not appear in the paper. But there was a crucial flaw in Jerne's hypothesis, as Burnet reminisced:

> Jerne postulated that gamma globulin molecules are continuously being synthesized in an enormous variety of different configurations. The origin of the diversity is left unexplained. When an antigen intrudes into the body, sooner or later globulin molecules of the appropriate natural pattern will become attached to the antigenic molecules or particles. The complex is then taken up by a phagocytic cell where, by hypothesis, the globulin can be released from the antigen. Such globulin molecules either in the macrophage or after transfer to another cell were said to serve as

119

a 'signal for the synthesis or reproduction of molecules identical to those introduced, i.e., of specific antibodies.' Even in 1955 this seemed wholly inadmissible. Most other aspects of the new theory were highly acceptable but the basic flaw seemed to be a fatal one. (Burnet 1967)

Around this time Burnet became interested in the Simonsen phenomenon, where the inoculation of adult allogenic spleen cells into fowl embryos and newborn mice leads to runting (Simonsen 1957). This graft-versus-host (GVH) disease was characterized by multiple lymphocyte lesions in the thymus and spleen, and the so-called Simonsen spots on the chorioallontotic membrane of the chick embryo were attributed to proliferating lymphocytes reacting immunologically. Burnet reasoned that autoantibodies might be produced by proliferating lymphocytes. In 1955 Carleton Gajdusek was working to identify the hepatitis virus in Burnet's laboratory. In seeking to isolate the virus, he used sera from normal and recovered patients, which he incubated with normal and damaged liver tissue. Some sera reacted against normal liver, the strongest reaction coming from a patient with a lymphoproliferative disorder (Burnet 1969c, p. 217). In conjunction, these two lines of inquiry (on lymphocyte-mediated GVH disease and autoimmune reactions) suggested that lymphocytes were key immunocompetent cells; that is, they were deployed in both cell-mediated immunity and antibody generation.

In fact, the evidence was growing that the lymphocyte plays a dual role, one, in immunity-producing antibody, and two, in mediating cellular reactions against transplantation grafts and certain microorganisms. Burnet would write in September 1955 that although plasma cells were known to produce antibody (these were later described as further developed lymphocytes), the function of lymphocytes was "not so clear" (Burnet 1956, p. 52). However, at the same time, he cited several recent studies supporting their capacity to generate antibody. In short, the assignment to the lymphocyte of a central role in immune function – both in recognition and effector arms – was quickly gaining support by the mid-1950s. The lymphocyte was becoming the Rosetta Stone that would finally fuse the humoral and cellular branches of immunity. Lymphocyte biology merges with the MHC story since the T-cell lymphocyte is responsible for the majority of rejection episodes, although the exact mechanism is still unknown. The interesting historical question is why the lymphocyte did not become the focus of interest before the mid-1950s, especially as Murphy had documented its role in graft rejection forty years earlier (Murphy 1926; Ford 1980). The implicit explanation in my account is that quite simply the thrust of immunology between 1910 and 1950 was so firmly committed to a reductionist immunochemistry that the broader tissues of immunopathology, although recognized, did not capture sufficient interest. Certainly Burnet's credentials as a proto-immunologist, that is, his research as a virologist, represents another research tradition. In any case, the pieces finally fell into place: "Rather suddenly 'the penny dropped.' If one replaced Jerne's natural antibodies by the cells which produced them and applied a selective process in a Darwinian sense to the antibody-producing cells, the whole picture fell into shape" (Burnet 1969, p. 205). With CST, Burnet and Talmage placed the selection process at an earlier stage, namely, at the level of the antibody-producing cell itself. This provision then accounted for both humoral and cellular-mediated immune reactions with one simple mechanism. The third component, random somatic mutation, was supplied by Lederberg (1959). CST is thus based on the notion that immune recognition selects lymphocyte clones capable of producing antibody or fulfilling cell-mediated immunity. Burnet viewed his presentation of CST as his most important scientific achievement (Burnet 1969c, p. 190) and believed that it warrented the Nobel Prize. He wrote Jerne in 1984: "I have often thought that you and I should have had a joint award for putting antibody production on the right track rather than the one I shared with Medawar [for tolerance in 1960]. Anyway we are both now on the list" (Burnet 1984).[8]

In a general sense it is important to recognize that the conception of the evolutionary mechanism of natural selection as the force shaping antibody fit – and its application to CST as selection of antibody-producing lymphocytes – is another example of selectionist theory that has been broadly and often metaphorically applied in several disciplines in the context of "evolutionary epistemology" (Hahlweg and Hooker 1989; Radnitzky and Bartley 1987; Callebaut and Pinxten 1987). Evolutionary epistemology (of which natural selection is one modality) has rich pastures in philosophy and history concerning such diverse issues as scientific progress (e.g., Hull 1988), scientific and general epistemology, problems of self-reference, and theories of mind, rationality, cognition, and learning (Hahlweg and Hooker 1989; Radnitzky and Bartley 1987). Note that CST was never conceived as a metaphorical extension of natural selection (as evolutionary epistemology generally is), but would be a literal application particularly in the consequence of survival value. In late nineteenth-century biology, natural selection was invoked by Wilhelm Roux to explain embryological development as "the struggle of the parts of organism," reflecting the differential adaptedness of organs within the body in their competition for nutritional substances (Roux 1881). (For the reasons that Roux's theory must not be confused with Metchnikoff's, see Tauber and Chernyak 1991, pp. 121–3). Before Mendel was rediscovered, August Weismann proposed "germinal selection," which invoked competition among the units of heredity prior to phenotypic expression; on this view, the most fit "germs" would thus yield the most fit organism (Weismann, 1895). Moreover, Edward Thorndike proposed a selective theory of the formation of neuron connections that has been revived in modern neuroscience theory (Thorndike 1911; Edelman 1987a; Changeux 1985). Thorndike's "law of effect" is closely analogous to the principle of natural selection, presuming that there is a population of stimulus–response pairs, more or less randomly mated, and from this pool reinforcers select the adaptive pairs mechanically. The law of effect has been used to explain all kinds of adaptive behavior and learning.[9] In the case of immunology, recall that Ehrlich first proposed a selectionist theory for antibody production. As he originally envisioned the situation, antigen randomly selected "receptors" by chemical affinity characteristics, which upon appropriate stimulation would secrete excess receptors (that is, antibodies) into the circulation. The theory was rejected when Landsteiner demonstrated that antibodies were generated against synthetic antigens, substances which had never existed in the evolutionary history of the organism (Landsteiner 1933, 1945; Silverstein 1989, pp. 107–12; Piatelli-Palmerini 1984, 1991b). The core assumption that underlay the instructionist theories was that nature is economical and would not produce thousands, not to say millions, of molecular species of little use; in short, Darwinian selection favors efficient genotypes.

In conclusion, if we step back and broadly survey the history of immunology in the 1940s, we discern a general reorientation from a chemically dominated discipline to one in which organismically based issues begin to challenge the dominant scheme.[10] This must be regarded as a retraction from a strict reductionist ideal, and Burnet must be credited with leading the vanguard in this regard. The writing of the *Production of Antibodies* (1941) was prompted by the inadequacy of the template theory to explain the *biological* nature of antibody generation. From that concern the question of tolerance became increasingly important as the fundamental question of immunity. And of course, the mechanism of natural selection had to be acknowledged as governing the operation of CST. This contribution is obviously due to Jerne's insight, which was intended principally to address the significance of natural antibodies. Natural antibodies were recognized in the 1890s and proved to be a vexing problem to early immunologists, both from theoretical and from practical perspectives. Ehrlich, of course, incorporated them into his side-chain theory, but this explanation soon fell into disfavor, and by the 1920s they were widely, though not entirely, dismissed as theoretical impossibilities (Keating and Ousman 1991). Although the conceptual antecedents of CST may be discerned in Ehrlich's side-chain hypothesis, neither Jerne nor Burnet make explicit reference to him in their refashioned proposal; Talmage did, and Lederberg carefully traces CST's history back to Ehrlich (Talmage 1957; Lederberg 1988).

In retrospect, Burnet might have invoked Jerne's reasoning, but his interests were directed toward other matters. Irrespective of the immediate scientific genesis of Jerne's contribution, CST did not *rise* directly from it, but rather from the more fundamental theoretical and experimental research on tolerance. The true conceptual basis of CST is inextricable from an explicit concern to define the self, and as described above, the roots of the orienting theory reach further back than Burnet's interpretation of tolerance, the acknowledgment for the need for self-markers, and the more general concern to integrate immunology with genetics and developmental biology. I believe it not unfair to suggest that Burnet was a worthy successor of Metchnikoff in their shared expansive view of biology: Both held an orientation to evolutionary processes and had an underlying understanding of the organism as a product of development, one whose programmed (or genetically encoded) potential is nonetheless deterministically realized through a dynamic setting. From this encompassing vantage, the organism in its entirety was again the focal subject of scientific inquiry. In this setting, the self was introduced.

123

4

From theory to metaphor

A. THE IMMUNE SYSTEM

The preceding three chapters have traced the conceptual and scientific history of immunology. There has been limited reference to the broader cultural setting and underlying philosophical orientations in conflict during the discipline's development. In a sense, I have attempted to set the stage for the ensuing discussion, which seeks to explore the interface between the science and its governing philosophy. To do so, the self concept will be carefully delineated, both as theory and as metaphor. So, first, let us briefly review the prevailing theory of immunology a century after the humoral–cellular debate was at its peak. As already summarized, by 1908, when Metchnikoff and Ehrlich shared the Nobel Prize, the basic character of immune reactions were known.

There are basically two forms of immune recognition: natural and acquired immunity. Natural immunity is characterized by effector cells and soluble factors that do not require specific or prolonged induction for their functions and more specifically do not require opsonization (coating) of their target. Such coating of the pathogen with antibody and other serum proteins offers "handles" for attachment and en- gulfment by specialized "eating cells" – phagocytes. The encounter with a phagocyte serves as the host's initial cellular defense action. This can result in the destruction of the microorganism by a blood or tissue phagocyte, and if the encounter is with an antigen-presenting cell (APC), a more complex interaction occurs, whereby the antigen (that constituent which elicits an immune response) is "processed," initiating the immune recognition (i.e., lymphocyte network) process. This so-called presentation of antigen results in a cascade of activation events resulting in the immune response of lymphocytes that "recognize" the antigen, which initiates their participation as primary effectors, or in other cases as recruiters in a more generalized inflammatory reaction. T-cell activation thus involves three elements: the T-cell receptor (Tcr), the antigen, and the class I or class II major histocompatability complex (MHC) molecules to which the antigen is bound. Although it is as yet unresolved how this interaction occurs, three possibilities are entertained: Tcr recognizes either an MHC–antigen complex, an antigen-altered MHC, or antigen alone (in this last case the MHC molecule merely serves as an anchor to present the antigen for recognition). The first model is generally favored as the most likely. In the simplest scenario, a helper T cell binds to the macrophage through its MHC protein (class II), which has already bound the processed antigen, and as a result, interleukins (a group of molecules involved in signaling functions between lymphocytes, APCs, and other cells) are elaborated to initiate the proliferate phase of the reaction. A clonal selection occurs, as only those lymphocytes from clones that recognize the processed antigen are stimulated to divide. These include not only the helper T cells, but the T cells involved with the actual encounter with the pathogen, the cytotoxic T lymphocyte (CTL). These two lymphocyte classes are representatives of a complex lymphocyte network; some classes augment immune responses, whereas others may suppress the reaction.

The T lymphocyte response is thus also based on the selection of particular lymphocytes, each of which has an immunoglobulin-related receptor of varying affinity, or fitness, for the antigen. To reiterate, the MHC complex serves as a crucial (albeit not exclusive) signature of the self, and when antigens bind to MHC, the MHC–antigen complex is seen as altered from the native (normal) state. This new MHC–antigen complex then functions as the docking site for effector immunocytes (cytotoxic, or killer, lymphocytes) and their ancillary force (helper lymphocytes). Antigen is thus "processed," lymphocytes are "sensitized," and soluble and cellular mediators "stimulate" antibody production from a potentially vast library of "receptors" that constitute a reasonable "fit" for the original antigen. The process of stimulation and proliferation of a particular lymphocyte that matches its receptor to antigen is referred to as clonal selection. Current theory holds that heterogenous immunocompetence resides in the unispecificity of surface receptors on antigen-binding cells, which when bound to antigen, initiate mitosis and development of a clone that expresses the same receptor specificity. Among the progeny of this clone are differentiated plasma cells (derived from B lymphocytes) that synthesize and secrete antibody molecules of the same specificity and affinity as the receptor molecules on the original antigen-binding cell.

For a regulatory system to function in a coordinated and meaningful fashion certain rules for lymphocyte–lymphocyte interactions must be followed: (1) Lymphocytes release their message only after an appropriate stimulatory signal, and thus resting antigen-specific cells do not constitutively produce lymphokines; (2) activation of cells to produce lymphokines must follow some recognition event; that is, a specific activation signal is required; (3) target cells must be able to decipher the message; that is, the signal initiates an appropriate response; (4) only the target lymphocytes are activated. In some cases the data supporting these particular criteria are compelling and suggest strong cooperation between lymphocyte subtypes. The testimony of the T cell (in a yes/no format of stimulation) is crucial in identifying the antigen as foreign, because by CST, T cells are "educated" in the thymus, where cells reactive to host constituents have been eliminated or silenced. Note that there are common features of antigen recognized by T cells and antibodies, but the mechanisms are clearly different, and the molecular basis for the commonality is not clear, for it is possible that B cells may act as APCs and re-present, in association with MHC class II, those regions of the antigen that their antibodies recognize. There are thus multitudinous arrays by which antigen is recognized, presented, and responded to by several limbs of the immune network.

If we couple these immune-sensitized pathways for self-protection with natural (presensitized) immune mechanisms of the phagocyte, we discern the extraordinary complexity of the immune response. The phagocytes encompass two classes of cells: the monocyte/macrophage and the granulocytes (neutrophil and eosinophil, so called because of their staining characteristics, initially described by Ehrlich). The macrophage, the mature version of the circulating blood monocye, is a key regulator of immunity, having astoundingly varied functional roles. Macrophages initiate and control acute inflammation that involves other immune cells (e.g., mast cells) and humoral factors (e.g., the production of complement). In addition, macrophages have direct toxic affects against parasites and tumors. As a first-line defender of host integrity, the macrophage is poised to engage the lymphocyte for further immune destruction. Thus, for this discussion, the macrophage's most crucial function is to present antigen (as discussed above) and elaborate soluble stimulatory factors for T and B lymphocytes. The macrophage in large measure fulfills Metchnikoff's original hypothesis that these cells are at the fulcrum of the immune reaction. The neutrophil until recently has been regarded as a short-lived cell whose functions were limited to simple search-and-destroy missions directed at bacteria. In fact, neutrophils, like macrophages, synthesize and release immune mediators (interleukins and tissue necrosis factor) and hence modulate lymphocyte activity (Lloyd and Oppenheim 1992). In addition, there is evidence that the MHC class II complex is induced in stimulated neutrophils, suggesting that these cells may play a significant role in the afferent limb of immunoregulation (Gosselin et al. 1993).

But phagocyte biology has not been viewed as essentially concerned with immune recognition, that is, MHC-coupled reactions. This problem of specific recognition is viewed by many as the true province of immunology, whereas natural immunity (viz. phagocytes) is understood as coming under some other rubric. In any case, immune processing is the basis of lymphocyte and humoral immunity, and serves as the memory system of the immunologic self. Formally the immune reaction begins here, at the level of specific recognition, a reaction that reflects the extraordinary ability of vertebrate immune systems to recognize, respond, and memorize a myriad number of antigens (on the order of 10^8–10^9). As described, acquired immunity is dependent on the generation of specific antibody, which arises from a mixed genome that arrays several components to make a particular protein. Recall that the molecular biology of immune selection is based on the discovery that a receptor for an antigen is generated on the surface of a lymphocyte, by the particular arrangement of a finite number of gene segments in a unique linear array. There are constant regions (that is, universal and homogenous) of the resultant protein that confer class recognition and function, and there are variable and hypervariable regions that accommodate a particular "fit" for the antigen. Generation of diversity (G.O.D.) is then explained by the shuffling of these genetic constituents and the selection from a source of hypermutated variable region genes to form distinct antibodies. The B lymphocytes with such antibody receptors are stimulated to secrete more antibodies, and the antigen is then neutralized by binding to a limited library of antibodies, each protein exhibiting different affinities for the antigen. The production of antibody is then a result of antigenic selection for those lymphocytes that will produce proteins that "recognize" that antigen. The selection is from a random population of potential antigenic receptors. The fitness (i.e., affinity) of the antibody receptor for antigen determines whether the biological response is then initiated: cell proliferation and antibody synthesis. In the response to an indefinite number of pathogens, exogenous antigen, and altered self (e.g., effete, cancerous, or damaged cells), the host creates an antibody library, a range of antibodies with varying affinities for the targets of immune attack. The immune reaction then arises from a spectrum of antibody-producing lymphocyte clones that are stimulated by antigen, each cell generating a single antibody. The "fitness" of antibodies actually increases during the immune response, in part due to mutation and in large measure because when the amount of antigen declines, only those antibody-producing B cells having the highest affinity for the antigen are able to be stimulated.

The discovery that immune reactions consisted of coordinated sub-populations of lymphocytes led to the designation of an immune *system* (Moulin 1989), but this was viewed in two different modalities. The immune system may be modeled as a complex array of antigen-driven interlacing cellular and humoral factors that regulate the immune response by feedback cycles. This model has already been sketched. A complementary hypothesis, in addition to the control mechanisms already described, postulates that a self-regulatory *network* is controlled by autoreactive antibodies, the latter control being the critical element that regulates the immune system. On this view the system is formally called the idiotypic network theory. The theory, first proposed by Jerne (1974), is based on the simple argument that if the 10^7 (or more) clones of the immune system are capable of recognizing any antigen, they should also be able to recognize one another; thus, he initially focused upon the notion that immunoglobulin idiotypes are organized as a network of complementary shapes. The network hypothesis postulates that the primary encounter with antigen results in the formation of a network of cross-stimulating antibodies and the creation of antibody-producing cells that are maintained even after the initial antigenic insult has been removed. In this model immune memory is a stable state, independent from the continual presence of antigen.[1]

The rudiments of an idiotypic network had already been suggested by Jerne earlier (1967a), but the network as a working hypothesis awaited finer definition (Jerne 1974). Before the network theory was formally proposed, a nomenclature for its components had already emerged (Jerne 1960): antigen carries several *epitopes* (immunogenic elements), and an antibody has, besides its own potential epitopes, *paratopes* (combining sites/receptors). Those epitopes that are carried by components of one individual are *idiotypes.* Collectively these comprise the *idiom* of the animal. Note that Jerne avoided *self,* using the more neutral *idiom.* In the network paper published fourteen years later he adopted a modified definition. He identified the idiotype as "a set of epitopes displayed by the variable regions of a set of antibody molecules. Each single idiotypic epitope I shall call *idiotope.* An idiotype then denotes a certain set of idiotopes" (Jerne 1974, p. 380). The definition reflects the discovery that immunoglobulin possesses unique antigenic determinants (Kunkel, Mannik, and Williams 1963; Oudin and Michel 1963). Idiotopy arises from the antigenicity of this variable region, which in turn generates anti–variable region antibodies, that is, designated anti-idiotypic antibodies as anti-idiotypes (Burdette and Schwartz 1987; Nisonoff 1991).

Jerne's theory postulates that the immune system is in a state of equilibrium, in that the original antibody with a particular paratopic specificity, Ab_1, is held in check by its anti-idiotypic antibody, Ab_2. Ab_2 is directed against the unique protein structure to the variable region, the idiotype of Ab_1. Introduction of antigen interrupts these interactions in a mutually interactive fashion: Ab_1-binding antigen removes its inhibitory effect on the "internal image" clone as well as its stimulatory effect on Ab_2. As Ab_1 production then ensues and concentration levels are enhanced, these effects are reversed, and equilibrium will be reestablished (Jerne 1974). In what Jerne described as Eigen behavior, the immune system may thus be viewed as a network comprised of antibodies and their regulatory anti-idiotypes. Homeostatic control of the immune reaction, specifically the production of antibody, depends on two mechanisms: (1) Antibody may interfere with antigen, binding to antigen-specific clones of antigen-binding cells and thus preventing further stimulation, or (2) a particular antibody may be suppressed by another antibody generated specifically to it. This hypothesis predicts that as antigen stimulates synthesis of antibody, the characteristic idiotype becomes more prevalent and in turn induces synthesis of anti-idiotypic antibodies in what is called the Id cascade. As the level of these anti-id antibodies (or cells) increases, they exert a specific suppressive effect on the further production of the original idiotypes. This cycle must continue if the anti-idiotypic antibodies themselves mimic antigens, but eventually the original stimulus is dampened.[2] A series of Id cascades that are generated to different epitopes associated with a multideterminant antigen may exhibit the ability to interact with one or another at both the B- and T-cell level. Thus, the model has been extended to T-cell regulation through the variable regions of the Tcr. The idiotypic network of B and T cells is viewed as possessing regulatory properties (e.g., Masaki and Irimajiri 1992) and thus may be understood as a complex array of responder and suppressor lymphocytes, where the introduction of antigen induces two distinct pathways, one leading to immunity, the other to its suppression (Bona 1987).

The immune network concept has been a fulcrum of wide experimental and theoretical interest (e.g., Bona and Kohler 1983; Kaufman et al. 1985; Bona 1987; Flood 1988; Lundkvist, et al. 1989; Atlan and Cohen 1989; Perelson 1989; Herbert et al. 1990; Parisi 1990) and not surprisingly has yielded many modifications (e.g., Perelson 1987). But there is also a strong lobby that regards the matrix of lymphocyte subgroups and regulatory cytokines as sufficient and sees an idiotypic network as simply an unnecessary theoretical appendage that has little direct experimental support (e.g., Cohn 1966; Langman and Cohn 1986; Klein 1990). Although the generation of anti-idiotypic antibodies has been demonstrated, the true functional relevance of these elements is still unclear (e.g., Klein 1990, pp. 386–7). Regarding our agenda, the important issue is that regulation became a key focus of immunology, and thus the *biology* of the immune response has regained center stage. The Metchnikovian challenge may now be addressed.

B. THE EMERGENCE OF THE IMMUNE SELF

Immunology as a discipline is generally viewed as emerging primarily from studies of infectious diseases at the end of the nineteenth century. But there is a second major concern of immunology beyond defining those processes that identify the pathogen and defend against it, namely, maintaining organismal integrity in normal physiological growth, repair, senility, and disease. This is what has been previously referred to as immune identity functions, and "the Metchnikovian challenge" is to decipher the basis of such activity. Thus, Metchnikoff's role in this second endeavor has been emphasized in the effort to highlight the dual aspects of immunological theory that originated in his formulation (Tauber and Chernyak 1991). So far I have traced the historical developments that must account for the distinction between the role of immunity as *defender* against the invader – maintaining host integrity against threats from the environment – and the function of establishing host *identity*. Ironically twentieth-century immunology developed in the context of focusing on the first function, even though the original theory of active immunity derived from studies of embryological development, where control mechanisms of growth and differentiation were assigned to phagocytic "immune" cells. As discussed in the preceding chapter, although questions concerning autoimmunity and transplantation were raised at the turn of the century, these problems were not seriously pursued until the 1940s. And it was only at midcentury, when the thematic concerns of immunology broadened to include the role of the immune system in preserving host integrity in an ever-challenging and evolving internal environment (i.e., dealing with endogenous insult), that the scope of the discipline was forced to incorporate issues of identity. No longer was it sufficient to examine how specificity was generated; that became a secondary question to the more fundamental problem of *what* was being defended. How was immune identity, the generative entity, established? Defense became subordinate to the issue of defining the entity itself. Not surprisingly the language of immunology reflected this transition, and here we endeavor to examine the appearance of the term *self.*

There exist a range of biological definitions of the organism, as was discussed in Chapter 2. Beyond the hegemony of any discipline's point of view, there are distinctly refracted visions due to the parochial prisms of assumptions, interest, methodologies, history, and language of the different subfields of biology. And thus self, a term borrowed from philosophy and psychology – where it has endured a complex etymology – has assumed various meanings and uses in biology. We discern such differences from the "lessons" applied by the Nazis and Lysenko or by Social Darwinists or by creationists, in their various causes (Lewontin, Rose, and Kamin 1984; Ruse 1988). We must be wary of misapplied or misconstrued theory; to resist and disavow allegiance to any sociopolitical program is my first concern. That caveat is important, because, at the same time, I will argue that the biological offering to our understanding of selfhood is embedded in a metaphysical foundation that underlies a broad cultural experience. Throughout this essay I explore that elusive relationship between a scientific construction and its wider intellectual milieu. Perhaps biology even offers a unique mirror of that *Zeitgeist.* But which biology?

For the various bioscience disciplines, each proffering its distinctive voice and viewpoint, the problem of how to define the organism is the fulcrum of interest and serves as the metaphysical boundary of its respective orientation. By discerning the view of organism endorsed, we potentially gain insight into the governing "rules" or the orientation of the biology being scrutinized. Certain biological disciplines would disdain employing *the self* as a useful term. It certainly is not used with the same meaning in diverse areas. For some, it immediately connotes anthropocentricity, perhaps self-awareness, which in turn implies self-consciousness. It is a term that conjures the image of a completed and bounded entity. *Organism* is generally the term preferred for the agent under discussion. For instance, evolutionists debate whether the unit of selection is the organism or another organizational level; the microbiologist might consider individual bacteria solely as components of a larger organizational conglomerate; the developmental biologist might debate the basis of identity that governs the maturation of the embryo (Tauber 1991a). Self in these contexts serves to define distinctness, biological unity, organismal identity. But as already discussed, competing visions of modern biology generate other conceptualizations of the self.

In immunology the use of self reflected a change in the vector of interests, that is, from immune reactivity directed against the outside world to immunity's source as endogenous activity. As the immune self increasingly became a central question after World War II, the metaphorical term *self* was enlisted to herald a discussion of that general issue. Such designations as Garrod's "chemical individuality" or Burgio's "biological ego" (Burgio 1990) cannot substitute for the rich evocation of *self*. *Self* is now commonly used in immunology and is even found in textbook titles, for example, in Klein's *Immunology: The Science of Self-non-Self Discrimination* (1982) and in a growing number of scientific reports and reviews. Using this coarse criterion of self appearing in titles and abstracts of the English immunological literature, since Burnet published *Self and not-Self* in 1969, a conservative estimate of the term's appearance indicates growing acceptance: Found once in 1973, by 1978 the number had grown to 77; in 1983, 192; in 1988, 383; and in 1991, 600 times.[3] *Self* is not a technical term, and when it was introduced, it was used hesitantly. And with good reason.

Returning to our central concern, consider how in Burnet's first publication on the CST, he notes with satisfaction that his modified theory views tolerance and the clonal selection hypothesis consistently: "Its advantages over Jerne's theory are its capacity to cover homograft and related types of immunity as well as the production of classical antibody" (Burnet, 1957, p. 68). But what is striking in this article – and consistently reappears in each of Burnet's more extensive treatises on CST, (i.e., *The Clonal Selection Theory* (1959) and *Cellular Immunology* (1969a) – is the surreptitious use of the term *self*. When the word was first introduced in the earliest publication, Burnet felt compelled to place it in quotations, for example, "response to '*self*' constituents and the related phenomena of immunological tolerance" (Burnet 1957, p. 67). However, twelve years later, when Burnet authored an immunology text explaining CST to a general science audience, he chose the title *Self and Not-Self* (1969b) ostensibly to appeal to this wider readership. (This work is basically the first part of his *Cellular Immunology* (1969a), which included in its second half technical documentation in support of his theoretical orientation.) *Self* barely makes its way past the title pages. The term is invoked, but often with scare-quotes, and never expanded upon (e.g., Burnet 1969b, p. 53). It is mentioned only rarely, in passing, and omitted from the index. The title, *Self and Not-Self,* invokes a generally graspable message, in its implicit pronouncement that immunology deals with distinguishing the organism from its world; but the study focuses on the mechanics of the entity, rather than the entity itself. The Metchnikovian challenge had yet to be met, but the stage was set.

Burnet was obviously not comfortable with *self* as a scientific term. If one surveys the literature, not *self,* but rather the terms *host, individual,* and *subject* are commonly employed to designate the responding organism. An examination of the key transplantation literature between Gorer (1930s) and Medawar (1950s) shows the term *self* to be absent. This is, of course, a limited survey, but clearly the use of *self–nonself* was not a common means to discuss immune phenomena prior to Burnet.[4] Interestingly, when the term *self* appears with Burnet, it is assumed to refer to the host or perhaps to the individual. But what is this self? Is not immunology the science of self-differentiation? Certainly not in the sense that Roux first invoked the concept for developmental processes. For him development was self-differentiation of the egg in relation to outer factors; that is, the determination of form (and its proper cause) are enclosed in the developing organism (Roux 1895, pp. 17, 44, 423, 777). Is it, then, the science of self-discrimination? But that provisional definition immediately shows inadequacies. The criteria of immunogenicity (the evocation of an immune response) are both vague and unreliable. Consider the important exceptions of immunologically ignored symbiotic/commensural/parasitic states (Chapter 2) or of nonantigenic "foreign" substances. The immunologist might object that immunity arises from a self that must be protected from the foreign or the diseased. However, when we ask the practical cogitator to define this self specifically, he or she most likely will attempt to refer to the given entity; assuming the genetic prerogative, he or she will cite the MHC as the signature of selfhood or even of the organism's *basic* identity. Then, in a circular turn, he will define immunology as those interactions based on the participation of the MHC complex, that is, characterized by immune recognition that is MHC restricted.

Obviously the most sophisticated immune recognition reactions defined by the MHC-associated foreign peptide as nonself is the basis of immunological memory and a wide spectrum of host definitional processes. However, the MHC system cannot be viewed as the *sole* basis of immunity. The so-called natural, preimmune, nonspecific recognition processes that account for "natural immunity" are not easily dismissable, and these reside well outside the processes dependent on MHC-mediated recognition. The MHC proponent simply designates these processes as "nonimmune." Although this argument provides a "solution," such a definition begs the question both by eliminating an entire mode of host discrimination mechanisms and more saliently by ignoring the alternative position represented by a biology that recognizes dynamic, developmental processes as a crucial component to any theory of organismic identity. From this perspective *self* is defined dynamically and might be viewed as a cognitive entity or operation (e.g., Varela et al. 1988) or even as a dialectical process (Chernyak and Tauber 1991). From this vantage immune reactions (of all types – and there *is* variety!) arise from encounters defined in a context of experience and complex interactive dynamics.

Concepts that invoke feedback loops, thresholds, connectivity states, and the like, create a hierarchical network that defies a single molecular definition. The self arises in process; that is, it is not a *given* entity. A definition in a current textbook of *self* as "everything constituting an integral part of a given individual" (Klein 1982, p. 5) or of *nonself* in another as "everything which is detectably different from an animal's own constituents" (Playfair 1984, p. 1) reflects simplifying assumptions that raise or beg more questions than they answer. These problems will command our further attention. Suffice it to note here that in either case the concept of self *does* serve an organizational function; the issue to explicate is the capacity in which it so serves. I will not endeavor to provide a rigorous scientific response, but will rather tackle the question of how *self,* the implicit foundation of immunology, functions as an organizing principle or as a metaphor in immunological thinking. The philosophical implications of such language have deep resonances in how the field is conceptually formulated and understood, both by the specialists and by its greater audience.

How is *self* employed by immunologists? It is important to precisely establish the ambit of use of the term *self* in this discussion. The first caveat is that the term falls within a continuum of meanings that can be seen to have two poles. On one end is the commonsensical notion of a bounded, autonomous entity; this is the reassured self-knowing being that we uncritically assume when we confidently refer to our self. Adapting Charles Taylor's (1989) usage, I will refer this view as the "punctual self," meaning the self as a stable entity, or perhaps metaphorically the self as occupying a locale in space. Taylor's punctual self is based upon what he calls the disengaged subject, which appeared in the objectification of Man in seventeenth-century thought. The modern figure searches for control, which requires taking ourselves out of our normal way of experiencing the world and ourselves:

> Radical reflexivity is central to this stance, because we have to focus on first-person experience in order so to transpose it. The point of the whole operation is to gain a kind of control. Instead of being swept along to error by the ordinary bent of our experience, we stand back from it, withdraw from it, reconstrue it objectively, and then learn to draw defensible conclusions from it. . . . We fix experience in order to deprive it of its power. (Taylor 1989, p. 163)

How John Locke empowers this position is discussed in the next section. For our purposes the "punctual self" is being employed to simply assert that there *is* a definable self to regard the world and itself so critically. But there is another strong sense of selfhood of which we are well aware, namely, the "elusive self" or, to follow the parallel metaphorical construction, an elusive space of indeterminate location and undefinable coordinates. This latter sense of the self is problematic and must be posited, affirmed, and ultimately discovered. When immunologists invoke the term *self*, they do so in different contexts and with diverse meanings, but I have endeavored to structure the discussion along this punctual–elusive continuum, which accommodates three categories of selfhood: ontological, cognitive, and embodied. Before we define these categories in detail, an explanatory summary will serve to orient the discussion.

First, *ontological* is admittedly obscure in the sense used here. In philosophy ontology is a branch of metaphysics that examines the nature of reality. (From the Greek *ont*, which is the present participle of *einai*, "to be.") I am deliberately drawing upon the rich connotative value as Martin Heidegger posed the question of selfhood in *Being and Time* (1927, 1962). He chose to ask not "What *is* the self?" but rather "What does it mean to ask, 'What is the self?'" The problems of personal identity, consciousness, and selfhood, when structured in this way, respecify the issue from an attempt to define an entity to an examination of the nature of the inquiry itself. The Heideggerian postulate in its full ontological scope situates the self with a unique comparity. Man (*Dasein*) is a being who can ask what is Being; to think on and of Being is the condition of an authentic personal life, and to recognize the significance of this ultimate metaphysical question as that which "authenticates" a true self is to discern the basic character of our selfhood. Thus, for Heidegger the authentic self is that which poses the basic ontological inquiry. As employed in this particular discussion, *ontological* refers to the quest for the source of immune activity, the identity from which the immune reaction arises to define the host organism.

Part of the power of the self metaphor is its ability to convey this elusive quest for what may be argued is an artifice, an approximation, a convenience by which the science structures its theory. In other words, self is a construct of a *question*. From this perspective *self* refers to the nebulous boundaries of some core essence that is demarcated by those immune processes that identify the other; the self, like a queen bee, remains guarded, the sequestered presence at the center of the buzzing hive. Obviously this referral to selfhood is closest to the elusive pole of meaning described above.

When the immune system is described as structured analogously to the brain or as functioning similarly to the mind, the self metaphor assumes a *cognitive* character. In this sense, then, *cognitive* is used loosely here to denote the metaphor of self as the mind entity. It plays on the pervasive concern of identifying mind functions by modeling of various sorts, and the immune system, in terms of both its structure and its function, readily lends itself to description by metaphors based on our conceptual understanding of the brain. This sense of self is more accessible to "practical" analysis and thus falls somewhere between the extreme meanings of punctual and elusive selfhood.

Finally, the *embodiment* metaphor is used in this discussion in part to complete the mind–body duality that arises whenever issues of identity emerge and also to reflect how the science of immune function has insinuated itself into our culture, assuming a cardinal role in asserting *who* we are. In other words, to a large extent today immunity defines ourselves as bodily entities, and common discourse employs the immune self to represent bodies – both individual and collective. I will endeavor to examine how this last instance employs *self* as closest to the everyday vernacular sense of a given entity. Any immune model may take elements from each category, but as seen below, dominance of one or another yields reasonably well-circumscribed domains for discussion.

Before describing how self is an integral construction of modern immunology, I must justify how (and why) I have chosen to frame this discussion in terms of citing self as metaphor. To do so requires, first, to define what I mean by metaphor. The topic of metaphor in the various discourses on language, literature, and social and natural sciences is immense (Shibles 1971; van Noppen, de Knop, and Jongen 1978), and to choose the boundaries of discussion, that is, the conceptual definition and appropriate use of metaphor, is to declare a particular philosophical orientation. The present discussion employs metaphor broadly, to encompass a wide sense of meaning. By *metaphor* I do not mean simply a comparison, either by abstraction or by homonymy. Although these aspects may be integral to metaphor, what is important in the context of this discussion is the elusive function that the metaphor serves, namely, that it may well be the only conceptual mode to organize experience in the absence of a comprehensive theory or model.

My position is consistent with that propounded by Max Black (1962) and extended by George Lakoff and Mark Johnson (1980), who regard metaphor as having an irreducible meaning and a distinct cognitive content; the metaphor in this view actually structures the very concept we describe. The development of this position dates to Aristotle, where, in the *Rhetoric, metaphora* meant a word used in a changed sense (Barnes 1984, p. 2250). The extension of metaphor to a more radical interaction theory may be traced to Samuel Taylor Coleridge, for whom language is not a conduit, but an expressive. medium for the imagination (Coleridge 1817, 1983, vol. 2, pp. 16–17). In his view metaphor becomes the linguistic realization of the creative unification of experience – fusing diverse thoughts and re-forming perceptions. This is a particular view on the matter and opposite to that argued by Donald Davidson (1978), who disallows assigning meaning to metaphors, that is, truth claims, a position that in turn has generated robust rebuttal (Black 1978). (For a succinct, but helpful history of metaphor's "stormy, tenuous, but tenacious affair with philosophy," see Kittay [1987, pp. 1–11].) I have adopted the more expansive orientation; i.e., concepts are not defined solely in terms of inherent properties or an implicit comparison, but rather emerge from their interactional properties, arising from experience in an open-ended fashion, with the important caveat that they may be based on similarities derived from cultural values. Thus, metaphor may create new meaning and serves as a means of structuring our conceptual system. In this regard the primary function of metaphor is to provide a partial understanding of one kind of experience in terms of another, which may involve the isolation of preexisting similarities or the creation of new ones. The endless complexity of perception is reflected in appreciating that we understand one kind of experience in terms of another.

Metaphor as commonly employed in science is "immensely practical" (Hoffmann 1985). As a tool for description and explanation, metaphorical language permeates the natural sciences, and biology, not surprisingly, richly embraces this rhetorical strategy (e.g., Temkin 1949; Bonner 1963; Canguilhem 1963; Rosen 1971; Young 1971; MacKay 1975; Salk 1975; Potes 1979; Phillips, Fillers, and Cohen 1980; Rather

1982; for a summary of the literature, see Paton 1992). Karush (1989) estimates that "immunology utilizes perhaps as many as three or four dozen metaphors which are distinctive to the field." Of these he lists several that evoke common human experiences, both biological and cultural in nature: memory, tolerance, recognition, surveillance, learning, responder, repertoire, helper or killer (T cells), virgin (B cells), allograft (rejection), (antigen) presentation, (lymphocyte) collaboration, restriction, and internal image. Interestingly *self* is not mentioned! Scientific metaphors essentially follow the same forms as metaphor in general: They may be broadly pervasive, that is, basic "root" metaphors or metaphor themes that form a worldview (e.g., the mechanical worldview regards the mind as a machine), or they may be more particular metaphor-based hypotheses or images (e.g., memory is like a "dictionary" or like "computer files") that generate functional relationships or abstract models (Hesse 1966; Wartofsky 1979). When scrutinized and understood, metaphors may be found to consolidate principles and assumptions of a theory or model based on their construction (Pepper 1942). Thus, metaphors may suggest new hypotheses, entities, or relations. Metaphors, whether with images or with models, may legitimately be regarded as "embryonic theories" that are part of scientific discourse and eventually contribute to the growth of more explicit knowledge (e.g., Hesse 1966; Orgony 1979), not to say problem solving (reviewed by Hoffmann 1985). Metaphor unites reason and imagination ("imaginative rationality" [Lakoff and Johnson 1980, p. 193]), offering a tool to comprehend partially what cannot be conceived by other means. I believe that this orientation to metaphor solidly straddles the territory between radical subjective and objective positions, in that understanding does not reside in a totally neutral conceptual system (contra objectivism), and imagination must be grounded within physical and cultural environments (contra subjectivism). Thus, conceptual metaphors are grounded by correlating different elements of experience, and they function to organize, cohere, and highlight that experience. In short, metaphors may create realities for us, and it is in this context that we now turn to explicate how self, a metaphorical construct of immense complexity, is used in immunology.

First, *self* appeared in the most unobstrusive manner, almost innocuously and quietly undefined. When Burnet introduced it in 1937 (Burnet 1940; and even later, Burnet and Fenner 1949), he left its meaning nebulous, as a vague gesture toward organismal identity – a fundamental property of even the simplest life. The context of self's critical appearance was discussed in Chapter 3; here we turn to how self appears in this largely implicit state. Note that self remained an undeclared entity for the first seventy years of immunology. Given the term's proliferation once it was introduced, it is mildly surprising that when we search for *self* prior to 1940, it simply does not appear in the literature. *Host, organism,* and *individual* are commonplace, and from our perspective they serve as reasonable substitutes in most contexts, but each is clearly deficient in transmitting our modern core concept: Immunity not only defends the host/organism/individuality, but *defines* that entity. And it is in the context of establishing organismal individuality that the host becomes a self. Thus, self is a rich metaphorical construction for all those immune activities responsible both for elaborating immune process and more fundamentally for pinpointing the *origin* of that function.

It is interesting to note that at the same time that Burnet and Fenner formally introduced the self/not-self terminology into immunology, there was active debate concerning the use of *self* in the philosophical literature: "For where but in the writings of philosophers do we even find "self" used as an independent word rather than as an assimilated reflexive suffix? The plain man may hurt himself: he never hurts his self. . . ." (Flew 1949). Antony Flew went on to protest the dangers of using "a new word" even though he understood that his adversary J. R. Jones was using *self* as synonymous with *person*, because "the phraseology employed leaves us 'softened up,' quite unready to resist the offensiveness" of faulty logic and the introduction of undefinable "metaphysical entities" (ibid.). He concluded with a firm admonition: "Get rid of some of the verbal rubbish which obstructs discussion [i.e., self]." Jones immediately responded with the justifications for using *self* (Jones 1950). In his original article on personal identity, *self* was used rather than *person* specifically because the former connotes continuity of identity as a result of its etymological derivation from the same root as that of the word *same.* He noted that the earliest forms were *I self, you self,* and *us self,* where *self* was an adjective meaning "same." In the thirteenth century, the genitive of the prefixed pronoun gave the forms *me self* and *your self,* which were then compounded into *myself, yourself,* etc. Meanwhile, the divided form, *my self* was also becoming an ordinary possessive phrase like "my nose," and *self* began to be regarded as a noun. But more important, Jones argued that *self* was more appropriate to convey a personal, introspective point of view. Flew's preference for *person/people* (which are "not-I's") conceals the introspective dimension that Jones wished to emphasize, and most pointedly the more general categories of person or people

> ... are terms of which by far the greater part of the denotation is covered pronominally by the third person of the personal pronoun. For this reason I would call them ... *third person expressions.* And my contention is that third person expressions tend to conceal that character of persons in which their peculiarity is that they are *all* "myselves." An alternative designation which was designed *not* to conceal this characteristic of persons I would call an "introspection world." (Ibid.)

Jones concluded by restating his need for a term to deal with the private self, as opposed to public object/person, in order to deal specifically with the issue of how personal consciousness confers identity. He readily drew upon John Locke's "modern" usage in *Human Understanding:* "Since consciousness always accompanies thinking, and it is that which makes everyone to be, what he calls self" (Locke 1689, 1934, p. 315). But this sharply defined sense and connotation of *self* invoked the ire of Jones's philosophical opponent.[5]

The fundamental controversies surrounding the use of *self* in the modern philosophical literature date from Descartes and represent the philosophical expression of the debate concerning the legitimacy of the punctual versus the elusive poles of meaning, outlined above. That debate remains unresolved. The term *self* itself was not taboo, and it was readily invoked by both analytical and less austere philosophers, even when attacked as lacking meaning. Wittgenstein might ironically proclaim: "The I, the I is what is deeply mysterious!" (1961, p. 80e), but the self (or "I") remains a crucial issue of twentieth-century philosophy. The message is quite clear: In 1949 *self* was charged with referring to a highly personalized, introspective, and subjective experience of personal identity. It was used against a rigorous analytic tradition that abhorred such ill-defined terms.

Burnet might well have been somewhat apprehensive about using *self,* but he needed to approximate an idea for which he could find no better language. He was forced to adopt this philosophical term and use it as a placeholder for a scientific language he did not possess. The extent to which Burnet was cognizant of the philosophical questions whirling about selfhood I cannot recount. But he must have been aware of the equivocality of the general senses to which *self* refers, that is the punctual versus elusive characteristics. Irrespective of his philosophical sophistication, it would do us well to review the issues that emerged from these debates and to examine the historical context on the basis of which polar meanings of selfhood developed. I take this admitted detour in order to place the analysis of the immune self more readily in its appropriate conceptual historical context. To do so without establishing the critical vocabulary would truly stymie that analysis.

C. THE PUNCTUAL VERSUS THE ELUSIVE SELF

The logical argument of this book might be summarized as follows: Immunology, like any science, models its data on borrowed schemata from common human experience. The self has emerged in the past half century as an operative metaphor for orienting immunity in terms of both the source of its activity and the object of its function. To understand the full scope of the metaphor, we must delve into how *self* has been used by humanistic and social disciplines in their particular endeavors. The meanings of *self* are, to a lesser or greater degree, at some fundamental level, generally shared throughout culture. More saliently for our purposes, the underlying metaphysical basis of selfhood, like vapor from a hidden chasm, pervades insidiously, and oftentimes overtly, the very fabric of the concepts that would tap into its evocation. In other words, if immunology uses the self to help define its conceptual apparatus, or even its particular operations, it does so with both the benefit of the rich meanings *self* evokes and also the liabilities that entail the use of such an anthropocentric notion. Metaphor is relied upon because a more explicit and carefully crafted image is unavailable. To discern, then, both the strengths and weaknesses of the self metaphor is an underlying concern of this study, but the dominant effort is in fact more modest: to identify how the characteristics of self are reciprocally reflected by immunology and philosophy.

I contend that it is useful to explore the wider intellectual boundaries of what the term *self* signifies and to attempt to define its contextual meaning beyond the confines of immunology proper. It is a central thesis of this study that the self is an elusive concept. By using the term *self*, immunology is treading on largely nebulous grounds. The self can hardly be viewed as a scientific concept. The term is borrowed from philosophical discourse to denote concerns about the source of immune activity, that is, the identity problem. In immunology self discrimination approximates the relation of a subject and a predicate – the immune processes, which defend, and the host, which is defended. There is a striking coherence with the term's resonance when we search for its use in philosophy. To appreciate what the self might confer to immunology as an implicit structuring of its scientific agenda, that is, of its underlying general metaphysics, a short historical survey of the term in philosophy is offered to illustrate simply the given versus the critical understanding. In immunology the average scientist uses *self* in its common, uncritical sense; however, as I will show, he or she in fact intuits its essential elusive character. The science inadvertently or perhaps even "subconsciously" understands this distinction. In no case do we expect exact correspondence between the science and philosophy, but there appear to be rich parallels in the epistemological discourse of each field. The benefit of such a study is a reexamination of subtle, perhaps, hidden, assumptions, prejudices, and insights that the self metaphor offers. The immunologist might not be aware of the philosophical richness of the concept, nor of the heavy burden its use might entail. But most tellingly, self remains problematic in philosophy, an unresolved conundrum dating back at least to Descartes.

Self is but one term among many of the vernacular language that deals with this elusive entity. There are the relatively neutral ones, such as *subject, inner man, person, ego, spirit,* and *soul,* and there are also those associated with the self as mind, including *consciousness, mental substance,* and *psyche;* finally, of course, there are terms constructed in particular philosophical contexts, such as *monad* (Leibniz), *transcendental unity of apperception* (Kant), *Dasein* (Heidegger), and *transcendental subject* (Husserl). Psychoanalysis also offers its special inventory of *id, ego, superego, libido,* and *sub-* and *unconscious.* The listings are quite open-ended, but any attempt to define our use of *self* with respect to an organism must fulfill two criteria, namely, the intuitions of embodiment and of interaction; arguably these are the two broad boundaries of an organismically based notion of selfhood. I turn first to the pre-Kantian philosophical evolution of the notion to highlight what I have referred to as the opposing punctual and elusive formulations.

Socrates exhorted his fellow citizens to *gnothi se auton,* literally "know thy soul or psyche." (Although there is a Greek pronoun for *self* in Plato, it is not used as a substantive; *psyche* functions as the noun corresponding to our *self* [Griswold 1986].) To know oneself is to know the psyche, which is best declared in its virtue and wisdom (*sophia*), upon which Socrates's entire ethic is erected. Psyche's complex and laden meanings may be simplified as that which is capable of attaining wisdom or, in Socratic terms, as the true self:

> The living man *is the psyche,* and the body . . . is only the set of tools or instruments of which he makes use in order to live. . . . Life can only be lived well if the *psyche* is in command of the body. It meant purely and

simply the intelligence, which in a properly ordered life is in complete control of the senses and emotions. (Guthrie 1971, pp. 149–50)

The roots of such an ethos may be discerned in Ionian scientific thought and in Pythagoreanism (Onians 1951), but they are best instantiated in Plato's *Phaedrus* and *First Alcibiades.*[6] The self proper, in its modern sense, first appears with Aquinas, but it does not become a focus of inquiry until the seventeenth century.

Many commentators have noted that the rise of modern science provoked a basic metaphysical reorientation. In Francis Bacon's definition science repudiates scholasticism and metaphysical analysis in favor of explaining how nature functions. The historical revolt was not a protest on behalf of reason, but the establishment of a *method* to study empirical facts: experimentation and induction. Through inductive reasoning, Bacon believed, nature's order would be revealed. To what extent he anticipated the necessity of reason to fulfill the general description of nature's reality may be debated, but it was the dual foundation of belief in a detailed order of nature and the severe rationalism of the Middle Ages that served as the foundation of the Baconian formulation (Whitehead 1925, pp. 62–72). Science, although achieving success with its pragmatic and experimental designs, requires more than simple induction of facts to constitute the "facts" of nature. Bacon underestimated the rationalist leap from data to theory. Observations alone do not reveal Order; abstraction is required. It is simultaneously the philosophical attitude of science and its empirical methodology that required the scientific method to be both abstract and inductive. The process of selecting data and then ordering them within a hypothesis, scheme, theory, or paradigm serves as the final pathway of the scientific method. Data collection and detached observation, although the primary means, are not the ultimate ends of science. It is the abstraction of the mind's observations, that fulfills the function of the scientific process; this is what requires a detached observer, a self that not only observes the universe, but situates itself within it. By the seventeenth century science had matured and proven its epistemological value to the extent of challenging the prevailing theism. Reality was to be newly discovered and ordered, for all was subject to scrutiny and skepticism.

The emergence of the modern self conception coincided with the rise of empirical science in the revolt against the Aristotelian science of the Middle Ages. Ordered by an anthropocentric perspective, the search for final causes and qualitatively differentiated levels of the cosmos were replaced by a view advocating impersonal observation to yield a world

143

of contingent correlations. Science, as the objective lens on the world, offered a means to define the self in relation to a newly self-discovered order. The basis of exploring this world presupposed self-presence and a meaningful order that would not achieve its epistemological synthesis between the competing radical positions of empiricism and rationalism until Kant. To achieve the personal detachment demanded by this new view of science, a revolution in the basic categories by which we understand the self was required. "The essential difference can perhaps be put this way: the modern subject is self-defining, where on previous views the subject is defined in relation to a cosmic order" (Taylor 1975, p. 6). Until the moderns, humanity oriented itself in the cosmic order, which it understood through reason. The Greeks bequeathed a human order as inseparable from the rational vision of the order of being. Descartes' *cogito* posits the existence of the self, whereas everything else (including God) is suspended in doubt. Modern science demands that the observer draw back from the world and concentrate on the processes of observation, whereby a new self-presence is asserted: "Aware of what we are and what we are doing in abstraction from the world we observe and judge" (ibid., p. 7). Not only new insights, but a new power to manipulate the world, confirmed and consolidated the underlying strength of this self-defining identity.

A key architect of this modern self-consciousness was René Descartes. He essentially dissociated the soul from a mechanistic universe, thereby constructing the self as a separate entity. "I am" depends on *what* I am, and in the *Second Meditation* Descartes casts aside the identification of this "I" with man, body, or soul and is left with the thinking "thing," the "I" of reason: "From this very fact, that I know I exist, and that meanwhile I notice nothing else to pertain to my nature or essence, except this alone that I am a thinking thing, I rightly conclude that my essence consists in this one [thing], that I am a thinking thing" (Haldane and Ross 1911, p. 190). It is fundamentally from this position that the dualism of mind and body takes form. (Descartes never uses the term *self*, but rather *le moi* or *je,* which are rendered "self" in the English version, a translation generally regarded as neutral and convenient [(Beck, 1965, p. 95].[7]) Descartes arrives from the terminus of metaphysical doubt to *"cogito, ergo sum"* – not explicitly found in *Meditations* (1648), but in *Discourse on Method* (1637) and *Principles of Philosophy* (1644) (Wilson 1978, p. 52) – essentially from an intuition, an "inspection of the spirit," where he surmises that self, subject, and object are identical, with no requirement for reference (Gibson 1932, p. 85). Thus, the post-Renaissance translation of *"gnothi se auton"* assumes new connotations from its classic origins but retains its essential mission. What is crucial to note from the perspective of our limited summary is that for Descartes, after all the tribulations of the argument are exhausted, there *is* a self that is firmly implanted in the ego, an ego of certainty and punctuality, to refer back to our original dipole. Descartes did not effectively determine how such an ego functioned or seek its definable boundaries, nor was the basis of its certainty to be universally accepted. But let there be no mistake – Descartes' posit of identity is hardly the end or even the beginning of the matter. He simply must be credited with effectively pronouncing the problem of selfhood in a post-Renaissance sense, namely, presenting the subject as the central nexus and riddle of knowledge. His "solution" was immediately attacked by other rationalists but most effectively by the British empiricists, who posed the issue quite differently. I will cite John Locke as offering a fecund approximation of the punctual self in answer to Descartes' challenge, a response to which "all subsequent writing has consisted merely of footnotes" (Noonan, 1989, p. 30).

Locke began his discussion in Book 2, Chapter 27, of *An Essay Concerning Human Understanding* (1689) by noting that the concept of identity must be joined to some substantive notion that can be assigned criteria that allow change and still let it maintain its identity. In theory no hidden substance is required, and in this sense the Cartesian dualism (body and mind as repository of soul) is addressed. When Locke deals with human identity, he acknowledges different criteria for "man" and "person," the former confined to a biological identity and the latter a thinking being, that is, a being with consciousness[8]:

> For I presume it is not the idea of a thinking or rational being alone that makes the idea of a man in most people's sense, but of a body, so and so shaped, joined to it; and if that be the idea of a man, the same successive body not shifted all at once must, as well as the same immaterial spirit, go to the making of the same man.
>
> *9. Personal identity* – This being premised, to find wherein personal identity consists, we must consider what *person* stands for; which I think, is a thinking intelligent being, that has reason and reflection, and can consider itself as itself, the same thinking thing, in different times and places; which it does only by that consciousness which is inseparable from thinking, and it seems to me essential to it: it being impossible for anyone to perceive, without perceiving that he does perceive. (Locke 1689, 1939, p. 315)

The thrust of Locke's concern is in this latter dimension, from which he defines a radical epistemological orientation based on what Taylor

145

calls the "punctual self" (Taylor, 1989, pp. 159–76). The subject assumes the power to objectify and remake his objects, and thereby distance himself "from all the particular features which are objects of potential change" (ibid., p. 171). A startling conclusion then results: Whatever is fixed by such identification is the work of the self, which is essentially the power to fix things as object. Locke shuns identifying the self with any material or immaterial substance and thereby disengages his view from the Cartesian polarity of body and mind. The Lockean self assumes a radical disengagement. But a fundamental dualism remains: The subjects's capacity for detachment is transfigured into an independent consciousness, ultimately leading to a radically subjectivist view of the self. Consciousness thus makes personal identity. The ideal of reason triumphs: rationality is a property of the process of thinking, rather than the substantive content of thought:

> For it being the same consciousness that makes a man be himself to himself, personal identity depends on that only, whether it be annexed solely to one individual substance, or can be continued in a succession of several substances. For as far as any intelligent being can repeat the idea of any past action with the same consciousness it had of it at first, and with the same consciousness it has of any present action; so far it is the same personal self. For it is by the consciousness it has of its present thoughts and actions that it is self to itself now, and so will be the same self. . . .
> (Locke 1689, 1939, p. 316)

Before discussing the philosophical ramifications of this statement, I note the centrality of the word *self.* Beyond its use as an adjective in the seventeenth century, it is already in wide pronominal use as a synonym of personal identity in philosophy, literature, and common speech (see *Oxford English Dictionary* 1971).

Locke assumes a radically reflexive posture. It essentially involves the first-person standpoint and demands the disengagement from unscrutinized beliefs. Each individual is personally responsible to so examine himself and his ideas. Locke's argument fits in well with the new optimism of Newtonian science – a radically objectified science witnessed by a detached observer examining a mechanical nature dispassionately and independently. These concepts are also wielded to build Locke's political doctrine, where autonomy becomes a value, limited only to the extent that its freedom infringes upon the freedom of others.[9] Thus, Locke dealt extensively with the question of the individual and the concept of identity to argue how society must guarantee both the rights of the individual and society's needs. Locke's influence as the great teacher of the Enlightenment emanated from the powerful account he gave of the new science as valid knowledge and a theory of rational control of the self, intertwined with the ideal of rational responsibility. The independent, scrutinizing self became the substantial pillar of modern identity, permeating psychological, social, and political theory. For instance, the Freudian ego, functioning as a steering mechanism, devoid of instinctual force, behaves as a Lockean rational (detached) self-controlling agent. As Taylor (1989) keenly observes, the modern ideal of disengagement requires a reflexive stance; by turning inward, we become aware of our own activity and the processes that form us. We will see how this search for the self was further developed by Rousseau from a very different perspective and radicalized by twentieth-century phenomenologists. Locke had already perceived that we must assume charge of constructing our own representation of the world, which otherwise goes on existing without discernment, without science. Self-objectification can only be carried out from the first-person perspective (a major credo of modern phenomonologists, as discussed in Chapter 6) and then mustered by the various agencies of self-control: economic, moral, and sexual (analyzed most famously by Foucault; see Chapter 8). The self that emerges from such analysis is fundamentally divorced from a given, taken-for-granted nature and cannot be identified as another piece of the natural world:

> It is hard for us simply to *list* souls or minds *alongside* whatever else there is. This is the source of a continuing philosophical discomfort in modern times. . . . Various solutions have been tried – reductionism, 'transcendental' theories, returns to dualism – but the problem continues to nag us as unresolved. . . . [T]his ungrounded 'extra-worldly' status of the objectifying subject accentuates the existing motivation to describe it as a self. All other applications seem to place it somewhere in the roster of things, as one among others. The punctual agent seems to be nothing else but a 'self', an 'I.' (Ibid., p. 175)

In the Cartesian world, the mind–body duality had become a vexing philosophical issue. The heritage of dualism pervades any discussion of self, or of what has been more traditionally referred to as the riddle of personal identity. Seeking to reunite the truncated body and mind has been an elusive quest bequeathed from seventeenth-century Western philosophy. At one extreme is the posited spiritual substance that preserves identity despite the ever-changing and unstable nature of thought and feeling. At the other end is complete skepticism concerning identity due to the inevitable failure of pinpointing this introspective or unknowable spiritual substance. David Hume offers one of the most famous denunciations of an evanescent and ephemeral selfhood in the section on personal identity in *A Treatise of Human Nature:*

> There are some philosophers who imagine we are every moment intimately conscious of what we call our *self;* that we feel its existence and its continuance in existence; and are certain, beyond the evidence of a demonstration, both of its perfect identity and simplicity. . . .
>
> Unluckily all these positive assertions are contrary to that very experience which pleaded for them; nor have we any idea of self, after the manner it is here explained. For from what impression could this idea be derived? This question it is impossible to answer without a manifest contradiction and absurdity; and yet it is a question which must necessarily be answered, if we would have the idea of self pass for clear and intelligible. It must be some one impression that gives rise to every real idea. But self or person is not any one impression, but that to which our several impressions and ideas are supposed to have a reference. If any impression gives rise to the idea of self, that impression must continue invariably the same, through the whole course of our lives; since self is supposed to exist after that manner. But there is no impression constant and invariable. Pain and pleasure, grief and joy, passions and sensations succeed each other, and never all exist at the same time. It cannot therefore be from any of these impressions, or from any other, that the idea of self is derived; and consequently there is no such idea. (Hume 1739, 1962, p. 300–1)

Hume further argues that we fail to distinguish properly between the idea of an invariable object (prototype of identity) and the idea of several objects existing in succession and closely related (i.e., diversity). Although change destroys identity, the mind overlooks the factor of temporal fluctuations in order to consolidate or impose unity. The self thus becomes an artifact, an invented metaphysical construction to account for this mental custom. Of course, most substantive concepts such as the self are designed to incorporate change to a certain latitude, and in this view there is no contradiction between acknowledging change and asserting sameness if such changes are characteristic of the thing or person.

We will not further explore this matter, but simply observe that for both Locke and Hume the problem of the unity of the self was problematic, whereas for Descartes the inviolate self was axiomatic. Locke believed he had demonstrated the basis of selfhood empirically, but his position was attacked by Hume, who argued that the self was but a convenient vehicle for maintaining essential psychological coherence. Hume vigorously argued that although the self is not experienced, we *do* experience changing impressions that supposedly refer to the self. The self is only a "bundle of impressions"; that is, we experience successive selves rather than one self. Furthermore, there is no concept of the unconscious for Hume; thus, the self is lost during sleep, and the sense of integration that remains even in such a dramatic disruption is simply an "error." The sense of self is achieved by *association* of various experiences in time, and to expand from the idea of succession to the idea of identity is a psychological conceit: "The self is but a necessary fiction. . . . [I]dentity is not in the different perceptions themselves, uniting them, 'but is merely a quality, which we attribute them'" (Langbaum 1982, p. 27). Thus, the self in Hume's view is a retrospective construction and an operative presence, constructed by association, imagination, and memory.

Already from this summary treatment of Descartes, Locke, and Hume, we discern that there are two overarching questions concerning personal identity that largely frame philosophical debate: (1) the genetic–psychological issue: What causes induce us to believe in unitary selves? and (2) the metaphysical–ontological question: What unites the perceptions within a single mind and distinguishes one mind from another? Hume is a central figure in forming the grounds of these issues; much subsequent discussion has been designed to respond to the troubling positions he assumed, namely, that identity is incompatible with change and that perceptions are ontologically independent entities; that is, perceptions are thought of as substances, and a self then has these perceptions, leaving the issue of what constitutes this receptacle (i.e., the self) as problematic.[10]

We might profitably further examine this skeptical position and trace it to our own period, for the analytical tradition begun by Hume is an important source from which the elusive self develops in modern philosophy, but I want to veer off into another tributary that has resulted in the postmodern notion of the doubtful self, namely, that origating with Rousseau. In this discussion of one of Romanticism's key architects, the role of the elusive self is clearly articulated in its full metaphysical significance. Man as self-realization is profoundly embedded in eighteenth-century thought; the very notion of freedom was based on self-definition. Reason objectified nature and compelled the subject to scrutinize himself as he did the world, and thereby this science not only ordered the world but also identified the knowing subject. But as a result, a schism between the subject and his object was established, one that of course, extended to the subject as other (e.g., his feelings and desires). Human nature was then viewed as a set of objectified facts. This rift between the personal I and nature (or I and other) was tantamount to sundering the unity of life. Although the objectified world served as testimony of the subject's self-possession (i.e., mastery and self-discipline), it also separated him from nature, and became the mark of a person's (self)-alienation. The division of soul from body, that of reason from feeling and imagination, and that of thought from senses were inevitably acknowledged as distorting the true nature of Man. The Romantic reaction to dualism, beginning with Rousseau, attempted to reverse this exile of the self from Nature, but in so doing, it was compelled to admit the essential elusiveness of the self.

Jean-Jacques Rousseau must be largely credited for declaring the elusive self that has dominated modern discourse concerning selfhood. He did so not in a rigorous analytic fashion, but through a paradoxical exposition in which psychological scrutiny led him to seek an elusive yet idealized self. His work may be divided into three stages: (1) *A Discourse on the Moral Effects of the Arts and Sciences* (1750), (2) *Discourse on the Origins of Inequality* (1755), and (3) *Émile* and *On the Social Contract* (1762) (Randall 1962, pp. 964–79). In the first he denounces society and contrasted the ideal of a free person without the encumbrances of civilized constraints: "One no longer dares to seem what one really is" (Rousseau 1750, 1987, p. 4). He thus turns to affirm self-reliance: "Are your principles not engraved in all hearts, and is it not enough, in order to learn your laws, to commune with oneself and, in the silence of the passions to listen to the voice of one's conscience?" (ibid., p. 21). At this stage Rousseau expounds Man's natural or original state as idealized. In his next work, the *Origins of Discourse on Inequality,* "nature" has become what is native in one's original endowment. Rousseau's analysis is based on a recognition that such a person, taken out of society into the pure state of nature, could never exist, yet to postulate that person's idealized nature, to acquire knowledge of uncorrupted human behavior, must serve to guide enlightened law and education. The issue for Rousseau was to explore how one is corrupted from natural goodness, that is, how society – more specifically education – transforms the innocent. To address the question, Rousseau dethrones the primacy of reason undirected to the collective good:

> Reason is what engenders egocentricism. . . . Reason is what turns man in upon himself. Reason is what separates him from all that troubles him and afflicts him. Philosophy is what isolates him and what moves him to say in secret, at the sight of a suffering man, "Perish if you will; I am safe and sound." (Rousseau 1755, 1987, p. 54)

We possess the two instincts of self-preservation and compassion, but society multiples our interests and desires, creating confusion and conflict. Even though we gain in serving others, competitive society rewards more selfish behavior. We must reeducate to develop the rational and intelligent cultivation of our native traits.

In the third stage, that of *Émile* and *On the Social Contract,* Rousseau views society as the only means to express human nature fully in its most aspiring humane goals. It is through rationally designed education that society will truly liberate one's full capacities of freedom, of a selfhood emancipated from artificial or conventional constraints on one's true *nature:*

> Man is born free, and everywhere he is in chains. He who believes himself the master of others does not escape being more of a slave then they. How did this change take place? I have no idea. What can render it legitimate? I believe I can answer this question. (Rousseau 1762b, 1987, p. 141)

Rousseau's answer, given in the *Social Contract* was the attainment of individual freedom through enlightened education and of collective freedom through a proper restructuring of society. The basic insight rests on the identification of liberty as autonomy, which makes the *Social Contract* arguably "the most influential book on political philosophy ever written" (Randall 1962, p. 973). For our purposes the Rousseaunian self is one derived from its broadest social context, a commitment "that whoever refuses to obey the general will will be forced to do so by the entire body. This means merely that he will be forced to be free. For this is the sort of condition that, by giving each citizen to the homeland, guarantees him against all personal dependence" (Rousseau 1762b, 1987, p. 150). Politics has thus been subordinated to the ethical imperative. According to Rousseau, autonomy is also thereby achieved, for one, in uniting oneself with all others, nevertheless is commanded only by oneself in this voluntary act:

> Since each person gives himself whole and entire, the condition is equal for everyone; and since the condition is equal for everyone, no one has an interest in making it burdensome for the others. . . . [I]n giving himself to all, each person gives himself to no one. . . . [H]e gains the equivalent of everything he loses, along with a greater amount of force to preserve what he has. (Ibid., p. 148)

The problem of government is hence to bring about conditions in which the majority will desire that which is best for all; as a result, education becomes Rousseau's central problem, and the very basis of his democracy, since the entire enterprise resides in the wise rule of the body politic. Underlying this structure of government, Rousseau asserts, is the fact that one must become one's own savior and, in the ethical sense, one's own creator. But it is society that, having inflicted the deepest insult, must both heal the wounds of individualism and assume responsibility for drawing forth a person's best nature. It is in the free act of subsuming the individual to society that the burden shifts to the collective. "Rousseau's solution of the theodicy problem, then, consisted in his removing the burden of responsibility from God and putting it on human society" (Cassirer 1954, 1989, p. 77). It is not this political aspect that is of central concern here, however, but an examination of the nature of selfhood such a theory presupposes.

Rousseau begins by postulating nature as good: Our depraved culture interposes obstacles to our natural instincts; we have essentially lost knowledge of our true self. This loss arises from the imposition of both corrupting societal mores and, more generally, the pressure of conformity to others. The independence of the self is thereby lost. The *Social Contract* allows escape from this other dependence through a true communing with nature, where reason and nature are aligned, and a person is thus unified in both the rational and the natural domains. The design of nature takes the place of reason as the constitutive good, which is at least in part recovered within our own nature by becoming attuned to our motivations and feelings. This inner voice eventually defines the good in Romantic expressivism, but Rousseau has played a crucial role in enlarging the scope of the inner voice. "The source of unity and wholeness . . . is now to be discovered within the self . . . [Rousseau] is the starting point of a transformation in modern culture towards a deeper inwardness and a radical autonomy" (Taylor 1989, pp. 362–3). However, peering within, the Cartesian clarity (that we are, in principle, transparent in ourselves) is lost. Rousseau suffered much self-doubt:

> We do not know ourselves, we know neither our nature nor the spirit that moves us; we scarcely know whether man is one or many; we are surrounded by impenetrable mysteries. These mysteries are beyond the region of sense, we think we can penetrate them by the light of reason, but we fall back on our imagination. Through this imagined world each forces a way for himself which he holds to be right; none can tell whether his path will lead him to the goal. Yet we long to know and understand it all. (Rousseau, 1762a, 1911, p. 230)

He concludes that only by introspection could he discover his selfhood:

> But who am I? What right have I to decide? What is it that determines my judgments? If they are inevitable, if they are the results of the impres-

sions I receive, I am wasting my strength in such inquiries; they would be made or not without any interference of mine. I must therefore first turn my eyes upon myself to acquaint myself with the instrument I desire to use, and to discover how far it is reliable. (Ibid., p. 232)

True knowledge of Man cannot be found in ethnography or ethnology. The only source is to be found in self-knowledge and genuine self-examination. It was this attempt to differentiate between *"l'homme naturel"* and *"l'homme artificiel"* that Rousseau himself regarded as his distinctive achievement (Cassirer 1954, 1989, pp. 50–1). He proclaims the self as underived and original:

No material creature is in itself active, and I am active. In vain do you argue this point with me; I feel it, and it is this feeling which speaks to me more forcibly than the reason which disputes it. I have a body which is acted upon by other bodies, and it acts in turn upon them; there is no doubt about this reciprocal action; but my will is independent of my senses. (Rousseau 1762a, 1911, p. 242–3)

Rousseau's *self* is thus not confined to the limits of the world of the senses. It emanates from the psyche's core function (feeling), from which all other psychological capacities are derived: "It no longer appears as a special faculty of the self but rather as its proper source – as the original power of the self" (Cassirer 1954, 1989, p. 112).

Such precepts are inherently unstable, even treacherous, to structure the self on. Rousseau, according to Jean Starobinski (1971), was left in an epistemological quandary because he recognized the fundamental "veil" placed between appearance and reality, between the self and other minds. Natural paradise is the state of transparent communication between minds, but to perceive our division from others must lead to an inevitable and immediate obstruction. The distance between individuals becomes the distance of individuals from the world, that is, the exile of the self from nature. Rousseau can only dream of a state of paradise, of perfect communion with society, with others, and in derivation, with our very selves. He concludes that we must live in opacity. Rousseau does not see the world, but the world reflected only in himself, and in so doing he loses his freedom to see. This universal sameness (i.e., himself reflected in the world) is the epiphany of a moral and ontological absolute, but if seen only in terms of self, the freedom can only be sought in freedom from otherness. The subtitle of Starobinski's study, *Transparency and Obstruction,* derives from the routes of possible escape: Either make one's own sameness function as a transcendent absolute (the world thereby becomes a transparent extension of the self), or avoid any encounter with the other by declaring that all that is

The immune self

different is an insurmountable obstacle (and thereby retreat into alienation). Starobinski's account is permeated by Husserlian phenomenology, which is discussed in Chapter 6. The underlying problematic for Rousseau (and Starobinski!) is how to deal with otherness. Difference is overcome by either of two strategies: unifying or maintaining autonomy. With each, freedom from difference or otherness is sought. When the transparency mode dominates, echoes of a Lockean dictate again are asserted: "The only truth accessible to us is in our ideas or sensations or sentiments, that is in consciousness" (Starobinski 1971, 1988, p. 75) But this encounter, at least by this modern interpretation, is phenomenological:

> Approaching an object, encountering a real situation, always blurs Jean-Jacque's vision. The mist or veil that comes between him and things in the outside world is dispelled only if he recovers the pure *sensation* of the thing. . . . Pure sensation involves the world's giving itself without opposition from us. . . . In experiencing pure sensation or exercising the imagination, consciousness does not confront an *object* distinct from itself. (Ibid., p. 221)

These themes, the self as absolute and the individual as alienated victim, are found throughout the nineteenth and twentieth centuries. Implied in, if not originating with, Rousseau's thought are the views of self as collectivity, the self as uniquely original, refuge in dreams as more "real" than the objective world, love as a remedy to alienation, renewal of history or submission to it, and reason as solution (Morrissey 1988, p. xxii.)

Rousseau failed to establish the self: "My self is something that I lack, something that constantly eludes my grasp. I am always someone else, someone without a stable identity. . . . The "self" is not the unattainable position of rest but the anxiety that makes tranquility impossible" (Starobinski 1971, 1988, p. 57).[11] Having sought a "natural state," Rousseau concluded that only an imaginary position could be envisioned. The difference arises over the great divide of the democratic and Darwinian revolutions, where Rousseau's probing of subjectivity could be resolved neither politically nor theologically, and there remained intractable problems of the individual unable to escape the opacity of subjectivity. Rousseau's significance resides in his struggle to emancipate the individual, but his failure lies in his inability to confront the other: "[I]n losing the Other, Rousseau loses himself" (Morrissey 1988, p. xxiv); as Starobinski concludes, there is no self without the other.

I am sympathetic with this interpretation and present it as part of the structure of the argument advanced in this essay concerning the

154

nature of the self and its definition in immunological discourse. (This essentially phenomenological position will be revisited in our discussion of William James and Edmund Husserl in Chapter 6.) The basis of claiming the legitimacy of one or the other view of the self – what I have referred to as the punctual and the elusive – has remained a major issue in philosophy. The responses of Kant, Hegel, Schopenhauer, and Nietzsche, which I will develop in Chapters 7 and 8, serve as the main background of the notion of selfhood as it emerged in Metchnikoff's formulation. And as we will see later, twentieth-century philosophy also contributes to the understanding of immune selfhood that reemerged after World War II. But here I simply wish to note the two poles of discourse regarding the nature of the self: one where the self remains essentially undefinable and the other where the assertive perspective of a self orders its world and thereby asserts itself. With this philosophical dipole established, we will return to examine how immunology in the latter half of the twentieth century developed its conceptual themes on the edifice of what the discipline called selfhood.

5

Immunology gropes for its theory

A. CONCERNING IMMUNITY AND COGNITION

Self, in its generic sense, is a complex metaphor used in immunology to signify those attempts to discern the source, the underlying root, of immunity. In scientific discourse self becomes the source from which immune activity arises to defend the organism against pathogens or endogenous deleterious or senile elements. But the self is not easily bounded and defined as an entity. As discussed in Chapter 4, I refer to one usage as ontological, to suggest the essential quality of the *inquiry,* not to define an *entity.* By the very nature of the question of selfhood and the evasive nature of the referred subject we are situated at the elusive pole of the self metaphor. It is in the self-seeking, or self-defining, process that the organism must emerge, and I use *ontological* broadly to refer to the admittedly nebulous quality of self-referential behavior. In this sense *ontological* refers not only to our view of the issue, but assumes the added dimension of the organism's *own* behavior. This usage, moreover, surreptitiously sets self outside reductionist discourse, fulfilling yet another aspect of my agenda. The transformation of self to a more experimentally fruitful and theoretically concrete usage (discussed in detail in this chapter) never completely absolves us from dealing with the elusive meaning of identity. In any case, as an articulated issue this is a relatively recent formulation. With respect to the organismic source, that is, the origin of immune identity, immunology was confronted with a novel problem. This is not to ignore the earliest observations concerning transplantation (see, e.g., Loeb 1901), but for complex reasons, transplantation and developmental biology were not important contributing disciplines to early immunology. Immunologists directed their earliest efforts to understanding those processes that defended the organism; the main concern was how immune

156

specificity is generated. As already detailed, the idea that immunity did more than protect was strictly implicit in pre–World War II immunology, and it was formulated as a specific issue by Burnet and Fenner as *self*/not-self discrimination only in 1949. Metchnikoff alone, because of his interest in developmental biology, truly formulated immunology in an ontological context, but his argument (as detailed in Chapter 1) was either misunderstood or, more often, ignored. Self was an ambivalent necessity for Burnet. He only used it with a certain unease. This discomfort was expressed when he referred to the self hypothesis of 1949:

> The self-marker hypothesis is obviously only a provisional and rather clumsy makeshift to draw together an important set of immunological and genetic phenomena that are neglected in orthodox immunological theory. Any attempt to follow its logical implications in any detail soon reaches a point where important questions must remain unanswered. (Burnet 1956, p. 104)

But beyond the hypothesis per se, the self metaphor never assumed a more elaborate description. As noted, in his first clonal selection theory (CST) paper (1957) Burnet did not expand upon what the self *is*. Even in his later theoretical work, Burnet almost scrupulously avoided the term. For instance, in his *Immunological Surveillance* (Burnet 1970) there are four references in the index to self, but only two refer to explicit usages, and those are simple truisms ("need for a more refined method of differentiating between self and not-self" and "the body can recognize self from not-self" [ibid., pp. 22 and 47]). *The Integrity of the Body* (Burnet 1962) follows a similar strategy, and an explicit discussion of selfhood (beyond describing immune reactivity to the foreign) is absent.

After the CST was proposed and actively discussed, Burnet turned to explore his understanding of the immune self more fully:

> Though I have spent my working life almost wholly at the bench, immunology has always seemed to me more a problem in philosophy than a practical science. My experimental contributions have been very few and in peripheral, rather than central, fields and I have been more interested in attempting to see the implications of other men's work than in providing new experimental material.
>
> My interest, for many years, has been concentrated on the immunological significance of self and not self at a highly academic level, yet it is fair to say that, at the present time, the most important unsolved practical problems of immunology are in just this field of self–not self discrimination. Most of the problems of immunization against microorganismal invaders have been solved, but organ transplantation and survival,

the immunological control of malignant disease, and the nature and cure of autoimmune disease are all immensely practical matters. I believe that it has been helpful to look at these from a broad biological view point and that there is continued justification to keep thinking about immunology at the theoretical level as well as at the level of experimental detail. (Burnet 1965)

He goes on to acknowledge his indebtedness to a Darwinian orientation in regards to CST and draws the broad conclusion that "in immunology, we deal with a microcosm that reflects vividly all the essential features of the biological cosmos." Human immune mechanisms have not only evolved over phylogenetic time, but represent the same evolutionary process within the individual through the *selection* of antibody fitness in the course of the organism's immune encounters. CST was fundamentally based on Jerne's selection theory, for Talmage, who, as already noted, suggested CST independently of Burnet, also drew a similar construct: "The term 'natural selection' implies that *adaptation of the individual* occurs by selection from a multitude of existing processes much as adaptation of the species occurs largely through a selection of existing individuals" (Talmage 1957, p. 244; emphasis added). Talmage emphasized the related issues of antibody specificity and the size of the antibody repertoire. Noting that immune sera contained antibodies of many different specificities, he argued that through shared affinity a limited number of different antibody specificities could distinguish a greater number of different antigenic determinants, because each cross-reacting antibody would appear as a distinct entity. This receptor site multispecificity reduced the number of different antibodies from millions by several orders of magnitude (Talmage 1959).

Immunity appears to be a specific application of a more general Darwinian observation. To quote Burnet again:

The most important is the general acceptance by biologists of the concept of organism in the sense that irrespective of how an organism has evolved, at any point in time the individual is a definitive functional mechanism, and that any potentiality for modification in the individual will be of definitive rather than of random-selective type. This, I believe, is now an inadmissible argument, not only in regard to the development of immunity, but also for the development of the functioning nervous system, particularly at its highest levels. One gathers that the just beginning development of computers that can *learn,* will require the insertion of some randomized form of flexibility of response which can be stabilized to the extent to which it is functionally successful. The enormous number of cells, and the immense variety of synaptic connections possible, provide an adequate material basis for accepting such a picture of how higher level functions develop in the mammalian and, particularly, the human brain. (Burnet 1965, p. 18)

This passage is important in several respects. First, Burnet regards the organism as an entity subject to selective pressures. An organism's identity arises in the course of its own life-history (encounters with the foreign) as a consequence of evolutionary selective mechanisms. He further elaborates:

> The situation in lymphoid cells . . . is much more directly comparable with a straightforward evolutionary one. If random differences in immunological reactivity can arise by mutation or otherwise within the population of lymphoid cells, then all the other requirements for selective modification of the population i.e., to all intents and purposes evolution, are present. (Ibid., p. 18)

Simply put, such an immune entity cannot be "preformed" in its entirety. There are obvious genetic (i.e., built-in) restrictions, but the organism is defined immunologically by a dynamic and everchanging *process* of immune selection. Burnet does not confront the genomic definition of the self, but implies that in the library of choices making up the immune repertoire of lymphoid cells (encompassing enormously diverse receptors), selection, by both serendipitous challenge and embryonic purging, will generate a self. That self has "soft" boundaries, compared to genomic definition, that is, major histocompability complex (MHC), but it is this open-ended process of choice and chance (with *resultant* change) that captured Burnet's interest and imagination. (Chapter 3 detailed Burnet's evolutionary orientation predating CST; here we only note the consistency of his thinking.) The second striking aspect of this passage is Burnet's application of the same biological principles governing the immune system to the brain. Antedating Edelman (Edelman, 1987a), according to Burnet the capacity to learn must indicate the presence of selectionist mechanisms inserting a certain degree of a randomized form of flexibility, which is ultimately stabilized by functionality. Aside from the scientific utility of such a proposed model, our concern must fasten upon the parallel between the immune and nervous systems, each being essentially a learning and perceptive process.[1] These are considered fundamental to any definition of the mind.

At this juncture we conclude that Burnet successfully launched the self metaphor, which soon expands in the more explicit constructions aligning or comparing immunity and brain functions, which will be discussed in detail. Suffice to note here that beyond Burnet's vague soundings for the immune self, he was aware of the broad biological issues in defining organism. Burnet *used* self closest to the pole of elusive identity. Understood as a metaphor, self expands the conception of organism from a pregiven host to an entity that arises from an evolutionary process, both phylogenetically and in the course of individual life history. Moreover, Burnet recognized the crucial aspect of randomness that placed the process of immune selection firmly within the domain of evolutionary, rather than developmental, problematics (Burnet 1957). Self in this regard becomes an interactive conception whose manifestation as immune processes fundamentally parallels matters of the mind. The various meanings of the self metaphor must be coalesced. Thus, when employing *self,* Burnet must also acknowledge the implicit notion of selfhood as an anthropocentric (and anthropomorphic) construction, whose very identity is based, if not exclusively, at least dominantly, on mind. There is a close correspondence between the mind and immune identity: Each orients and mediates us with our world and essentially defines us. In this context, an allusion to the brain's possessing a parallel function to that of the immune system arises naturally from potentially shared biological properties that, on an abstract level, are essentially the same properties of selfness: the ability to perceive, react, learn, and cognate. Thus, the meaning of self begins to move toward the punctual pole as it assumes affinity with function of the brain.

A second parallel to neural function is found in the hypothetically shared mechanism of biological operation. There is a certain scientific *and* philosophical coherence in the application of selectionist mechanisms to both immune and neurological functions (Piatelli-Palmerini 1991a). The paradigm is similar in conception to the insight into evolutionary mechanisms whereby, beyond random mutation of point nucleotides alone, genetic shuffling of entire chromosomal segments apparently also gives rise to speciation (Mourant 1971; McClintock 1987). The issue at hand is to establish "new structures," in the sense of new functional opportunities, by employing available units in novel ways. This is the type of mechanism by which diverse immunoglobulins are generated, and regarding CST, the selection of a reactive repertoire follows the same logic. Further, this is the basic concept of Chomsky's generative grammar, which has been applied broadly to linguistics and to cognitive science. Such selectionist theories share a similar conceptual structure: A matrix of units (genetic segments or neurons or immunoglobulins) can be sorted by adaptive stress to allow functional response that is to a certain degree open ended. Recent data suggest adaptation at the genetic level itself (Foster 1991; Sarkar 1991a); it is clear that even without invoking neo-Lamarckian mechanisms, selectionist processes at an ontogenetic level maximize biological diversity by dynamic sorting, amplifying, and purging operations. The permutations of choice offered by such strategies far exceed the level of complexity predicted by some fixed (or linear) logic. By positing "architectural modeling," with flexible units building biological responses, learning, whether cognitive or immunological, becomes connected to an operation of selecting basic units to compose the perceptive apparatus. This is obviously a novel twist to the neo-Darwinism of the mid-1950s, but it closely follows the fundamental understanding of selection, something which both Jerne and Burnet apparently accepted.

Burnet brought a term with psychological and philosophical connotations to his scientific discussion because he recognized the significant role of identities that could not be more explicitly defined. He was forced at first to employ *self* in its more elusive sense, but the notion quickly assumed firmer demarcation in its association with cognition, to which we now turn. When Burnett formally introduced the self metaphor in 1949, he prepared the critical context for other terms that were consistent with that notion of personal identity, or what would soon become explicit, a *knowing* entity. For instance, immunological "training," "recognition," "learning," and "memory" are terms that rely on a cognating creature. Thus, in erecting a new theoretical edifice on the foundation of the self, the appearance of *information* finds a conducive environment for evoking a rich metaphorical potential. Information is certainly a tool that a mind uses; thus, its appearance might be regarded as a natural member of this new immunological lexicon. Review of the historical development of this cognitive perspective reveals that the ready analogy of the immune system to the nervous system awaited a complex metaphorical approximation of the self. Only when the self was in place, could a cognitive apparatus be added to develop fully the model's inherent potential. Not surprisingly, such a formulation was articulated by Burnet shortly after he presented the self metaphor. I believe the faint outline of what was to become a novel conception of immune function can be discerned in Burnet's earliest use of the word *information.*

Information, its manipulation and control, has had a complex and important history in biology. The novelty of information theory in the mid-1940s was not concerned with its measurement or transmission, which date to the nineteenth century, but with new notions regarding the logistics of information storage. With the development of the modern computer, information technologies and their powerful epistemic influences began to permeate other academic disciplines and the broader cultural sphere. The impact of automated control systems on modeling biological processes was anticipated and promoted by Norbert Wiener and his colleagues during the early 1940s, and with John von Neumann, grandiose schemes for applying complex field control and communication engineering to several areas of biomedicine were planned. Their interest ranged from modeling neural function to viral replication. Just as Burnet and Fenner published *The Production of Antibodies* (1949), Wiener's *Cybernetics* captured attention in diverse fields, as information theory was offered as a new paradigm. Von Neumann's ideas concerning self-duplicating automata soon followed and similarly enjoyed wide popular and scientific interest (Heims 1980). His reproducing machines copied via coded instructions, and thus such "copiers" (or cybergs) first appeared to serve as models for genes.

There was much speculation concerning the application of cybernetic theory to biological systems in general and to genetics in particular. For example, despite a lack of experimental support, J. B. S. Haldane, in an unpublished paper written in 1948, attempted a calculation of the "total amount of control (information = instruction) in a fertilized egg, and various other similar points" (Kay 1994). H. Kalmus made similar attempts and concluded that genes could be described as messages or sources of messages and thus be regarded as the basic elements of biological control (Kalmus 1950). The most aggressive proponent of cybernetic theory application to biology was Henry Quastler (Kay 1994), who, in 1952, organized a symposium to integrate information theory and biology (Quastler 1953). Of particular note, the general proposal that proteins might be regarded as "message" and amino acid residues as an "alphabet" was applied at this meeting by Haurowitz to the specificity of antibodies (Haurowitz 1953). (Burnet specifically cites Haurowitz's theory in *Enzyme Antigen and Virus* [pp. 65–7, 1956]). This history is important inasmuch as we should recognize that cybernetics was considered as a new means of understanding genetic transmission before the Watson–Crick paper of 1953, and immunologists were keenly aware of its possible application to the production of antibodies. In fact, the generation of diverse proteins with directed immune specificity was viewed as an important problem of protein synthesis, and thus it is not surprising that Burnet would also comment on the possible ramifications of information theory.

In 1955, shortly after the self-marker theory was proposed, Burnet, like most of his contemporaries believed that protein formed in some nebulous fashion as a template on some preexisting molecular structure. In *Enzyme Antigen Virus* (1956) Burnet was struggling to account for protein synthesis, specifically antibody production, as in his earlier monographs, but now further modified to account for the information contained in DNA. It is not necessary to summarize the particular issues raised in this short monograph to appreciate the appearance of *information* in the immunological literature. Consider how *information* is introduced in the overview printed on the book's jacket:

> The problem [protein, virus, and antibody production] is discussed from the point of view of the ways in which 'pattern' can be manifested in protein or other macromolecules and an attempt is made to sketch the outline of something like an 'information theory' of the cell in which macromolecular pattern serves as a means of conveying 'information' within the organism.

Burnet, as in his earlier attempts, sought a common ground to explain antibody production within the broader concerns of protein synthesis and viral replication. He did so on a persistent template hypothesis that built on a vague notion of biological patterning. This was an intermediate stage in acknowledging the basic information flow from DNA to protein, which was in the air, so to speak, but had no experimental basis. As he wrote:

> It is the thesis of this monograph that, where biological matters above a certain level of complexity are concerned, most interpretations must be in terms of macromolecular pattern which, by interacting with complementary or near complementary pattern in some other functional situation, can induce action or, if it is more convenient so to express it, convey information or instructions. (Burnet 1956, p. 58)

The details of how Burnet accounted for antibody synthesis are not of concern here; I wish only to acknowledge the early appearance of "information" as a governing principle. He was well aware of the theories expounded by Haldane and Haurowitz (ibid., p. 17), and in a section entitled "Information Theory in Biology" Burnet acknowledged the profound impact cybernetics made on his own theorizing:

> Since 1945 there has been widespread recognition amongst scientists and the public generally of the importance of the principles which have emerged from experience in the development of electronic communications and control systems. . . . In our field, the relation is a good deal more distant, but one cannot escape the attraction of the general approach. This monograph was originally conceived as an attempt to de-

velop something analogous to a communications theory that would be applicable to the concepts of general biology. However, it has not been found possible to make any serious use of the already extensively developed concepts of information theory in the strict sense. In part this is due to . . . only the most generalized sketch of an outline [that] has yet to be given of how information theory at the strict level can be applied to biology. The only extended account of such an approach that I have been able to find is the symposium edited by Quastler (1953). (Ibid., pp. 164–5)

Although Burnet felt confident that macromolecular pattern replication was the basis of information flow in the cell, that hypothesis, and thus the thesis of the book, was weakened by lack of experimental support. Burnet sensed the potential power of information theory, but its application as a research program was highly problematic at this time.

In *The Clonal Selection Theory of Acquired Immunity* (Burnet 1959) information is again used, but now in a completely different context. No longer directly concerned with transfer of genetic instructions, information appears as a library of ready-to-serve lymphocytes. Information would have "to be stored, either individually or collectively, in the cells of the antibody-producing system," and "long-lasting retention of information" was required to account for immunological memory (p. 47). The clonal selection theory placed *information* firmly within immune theory, setting the stage for the full expansion of the self metaphor to the cognitive domain. Although Burnet first uses "cognate" in passing in 1959 (p. 70), by 1963, in *The Integrity of the Body,* a full-blown language metaphor is presented. In the section where Burnet is explaining the clonal selection theory, he illustrates how "the body acquires or generates the information which allows it to differentiate immunologically between what is self and what is not self" (1962, p. 94). He draws an analogy with another information transfer example – words:

If it is true that around four amino acids units are responsible for each specific immune pattern . . . we are in a position to make use of the analogy between 20 common biological amino acids and the letters of the alphabet. Each pattern could be represented by a four-letter combination. . . . [W]e can imagine an electronic computer set to produce at random four-letter groups from a 26 letter alphabet. If 10^7 words are asked for, we should have 99-percent probability of getting at least one example of every possible four-letter word. . . . Our computer has another characteristic. Once the selection has been completed [to create the reactive library], all the remaining "words" are stored in the memory and when any combination is asked for it can be produced in unlimited numbers, but only if it is in the memory. (Ibid., pp. 94–5)

Here Burnet utilizes analogies from language, information processing, and the seductive power of cybernetics to explain immune function. Jerne was to develop the cognitive model further.

My concern is to focus on how the immune network – formulated in cognitive terms – became part of the self metaphor. In 1960 Jerne had regarded the antibody-forming system as "comparable to a typewriter" (Jerne 1960, p. 348) or "analogous to an electronic translation machine" (ibid., p. 341), thus implicitly appealing to models of mind derived from cybernetics:

> The antibody-forming system in an animal is analogous to an electronic translation machine, the parts of which are scattered among other devices in a factory exercising a large variety of functions. In trying to understand the design and structure of this translating machine, we might obtain useful clues by studying parts such as bits of magnetic tape and photographic film. But it would be equally important to analyse the general performance of the machine, in order to deduce what essential operations have to be provided for, and to try to induce the machine to make mistakes that would reveal limitations of its complexity. Having observed that the machine translates a foreign language into English, we might say: The machine produces English, but recognizes only Foreign. We would realize that both English and the foreign language are composed of the same alphabet, and that single letters cannot be the units that are recognized by the machine. We might also feel sure that a sentence cannot be recognized by the machine. We might also feel sure that a sentence cannot be recognized as a whole because the number of possible sentences would surpass any tolerable complexity. We would probably conclude that an important feature of the machine would have to be an ability to recognize single foreign words, and that the mechanism by which it functions must, in some form or other, include the consultation of a Foreign–English dictionary. (Ibid., p. 341)

In the mid-1960s Jerne dealt explicitly with the metaphorical meaning of immunological "memory" and "learning." In an anthology devoted to diverse systems of learning behavior, he noted how immunologists used metaphors, such as memory and recognition, that were obviously derived from brain function, but he considered "the analogies between the immune system and the central nervous system . . . to be quite superficial" (Jerne 1966b, p. 151). Citing his selection theory, however, he tentatively suggested that "learning" shared the same tenets in both contexts:

> . . . "selective" theories of antibody formation . . . are based on the idea that the organism acquires no *new* knowledge by the exposure to antigen

165

but has already spontaneously created samples of antibodies of all possible specificities in order that the "fitting" ones can be selected as circumstances may dictate. In considering learning by the central nervous system, this would correspond to the idea that we can never truly learn anything new. As pointed out by Socrates (375 BC) and Kierkegaard (1844), understanding must be preceded by recognition, and all learning, therefore, must consist of a "recollection" of knowledge already present in the soul. (Ibid., p. 157)

In another article Jerne went on to draw more explicit comparisons and contrasts with the nervous system (Jerne 1967b). Each system has a history of encounters with the world that remain present both in the form of irreversible changes (with the alteration of each preencounter state into a new one) and in the forms of a *memory* of that encounter (always affecting the next response). In the latter sense both systems *learn* from experience. Although these are somatic events, and no Lamarckian transfer to progeny of such cognitive experiences is known to be possible, Jerne postulated that there is a certain plasticity of learning from a species-wide capacity delimited by genetic factors. Jerne further speculated that the nervous system might well share similar molecular mechanisms to those found to generate antibody diversity. Solely on the basis of knowing that immunoglobulin had variable and constant regions he suggested the following:

> Analogous to the utilization of the diversity of the variable part of the antibody light chain in the immune system, learning from experience is based on a diversity in certain parts of the DNA, or to a plasticity of its translation into protein, which then controls the effective synaptic network underlying the learning process. I would, therefore, find it surprising if DNA were not involved in learning, and envisage that the production by a neuronal cell of certain proteins, which I might call "synaptobodies," would permit that cell to enhance or depress certain of its synapses, or to develop others. (Ibid., p. 204)

He went on to propose that nervous system learning may well operate on the same selection basis found in the immune system.

By 1974 Jerne's inferences became bolder: The brain is likely to reflect the same basic structure proposed for the immune system:

> . . . the immune system, when viewed as a functional network dominated by a mainly suppressive Eigen-behavior, but open to stimuli from the outside, bears a striking resemblance to the nervous system. These two systems stand out among all other organs of our body by their ability to respond adequately to an enormous variety of signals. Both systems display dichotomies and dualisms. The cells of both systems can receive as

well as transmit signals. In both systems the signals can be either excit-
atory or inhibitory. The two systems penetrate most other tissues of our
body, but they seem to be kept separate from each other by the so-called
blood-brain barrier. The nervous system is a network of neurons in which
the axon and the dendrites of one nerve cell form synaptic connections
with sets of other nerve cells. In the human body there are about 10^{12}
lymphocytes as compared to 10^{10} nerve cells. Lymphocytes are thus a
hundred times more numerous than nerve cells. They do not need connections by fibres in order to form a network. As lymphocytes can move
about freely, they can interact either by direct encounters or through the
antibody molecules they release. The network resides in the ability of
these elements to recognize as well as to be recognized. Like for the nervous system, the modulation of the network by foreign signals represents
its adaptation to the outside world. Early imprints leave the deepest
traces. Both systems thereby learn from experience and build up a memory that is sustained by reinforcement and that is deposited in persistent
network modifications, which cannot be transmitted to our offspring.
These striking phenotypic analogies between the immune system and the
nervous system may result from similarities in the sets of genes that govern their expression and regulation. (Jerne 1974, p. 387)

By around 1985 Jerne had fully developed this position, and the formal comparison was well under way. Addressing the theme of self–
nonself explicitly, Jerne argued that a decisive mechanism to overrule
antiself was the encoding of antibodies that recognized the self. According to Jerne, these self antibodies are then controlled by the production of anti–self antibodies during ontogeny and normal immunoglobulin is essentially a "stable population of molecules that are anti-
idiotypic to the anti-self molecules likewise encoded in the germline"
(Jerne, 1984, p. 18). Based on rather simple calculations, Jerne surmised
"in its dynamic state our immune system is mainly self-centered, generating anti-idiotypic antibodies to its own antibodies, which constitute
the overwhelming majority of antigens present in the body" (Jerne
1985, p. 851). From this position, Jerne clearly draws the parallel of the
immune system with the brain:

> . . . the immune system (like the brain) reflects first ourselves, then pro-
> duces a reflection of this reflection, and then subsequently it reflects the
> outside world: a hall of mirrors. The second mirror images (i.e., stable
> anti-idiotypic elements) may well be more complex than the first images
> (i.e., anti-self). Both give rise to distortions (e.g., mutations, gene re-
> arrangements) permitting the recognition of nonself. The mirror images
> of the outside world, however, do not have permanency in the genome.
> Every individual must start with self. (Jerne 1984, pp. 19–20)

167

The immune self

There is a striking "bridge" of the self as *entity* and the self as *process* in these few short lines. Jerne invokes the relational basis of Kierkegaard's formulation of the self (see the opening lines of *The Sickness Unto Death* [1849]) in the image of the mirrors endlessly reflecting the self's image, and at the same time he notes the genetic basis of the self as the "starting point" of this dialectical identity. In a curious way, Jerne would have it both ways, seeming to acknowledge the inner tension of resolving what is fundamentally a philosophical conundrum.

In a paper published the next year Jerne drew a more explicit parallel with language. The immune repetoire of specificity (exceeding ten million different proteins, that is, over a thousandfold more than all other body proteins taken together) resides in the variable region of the immunoglobulin molecule, which may be regarded as analogous to a lexicon of sentences, rather than to a vocabulary of words (Jerne 1985, p. 850). In this case different amino acids (i.e, words) forming diverse proteins (i.e., sentences) still possess the mirror-image structure required to bind the corresponding antigen. Thus, although the antigen may be constructed with different words, these possess the essential "sentence" structure required for recognition. Jerne then turned to Noam Chomsky's theory of generative grammar, specifically the hypothesis that certain deep, universal features of language competence are innate characteristics of the human brain, which in turn suggests that language acquisition is dependent upon a DNA-encoded function. Viewing generative grammar as a possible model, Jerne wrote: "I find it astonishing that the immune system embodies a degree of complexity which suggests some more or less superficial though striking analogies with human language" (1985, p. 852). The flexibility and fine specificity of the immune response dwells in the ability of genetic components to sort, select, and assemble into diverse immunoglobulins, but why Jerne chose to parallel such a process to language reveals a metaphor of great complexity. Heavily indebted to Chomsky's generative grammar, Jerne attempted to crystallize the analogy between linguistics and immunology (Jerne 1984; 1985). He used the linguistic theory as a model for explaining how the immune system reacts to the universe of antigens with such adaptive, versatile precision. But in formulating this analogy, Jerne had to invoke the role of the brain and thereby found himself addressing the entire issue of mind–body modeling. In this construction he alludes to characteristics of body as entity and of mind as process. The immune system parallels this structure: Immunoglobulin is the entity, and network function its process.

168

Jerne placed himself in a classical predicament: Is language the *voice* of the mind, or the mind itself? The position that language essentially constitutes what we know about the world and represents the epistemological model for understanding every form of human endeavor was strongly advocated by Whorf (1956). This so-called internalist conception of language is countered by the claim that no private language exists, because meaning arises only in relation to a complex network of other communication in public interactions. (This position is expounded by Bakhtin [1986] but also argued by Dewey, Heidegger, Wittgenstein, and Quine [Kent 1991; Jameson 1972; Rorty 1979]). An attempt to offer a synthesis is given by Schultz [1990].) Those who have used language as a source of ideas about immunology have assumed that language and cognition have more to them than can be subsumed under behaviorism, that is, that there are mental processes that do not fit the physical stimulus–response model. In drawing a parallel between the immune system and language, the shared implicit characteristic is the generation of *meaning*. It has been suggested that semiotics, the study of signs, might be applicable to immunology with respect to the kind of phenomena both study; the parallel, then, between language and the immune system has been addressed under the auspices of whether the latter qualifies as a semiotic system (Sercarz et al. 1988). Ferdinand de Saussure's *Course in General Linguistics* (1916, 1983) has had a decisive influence in semiotics. He argued that the signified is inseparable from the signifier, that they are two aspects of the same unity. From this perspective language is a system of signs in which the relation between signs and what they signify (words of "things," expression of "ideas") is arbitrary or conventional (as opposed to "natural"). Thus, words (signs) embedded in a given language have meaning only within that system. Signs, then, are "determined positively not by what they are but negatively, by what they are not, by their differences from the other elements of the system. They are what the others are not. Language, in this respect, is not a system of identities. It is a system of differences" (Behler 1991, p. 66). One can easily discern the attraction of such a formulation for theorists of immunity.

The first matter to establish is what qualifies as a "sign" in immunity. Bona (1987) and Golub (1988) make an important distinction in applying semiotic control criteria to immune phenomena, by noting that the immune response and the immune system should be distinguished in such a discussion. A stimulus–response reaction fails to become semiotic if no signs are used. The immune response regarded merely as antibody–antigen interaction does not meet the standard of

169

semiosis. But Golub argues that the immune system does qualify as semiotic if one accepts the Jerne internal idiotypic image system. Because the elements of the immune system communicate with one another through idiotypic recognition in the absence of exogenous antigen, the immune reaction, which results in the clonal expansion of a particular idiotype, is responding to its already present internal image. In this sense "the same receptor recognizes and responds to both the antigen and the 'sign' of the antigen, the internal image. From this it can be seen *that the network is its own sign system* which 'sees' in itself the sign of the antigen" (ibid.). Whether one accepts such a formulation of a sign is problematic, but of even more salience is to elucidate how – on the model of semiotics – "meaning" is generated within the immune system. Meaning in language is based on the word as sign (Saussure, 1916, 1983), which consists of the signifier and the signified; the sign conveys meaning by indicating something else – its referent. At this level it is not apparent how semiotic analysis has been helpful. Semiotics has only classified signs and their systems and has not established generative mechanisms to show how signs become signifiers: "[T]here is no explicit mechanism for the origin of meaning. It is somewhat like reading a taxonomy text before natural selection was proposed as a mechanism for the origin of diversity" (Varela 1988). How to choose between a taxonomic (or structuralist) versus a generative explanation for meaning is the key impasse, and a strong case can be made that analysis of the immune system might offer semiotics a useful model for language, rather than vice versa (e.g., Celada 1988; Golub 1988; Prodi 1988; Varela 1988; Violi 1988).

It is of interest in regard to the general structure of our analysis to situate the self concept along the punctual–elusive axis. Evident in the discussion concerning immune function as analogous to language is an underlying effort to concretize further the self metaphor. If Burnet began with an elusive self (what I dubbed ontological), the program to establish congruence with language – and the later semiotic endeavor – must be regarded as efforts to move the self of immunity toward the punctual role. As science, the immune self has an epistemological goal to concretize itself – or in earlier terms, localize its phenomena into definable coordinates. The "core of identity" and "the source of immunity" are vague and clumsily abstract from the point of view of an experimental science. As elusive as the language analogy might be, it still represents movement toward a better punctuated position. But the application of shared principles between language and the immune system must be viewed with caution. Although each may be viewed as having grammar, inasmuch as they have a limited set of elements and general principles (rules) to combine them, the analogy has also been met with a strong sense of skepticism:

> What makes a natural language a semiotic system is the existence at the same time of a grammar and a semantics, not the mere capability of combining the elements of the system according to given rules. But it is precisely the existence of something like an autonomous semantics that seems problematic in the immune system. (Violi 1988)

The issue revolves around the previously mentioned notion of whether T and B cell recognition constitutes "cellular communication" analogous to *meaning* as defined by language discourse or, to the contrary, such recognition is more properly viewed as but a sophisticated stimulus–response pattern, with "communication" simply an alternative metaphor. Immune theory as present cannot truly resolve this question, for how meaning can be conceptualized as generated by the immune system is still incompletely understood.

The recognition of molecular forms by the antibody repertoire is suggestive of an open-endedness analogous to language. Violi (1988) brings up one crucial caveat:

> In the linguistic production, not only do we recognize a given form (for example the sounds of language) but we also "grasp its meaning," i.e., we attach to its formal components an interpretation which is not causally linked to them. Does the immune system do something like that, or does it merely react to new molecules in the same way as, for example, the nervous system would react to a new stimulus never experienced before? (Ibid.)

In short, the question is whether processes of the immune system are causally determine or there is processing (*qua* semiosis) that can be regarded as "interpretive," involving what Eco calls the C space.[2] At the level of single reactions it is quite apparent that simple stimulus–response patterns are operant, but if the entire system functions as more than the sum of its parts, possessing emergent properties, then meaning as an emergent quality of language may be understood as analogously present within the immune system. These issues are increasingly being addressed with the growing appreciation of the multidirectionality of the flow of information within the immune network in the steady state. It has become clear, for instance, that the same idiotope can be shared by antibodies with various specificities, and antibodies in the steady state can exhibit multispecific binding properties (Bona 1987). How control ensues both in resting and evoked responses is the primary experimental concern. In pursuing the analogy with respect to the connection between language and mind, we may ponder how cognitive operations may be expressed within the immune system, if it functions according to rules similar to those governing a generative grammar.

The metaphorical complexity of Jerne's model, irrespective of its scientific utility, presents a most interesting question for any theorist who ponders the philosophical basis of selfhood. In this regard, immunologists follow their linguist counterparts in posing essentially parallel questions. Jerne implicitly addressed the shared general issue of how to place deterministically caused elements into the broader context of a semiotic, process-oriented model. This is the same fundamental question facing theorists of language generation. The "punctualists" might be more at ease with viewing language – and by analogy, immune functions – as an expression of the self, that is, somehow a "tool" of the ghost in the machine. Language/immune *process,* working on its structural hardware, serves an alternative formulation: Immune function (*qua* process) is not the "voice" of the underlying self, but must be viewed as *the* self incarnate. Jerne did not explicitly define the issue in such terms, choosing only suggestive allusions and forays, but with the collaboration of immunologists and cognitive scientists more explicit models soon emerged.

From the model of idiotypic controls, an active endogenous system of immune activity was discerned. The normal processes of immune surveillance, such as those found in germ-free animals (Benner et al. 1982) and described for destruction of senescent erythrocytes (Singer et al. 1986), require active antibody production that probably exceeds that demanded for action against pathogens:

> I see the immune system as continuously seeking a dynamic equilibrium – and by 'dynamic' I mean that a vast number of immune responses are going on all the time, even in the absence of foreign antigen. The old term "immune response" suggests that the system is "at rest," waiting to "respond," whereas I think it is continuously active, interacting with self-antigen, idiotopes, factors, etc. (Jerne, personal communication, quoted in Golub and Green 1991, p. 486)

On the basis of this realization the hypothesis has been proposed that the self is not negatively defined (by ontogenetic thymic purging of self-reactive clones); rather, fundamentally a positive identification of the self configures the immune system (Pereira et al. 1989; Kocks and Rajewsky 1989). The Paris School (referring to Antonio Coutinho, John Stewart, Francisco Varela, Nelson Vaz, and their collaborators) has actively pursued such an ambitious theoretical and experimental program to critically revise CST, and challenge that negatively defined notion of self, that is, of the self as that entity that evokes no immune challenge (reviewed by Varela et al. 1988; Varela and Coutinho 1991). The issues they address explicitly are those activities operating in the unchallenged state, where the ongoing immune network functions undisturbed by foreign insult, as for example, internal lymphocyte activities and natural antibody production in unimmunized animals, pre–immune repertoire selection, tolerance, and immune memory. Theorists of this school argue that the global properties of the immune system cannot be understood from analysis of component parts alone; "emergent properties," "nonlinear network or complex systems," "global cooperativity," and other terms borrowed from diverse scientific disciplines similarly concerned with dynamic interactions of complex systems are invoked to describe the immune system as a whole (Varela and Coutinho 1991).

The Paris School contrasts itself with CST in offering a radical alternate immune paradigm (Coutinho and Stewart 1991). The first principle of this so-called autonomous network theory (ANT) is that the immune system has its own endogenous activity that does not require stimulation from antigenic insult. It is regulated by *self*-recognition, as opposed to being driven by nonself elements. In the classical formulation of CST self is definable only from an external perspective as the total molecular constitution of the organism that is ignored by the immune system. In the ANT schema the self is operationally defined by the immune system itself; that is, immunity constitutes molecular identity and recognizes nonself in the context (viz. MHC) of its selfhood. Thus, the self for CST and ANT is not the same entity, nor does the term mean the same thing. From their divergent understanding of the self, striking contrasts ensue in the interpretation of tolerance, autoimmunity, and immune responsiveness to the foreign. What from the perspective of CST are anomalies, such as the high prevalence of autoantibodies or the poor correlation between such factors and "autoimmune" disease, are explained by ANT as natural outcomes of normal immune activity with regulatory factors either present or aberrant (accounting for these innocuous anomalies or disease, as the case may be). Strong corroboration of the argument for such autonomous activity is gleaned from studies with germ-free animals that demonstrate ongoing lymphocyte activation (Pereira et al. 1989). In the ANT model autoimmune disease would arise from unregulated autoimmune reactions, which might be treated either by immune suppression of bandit clones or alternatively by stimulation of the immune system to reintegrate these uncontrolled clones, as possibly demonstrated by the infusion of massive quantities of immunoglobulin in certain autoimmune diseases (Sacker 1992). A similar rationale could be invoked for immune stimulation in the case of transplantation recipients, who, when exposed to allogeneic antigens in the form of blood transfusions, experience heretofore unexplained beneficial effects (Churchill and Kurtz 1988, pp. 279–80). In addition, immune stimulation to provoke immune integration of heterologous components into the self network is ANT's hypothetical account of how spontaneous abortions may be corrected by active immunization of the mother with paternal transplantation antigens (Coutinho and Stewart 1991).

Regarding immune reactivity to nonself, the Paris School attempts to address the unexplained nonspecificity of antigenic challenge, where 90 percent of all antibodies so stimulated are actually unreactive with the immunogen (ibid.). Beyond questions of sensitivity, clearly much of the immune response is "selected" by factors other than the specific antigen. In addition, upon infection more than half of all the lymphocytes appear activated, raising another fundamental concern about specificity. There are reasonable CST-based explanations to these findings, but Coutinho and his co-workers have raised interesting problems of interpretation and thus fundamentally challenged CST's understanding of selfhood and how it is derived. They have thereby reopened the question of a model of the self and its very definition in immunology.

From this perspective of a revisited and revised notion of the self, the Paris School makes explicit reference to the nervous system to illustrate its hypothesis; even the nomenclature is appropriated. The "self-related immune network" is designated the "central immune system" (CIS), and the reactive, rapidly turning over lymphocytes posed to assail foreign intrusion are the "peripheral immune system" (PIS) (Varela and Coutinho 1991). The CIS, as the true center of immune identity whose ongoing activity comprises the core immune function, is obviously analogized to the brain and contrasted with the peripheral system, which, although crucial for reactions to the outside world, is subordinate to the central faculty. According to the Paris School, this reacting faculty, that is, the PIS, has erroneously been the conceptual focus of immunology; it is the CIS, the *self-related* immune network that must be characterized in order to understand the functional structure of immunity. CIS and central nervous system (CNS) are tied together as the focus of selfhood, and to understand the CIS, insight must be drawn from what is understood about neural cognition. In "Cognitive Networks: Immune, Neural, and Otherwise," the Paris School clearly draws on the shared metaphor of the CIS and CNS:

There is a strong intuitive sense in which immune systems are *cognitive:* they recognize molecular shapes, remember the history of encounters of an individual organism, define the boundaries of a molecular "self," and make inferences about molecular species likely to be encountered. By and large immunology has left these admittedly cognitive terms undefined or at a metaphorical level and has concentrated, instead, on the molecular details of immune components. (Varela et al. 1988, p. 359)

The nature of immunity must be addressed in terms of its cognitive mechanisms, which assume analogy to neural networks (even as they are admittedly different). Immune identity arises out of the network's activity, in toto and in everchanging dynamics:

Clearly, one can define "self" from a biochemical or genetic or even a priori basis. But from our vantage point, the only valid sense of immunological self is the one defined by the dynamics of the network itself. What does not enter into its cognitive domain is ignored (i.e., it is non-sense). This is in clear contrast to the traditional notion that IS [immune system] sets a boundary between self in contradistinction to a supposed non-self. From our perspective, there is only self and its slight variations. That which is foreign is only so because it is similar to (or only slightly different from) self: the Unheimlich [sinister] of that considered as foreign can only come from this proximity (*as Freud pointed out long ago*). (Ibid., p. 365; emphasis added)

The self has no firm genetic boundaries, but arises from experience and self-creative encounter:

It must be stressed that the self is in no way a well-defined (neither predefined) repertoire, a list of authorized molecules, but rather a set of viable states, of mutually compatible groupings, of dynamical patterns. In effect, a molecule is neither self nor antiself, as a musical note does not belong more to a composer than to another one. The self is not just a static border in the shape space, delineating friend from foe. Moreover, the self is not a genetic constant. It bears the genetic make-up of the individual and of its past history, while shaping itself along an unforeseen path. (Ibid., p. 363)

The model of the immune system as cognitive has gathered other supporters. For instance, dissatisfaction with CST has led Irun Cohen to conclude:

. . . progress in immunology appears to have rendered the clonal selection paradigm incomplete, if not obsolete; true, it accounts for the importance of clonal activation, but it fails to encompass, require, or explain most of the subjects being studied by immunologists today. . . . More importantly, what we have learned about autoimmunity directly contradicts a major corollary of the clonal selection paradigm: auto-immunity is not

175

The immune self

an aberration, but is a property of all healthy immune systems. (Cohen 1992a)

Cohen argues that instead of a science focused on the discrimination between self and nonself, the task of immunology is to understand the true, more basic function of how immunity enhances fitness. Cohen, like his Parisian colleagues, argues for a paradigm shift: CST has been eclipsed by "the cognitive paradigm." After schematizing the enhanced fitness problem (survival against pathogens) common to adaptive, cognitive systems, that is, the problems of maximizing sensitivity to differences between signal and noise, defining the context of the signal, and finally, controlling the response, he explicitly throws down the gauntlet:

> Cognitive paradigms are founded on the idea that any system which collects and processes information will do its job most efficiently by having an internal representation of its subject. Simply put, a cognitive system is a system that extracts information and fashions experience out of raw input by deploying information already built into the system; in a sense, a cognitive system is one that knows what it should be looking for. This internal information, which precedes and imposes order on experience, can be seen conceptually as a blueprint for dealing with the world. In the abstract, cognitive systems can be said to behave with a sense of direction; their internal organization endows them with a kind of intentionality. Cognitive systems, then, are not passive processors or recorders of information; they are designed to seek very particular information from the domain in which they operate. (Ibid.)

Cohen proposes a particular view of neural function that posits that internal images are prewired in the structure of the nervous system. Thus, for instance, infants recognize human faces because the human brain is preprogrammed for such perception, and bees are hard-wired with the internal image of an "essential" flower. Encoding important features of the environment, such images facilitate the system's interaction with the environment by providing patterns to which the system can selectively respond. The key property of the cognitive paradigm is "the definition and creation of information by internal representations, *the intentionality of a system* (ibid.; emphasis added).

Cohen, along with the Paris School, composes a metaphor of the self closely aligned with our intuitive understanding of the mind. The comparison of the architecture of the immune system to models of neural networks – complete with analogous computer program simulations (Stewart, Varela, and Coutinho 1989) – offers not only a scientific program, but a firm foundation for altering our orientation toward immu-

176

nity. As already mentioned, various mathematical models of the immune system have been proposed following the idiotypic network proposal of Jerne. Most models set up systems of differential equations that serve to represent the behavior of the system (see, e.g., Perelson 1987; 1989). An alternative approach is the cellular automaton, first utilized by Kaufman, Urbain, and Thomas (1985), which attempts to model more "biologically." It is a dynamic model whose evolution is described by local interactions. Complexity is introduced with no need for additional qualitative difficulties, but the simulations are of restricted size. It is this kind of modeling that the Paris School has employed with some success. (See Celada and Seiden 1992 for a review.)

The cognitive model argues that immune theory must take account of why foreign proteins, with a potential to paralyze specific immune reactions by polyclonal activation, evoke a largely discriminating response. This highly targeted focus suggests a selective or even intentional property. And, of course, the alternate context of the antigen as either benign or deleterious also suggests an ability for cognitive assessment; as it cannot be readily understood by simple encounter, processing, characteristic of hierarchical systems, is implicated. Such complexity regarding the responsive portion of the immune reaction is also highly dependent on context, another determinant defined most readily by systems analysis. All these matters seem more appropriately incorporated into models where the self is defined in a positive fashion. Reaction to the foreign becomes secondary or a by-product of the central self-defining function, as the pathogen is *known* only in the context of its association with self; that is, foreign peptide binds to MHC for T cell recognition (Kourilsky and Claverie 1989). These two concepts, namely, the self as positively identified and the *other* perceived in the self's context (viz. MHC), together potentially herald a major shift in immunology's theoretical structure. And in the more specific context of this discussion, namely, the development of the self concept in immunology, note how the underlying sense of the language has radically shifted.

Before describing the third major metaphorical category of the immune self, the body metaphor, I must note that there is an inherent ambiguity in viewing the mind metaphor along the elusive-punctual axis: The mind as an *entity* is "punctual" by our earlier Lockean definition only under certain conditions, and more generally connotes an elusive and hardly definable character. Before further elaboration of this issue (discussed in Chapter 6), we might argue here that the self may be regarded as both an *entity* and a *process* to capture the bivalent character of the dynamic subject. The difficulty of conceptually grasping the self resides in part within the very foundations of our subject–verb structured language. From the punctualist perspective, only a subject can act. In this regard, then, how can there be immune activity without a definable entity – a self that acts? Perhaps we require a new grammar, for the self is neither subject, nor object, but is actualized in action (Whorf 1956). The self becomes, on this view, a subject-less verb, or perhaps predicate. These musings reflect a general concern with discerning how properties arise from structure – for instance, how the mind emerges from the brain. The approach to this general issue quite dramatically separates the anatomists, whether gross or molecular, from their colleagues concerned with emergent process phenomena. It is in this sense that I believe it fair to divide immunologists primarily concerned with the system character of immune function from those content with a genetic prescription of identity. I am specifically referring to those attempts to understand the dynamic regulatory mechanisms that govern how the immune molecular and cellular profile is altered in response to new challenges. The cognitive metaphor reflects these latter concerns for, like the mind, the immune system is in flux. Always changing in composition (the immune profile of clonal activity and memory), immune selfhood is characterized by a certain "indeterminateness"; a focused center, a postulated organizing principle, remains elusive and undefined.

Perhaps the most explicit conception of a decentered self in current immunological theorizing is that of Irun Cohen's immune homunculus (Cohen 1989; Cohen and Young 1991; Cohen 1992a, 1992b). He has proposed, in contrast to the dominant paradigm of the clonal selection theory, that the immune system is in fact organized to recognize particular host constituents. On this view, immune cells are keyed to a library of targeted self antigens that represent an "outline" of the normal body analogous to the mapping of the homunculus of the cerebral cortex or late-seventeenth-century embryological depictions of the preformed miniature adult in the head of the sperm (Gasking 1967, pp. 53–4; Farley 1982, pp. 20–2).

Cohen adopted the homunculus image to depict his theory of autoimmunity, where immunological dominance of selected self antigens is explained by network theory (Cohen 1989). In his thesis major autoantigens are dominant because they are encoded in the organizational structure of the immune system through a balanced idiotypic/anti-

178

idiotypic network of interconnected lymphocytes. A functional representation of self antigens is thus encoded within a cohort of lymphocytes. Accordingly the composite picture of the immune system's self is assembled by this set of dominant self antigens and their respective lymphocyte network:

> The neurological homunculus encodes neurological dominance. Likewise, the immunological homunculus, composed of immune networks centered around a relatively selected few self antigens, encodes the dominance of these antigens. These antigens are dominant, therefore, because the response to them is already anticipated by preformed lymphocyte networks, a distributed picture of the immunological self. (Cohen and Young 1991, p. 107)

Cohen's theory, closely aligned with the Paris School, addresses the problem of errant autoimmune responses. They argue that a limited set of epitopes induces tolerance, and with normal tissue injury limited autoreactivity occurs. A central postulate is that T- and B-cell reactivity is enhanced against microbial antigens that exhibit selflike epitopes, because the system is already primed by their "embodiment" in the homunculus. In addition, heat shock proteins – highly conserved in phylogeny, so host and parasite essentially share them – signify invasion, inflammation, and tissue damage: "Their homunculus-encoded dominance directs the effector immune response to the parasite as long as the stress persists. . . . Thus the host can count on exploiting the very same homunculus for self tolerance and for microbial intolerance" (ibid., p. 109).

Foreign antigens then are recognized as "other" not by their intrinsic foreignness (i.e., their novelty), but because they are presented in a *context* that changes their shared "selfness" to a representation that declares their pathology. In other words, the foreign is destroyed not so much due to differences in the molecular structure that might distinguish a foreign substance from host constituents, but because, for instance, shared antigens do not appear "normal" in the context of infection. Cohen uses the analogy of Maurits Escher's ambiguous ground/ representation etchings, where diverse figures emerge depending on what the viewer perceives as the background context (Cohen 1994). The trope is the alternate way one views what is figure and what is its context, so that what first appeared as ground may assume a new form as a figure. When visual cognition perceives their interplay, representation – that is, the figure – is shown to be constructed from an ambiguous universe, where background and image are exchanged at liberty. From this perspective differentiation of self and nonself may become ambiguous. Meaning is thus actively sought by an immune system that defines selfhood in a contextual dynamic. He argues that self antigens in their "normal" setting are made tolerant by the animal's control mechanisms, which are loosened or abandoned when the context of these "self" antigens is altered in the setting of microbial invasion. In this scheme, the self is operationally defined as that which elicits immune recognition in a positive fashion, albeit at a level below that evoked by the appearance of recognizable (i.e., still "self") antigen in the context of infection. In other words, autoimmunity is a normal characteristic of the immune system, which constantly seeks to identify and monitor key host constiuents. If these self antigens are altered in a contextual sense, their "meaning" changes and an active immune reponse is initiated.

Cohen's theory is highly controversial but, irrespective of its scientific merits, our concern is to ouline the breadth of the cognitive metaphor. He has persistently expanded what he perceives as the strong parallels between immune and brain functions. Ignoring the differences between Cohen, the Paris School, and others who embrace this cognitive orientation, the immune self is regarded by these immunologists as dynamically emerging in a process of self-identification, which presumably changes and adapts continuously throughout the life of the individual. Not only are the dynamic elements emphasized, but how the immune system is structured as self-defining challenges other "genetic" conceptions of the immune selfhood as a given entity. This cognitive view of immune function is fundamentally formulated on self-seeking and self-organizing activity, whose structure is decentered from any bounded self. Immune identity can be defined only in particular contexts and from such histories selfhood emerges. This so-called cognitive paradigm thus fully embraces the mind metaphor, and in this rendition the contingent character of immune selfhood is emphasized. Note, however, that Cohen "embodies" the self in the homunculus, and thus the metaphor reaches a somatic representation. Thus again, as we saw in Jerne's attempt to model immune function (see p. 167), a complex interplay of images based on *entity* and *process* metaphors are required by these theorists to depict the phenomena. The ambiguity, and ambivalence, of Cohen's metaphors may perhaps be best discerned in contrasting his construction with others that more explicitly embody the immune self so that identity assumes a more concrete, if not punctual, state.

B. THE BODY METAPHOR

The meaning of *self* in immunology since Burnet's initial use has assumed additional connotations from general public usage. These newer meanings effectively transmit highly technical, if not abstruse, concepts to an uninitiated lay audience. At the same time, the language, whether inadvertently or deliberately, perpetuates an imperialistic effort to convince that modern immunology promises truly extraordinary therapeutic applications, as well as revolutionary insight into our biological nature. (I would not argue to the contrary, but I am taking a critical stance regarding the rhetoric and the conceptual apparatus it conveys). As a means of effectively transmitting such a promise and engaging its audience, the rhetoric of modern immunology invokes the immediacy of ourselves *as selves*. In this context the self metaphor is conceptually tightly allied to the vernacular usage of the notion. It is the punctuated self, unconfused with philosophical doubt, that is being discussed. Immunology seeks no less than to play a central role in defining us, effectively competing with the neural sciences, with the tacit claim that the secret of discerning our fundamental biological character resides in understanding our immune function. So, following the ontological and cognitive senses in which immunologists have employed *self,* a third metaphorical category of the immune self must be discussed. I will refer to this sense as *embodiment.* In many respects it is the least abstract and most accessible image of those already considered.

The transition from the mind to the body metaphor assumes sophisticated theoretical constructions and necessitates caution for its dangers of misplaced generalization and misrepresentation to the general public. Examples of these metaphorical constructions abound from the trivial, such as an illustration of the B cell receptor as schematized hands reaching out to grasp (Hynes 1991), to the adaptation of "homunculus" to explain a sophisticated theory of autoimmunity (Cohen and Young 1991). There is also a strong physiological linkage between the immune and neuroendocrine systems. Already in 1896 J. N. Mackenzie had shown that an immune reaction might be precipitated by suggestion (inducing rhinitis with an artificial rose) (Mackenzie 1896) and beyond suggestion, Serge Metal'nikov in 1926 showed immunity to be susceptible to Pavlovian conditioning (Metal'nikov and Chorine 1926). From these early clinical origins the hybrid discipline of psychoneuroimmunology has emerged to examine the interconnections linking the immune and neuroendocrine systems. Bidirectional pathways of

The immune self

communication between the systems, as well as the influence of psychosocial factors on immunologically mediated disease, have been discovered, and an understanding of the basis by which each system influences the other is rapidly growing (Ader, Felten, and Cohen 1991). The immune system shares multiple mediators and receptors for neurotransmitters and hormones. Various immunocytes are able to synthesize these factors and metabolize their own products, as well as mediators first identified from neuroendocrine sources. In turn, these neuroendocrine hormones and autonomic pathways directly affect thymus and bone marrow function and modulate macrophage and lymphocyte behavior. Studies of Pavlovian-like conditioning, anatomic ablation experiments, and pharmacological manipulations have extended the biochemical and molecular biological approaches to map the linkages that are based on shared anatomic (the immune system is innervated), ontogenetic (part of the thymus derives from neurocrest ectoderm), and phylogenetic histories. Finding the brain and immune functions interconnected represents a striking concretization of immunology as cognitive – in a different (and conventional) sense from that considered above.

The cognitive faculty takes on more than simply perceptive functions, for intrinsic to a *knowing* entity is *knower*. The self metaphor projects deeply into the very notion of the subject. Thus, it is not surprising that the self becomes incarnate, with powerful embodiment metaphors being enlisted to propel this conceptual argument. Before we proceed with specific examples, it is worth noting that such metaphors used within science rest upon a more general phenomenological process by which we organize our experience. Most of our fundamental concepts are organized in terms of one or more spatialization metaphors (e.g., good is *up;* bad is *down:* "Things are looking up; things are at an all time low"). A scientific theory may also rely on metaphors with these physical and/or cultural biases, for example, "high-level functions" of physiological psychology is based on RATIONAL IS UP, whereas "low level phrenology" is MUNDANE REALITY IS DOWN (Lakoff and Johnson 1990). Metaphors may actually correlate with our physicality; for example, happy is *wide* ("I'm feeling expansive") or *up* ("I'm feeling elated"). This feature of our language essentially arises from the phenomenological insight that the world is known and structured by our perceptions, that is, by our embodiment of experience (Sallis 1981; Gill 1991). Therefore, by extension experience is structured metaphorically through such allusions to forms of embodiment. Understanding emerges from interaction and negotiation with our environment, which structures our experience in terms of natural dimensions (i.e., personal,

182

embodied, emotional [Lakoff and Johnson 1980, pp. 90–4, 230ff]). Spatialization metaphors are neither random nor universal, but rooted in physical and cultural experiences. For instance, *up* is viewed as active in our culture and connotes a positive attribute, whereas passive (and *down*) may be viewed as preferable in others.

Spatialization is particularly pertinent to our inquiry regarding the metaphorical embodiment of the immune self. The metaphor is readily applied: Our body contours bound and mark off the rest of the world, for the environment is experienced as "outside." Thus, each of us is a container with an in–out orientation, which we project in our experience of the world. Defining a territory and putting a boundary around it is not simply an act of quantification, but extends from containers and substances to our own self-image. Relying upon the power of such metaphorical thinking, the embodiment of immunity may have enormous influence on how we conceptualize immune structure and function.

Not surprisingly the embodiment metaphor, because of its seductive power, has been extended well beyond professional discourse. Critiques of the reduction of the immune self to the body have been advanced by sociological interpreters of immunology. For instance, Emily Martin, an anthropologist, having surveyed the popular press, interviewed the general public, and examined the scientific and textbook literature, concludes:

> *The self has retreated inside the body,* is a witness to itself, a tiny figure in a cosmic landscape, which is the body. This scene is one that is both greatly exciting and greatly bewildering. Dramatic forms of spatial disorientation are particularly apparent in the large numbers of people who interpret scientific images as visions of something colossal and distant inside us. (Martin 1992, p.125)

Her examination has assembled much interesting material, but she seemingly cannot separate metaphor from more narrowly construed scientific intent and freely erects concordant comparisons across several levels of discourse. Donna Haraway, like Martin, has her own position to stake out and is not shy to use immunology to fashion her ideology. Haraway sees immunology as portraying the body as a nation-state, organized around hidden issues of gender, race, and class, or what she calls the "hierarchical, localized organic body" (Haraway 1989). Martin believes "the science of immunology is helping to render a kind of aesthetic or architecture for our bodies that captures some of the essential features of flexible accumulation" (Harvey 1989, pp. 141–72; Martin 1992). As in so many such ideological discussions, to suggest that these uses of science are mainly political and thus may be easily dismissed as specious – or even obviously false – is to ignore their potency. Regardless of what one may think of the particular legitimacy of the inference, science as metaphor evokes a certain authority in the public domain. And it is the use of metaphor that relates the technical issues to broadly appreciated conceptual images.

Since the self/nonself designation has been adopted as the defining axis of immunology, the imagery of host defense has become more personalized. For instance, the metaphor of the body as nation is widespread: "It can be as difficult for our immune system to detect foreignness as it would be for a Caucasian to pick out a particular Chinese interloper at a crowded ceremony in Peking's main square" (Dwyer 1988, p. 29); the notion that the immune system "has developed to function as a kind of biologic democracy" (Jaret 1986, p. 709) is readily extended to warfare imagery, with killer T cells bestowing "poisonous kisses" (Nilsson 1987, p. 105), assuming deadly embraces (ibid., p. 25), or becoming a David killing Goliath (ibid., p. 100). Such images written for a general audience are derived from the scientific literature and serve as powerful metaphorical statements, irrespective of how we might view their interpretation (Martin 1990). Their power resides in the immediacy and pervasive sense of embodiment as mediating perceptive and cognitive structures. What is so intriguing about the critiques of Martin and Haraway is the dramatic bidirectionality of the metaphors (which also reveals why their arguments are circular): The immune system is cast in terms of images from the everyday world that are in turn refashioned to be utilized as material evidence for ideological analyses and sociological constructions. The extended self metaphor thus assumes several layers of complexity functioning in the realms of the individual, gender, state, and history.

The evolution of the self metaphor reveals changing concepts of immunity. There has been a concretization of the image, from a vague intuition concerning identity to specific research programs with cognitive functions (both intrinsic to the immune system and interactive with the neuroendocrine system) that define mechanisms of tolerance and autoimmunity. As immunology has grown in scope and become more sophisticated, the metaphors employed within it have changed accordingly, following the overall growth and change of a scientific language. This recognition, however, does not address a pivotal issue, namely, to what extent does the metaphor *drive* conceptual change within the science? In other words, to what extent does the notion of self implicate and impart meanings beyond the pale of bona fide scientific vocabulary? Clearly in Burnet's use of the term a vague sense of origin is evoked, and his audience might descry the sense in which he wished to alter the manner immunologists viewed the questions of that period. Today the use of *self* has assumed other connotations that may fairly be called imperialistic. At least this is the interpretation of certain critics, and it seems to be a shared perception of some significant segment of the general public. Let us briefly review the case.

If immunity is to be portrayed as a metaphor for us ourselves, if the discipline would encompass images that evoke cognitive and bodily metaphors, the evidence is strong that the science would lay claim to be a defining parameter of our very *selves*. Immunology has, or perhaps is only attempting to appropriate, identity. This was Metchnikoff's initial agenda; it was reserved as a central tenet of post–World War II immunology in Burnet's theorizing, and the immune self in all of its current metaphorical guises must be viewed as successful to the extent the science has fulfilled this programatic mandate. When Avrameas (1986) entitled an article on autoimmunity "Natural Antibodies: From 'Horror autotoxicus' to 'Gnothi Seauton,'" he offered us what may well be the most provocative identification of the immune system with the self. The Delphic command is invoked in his dictum that "since the immune system can recognize self constituents, it should possess the 'know thyself'" ability. He does not explicitly address the contrast between Ehrlich's position and his own, but from the general argument it is evident that he considers autoreactivity to be a homeostatic mechanism, and contrary to viewing autoimmunity as an aberration, he regards it as a central characteristic of the immune system – a position closely aligned to that of Cohen and Coutinho (Avrameas 1991). Irrespective of the validity of this position, the metaphorical use of the self has unabashedly assumed a new intellectual orientation, tapping into the well-spring of Western philosophy. Beyond the similies of the body or the nation, immune function here draws upon the very fabric of our cultural identity and ethos, and one may now ponder to what extent the immune self has presumed such identification at the very foundations of purportedly "scientific" analyses. "*Gnothi se auton*" goes well beyond invoking an allusion to the mind or embodiment; it evokes a Western psychological conception of a bounded, autonomous, 'discoverable' subject.

The self concept has thus fully emerged in immunology within a short, forty-year history. First, the discipline sought to concretize the self metaphor that had initially been used to summon a vague notion of identity or of the origin of immune activity or of that "entity" to be defended. The movement toward a punctual self in part reflects the progress of the science of immunology, and we briefly sketched how theorists of immunology borrowed from the vocabulary of mind, language, and cognition to construct their own models. As the science has assumed growing dominance in medicine, it has effectively advanced the portrayal of immune function as essentially constituting ourselves. In this sense the language does not seek to discern what we are, but simply assumes and rests upon personal identity as given. This is the "punctual self," the version of the self metaphor that has captured the public's understanding. This punctual self mobilizes the generally comprehended and culturally potent concept of selfhood. No longer concerned with language, mind, or cognition, the punctual self is we ourselves, unfettered with problematic or demurring definitions. From this perspective the self metaphor is both wielded with facility and readily grasped.

In immunology the self metaphor assumes new complexities when it explicitly appeals to traditional images of the self in discourses outside its discipline. There is strong evidence that the immune self now extends well beyond the original attempt to elucidate the source of immune activity or to invoke a parallel to cognitive function or a simile to gross embodiment – the immune self has now *become* us ourselves. The immune system, the master epistemologist, *knows* what belongs, defines that entity, and designates it as self. Moreover, this view is not idiosyncratic to Avrameas. It represents a popularized concept of immune function. Both the lay press and the general public increasingly support the notion that "the immune system is the whole body" and even more that "it's the whole thing" (Martin 1992). The immune system is conceived as being somehow autonomous: "These cells [immunocytes] act on their own" (ibid.). When the immune system is destroyed, the self, in this broadened sense, is similarly shattered. Our best illustration of that conclusion is found in the testimony regarding the AIDS epidemic.

Because of the devasting medical and social impact of the human immunodeficiency virus (HIV), the image of its capacity to destroy the immune self is particularly powerful. Whereas cancer may be "regarded with irrational revulsion, as a diminution of the self" (Sontag 1989, p.12), HIV, with its immediate insult on immune identity, is regarded with horror as a devastation of self. HIV is a "plague" that "invades," "pollutes," "attacks," "annihilates." Macrophages, normally responsible for processing foreign antigen, harbor the virus, and helper (i.e., noble) T cells (ironically through HIV docking to a self-recognition receptor,

CD4) are targeted for destruction. The immune network disassembles, the self is dissolute, and the "war" is lost. The biology of HIV is understood metaphorically through this "besieged fortress" image: The immune system is attacked, the cellular soldiers of the self are destroyed, integrity (self-defensiveness) is violated by the unchecked growth of microorganisms (normally mundane opportunistic organisms), and the HIV-infected designation changes to the advanced stage of AIDS, with death ultimately triumphing. The military imagery is potent and critics have justly warned against its power:

> The age-old, seemingly inexorable process whereby diseases acquire meanings (by coming to stand for the deepest fears) and inflict stigma is always worth challenging. . . . Not all metaphors applied to illnesses and their treatment are equally unsavory and distorting. The one I am most eager to see retired – more than ever since the emergence of AIDS – is the military metaphor. . . . [T]he effect of the military imagery on thinking about sickness and health is far from inconsequential. It overmobilizes, it overdescribes, and it powerfully contributes to the excommunicating and stigmatizing of the ill. . . . We are not being invaded. The body is not a battlefield. The ill are neither unavoidable casualties nor the enemy. (Ibid., pp. 94–95)

The pervasiveness of metaphorical thinking in the AIDS epidemic clearly illustrates the potency of such images. Treichler, in showing the cultural values enlisted to describe AIDS, lists several antonyms. Of the coupled oppositions, self/not self is the fulcrum that largely structures the others, namely, heterosexual/homosexual, safe sex/bad sex, knowledge/ignorance, innocent/guilty, virtue/vice, vagina/anus, love/death, doctor/patient, natural/alien, normal/abnormal, wife/prostitute. All these antipodeans share the structure of self-identification versus the foreign, the other, the rejected, and have high emotive content precisely because of their personalization. But in addition, there are metaphorical references to immunology in particular, with propositions such as "An infectious agent that has suppressed our immunity from guilt" or "The crucible in which the field of immunology will be tested" (Treichler 1988, p. 32–3). The first quote, which echoes and amplifies the oppositions listed above, signals a clear message to a particular cultural context of AIDS, whereas the second (quoted from Robert Gallo, a leading virologist) concisely projects a specific challenge to immunology: The "fight" against AIDS is the crucial battlefield where scientific expertise and credibility are at stake. It is a calling to immunologists to come forth from their cloistered laboratories, bring their

187

alchemical and healing wisdom, and minister to the besieged. "Crucible" evokes magical powers, medieval sorcery, and secret knowledge to combat a plague of fierce and mysterious force. The alchemy must now be applied to "a golden opportunity for science and medicine" (ibid.). The challenge could be made no more explicit.

The intimate cultural links between the domains of HIV virology, immunology, and pathophysiology, on the one hand, and the constellation of associated social, moral, and psychological reactions to the disease, on the other, compose a complex image of the self, whose reliance on concepts borrowed from immunology can scarcely be overestimated: "AIDS is not simply a physical malady; it is also an artefact of social and sexual transgression violated taboo, *fractured identity – political and personal projections*" (Grover 1988, p. 18; emphasis added). More specific to our purpose:

> Of course, AIDS is a real disease syndrome, damaging and killing real human beings. Because of this, it is tempting – perhaps in some instances imperative – to view science and medicine as providing a discourse about AIDS closer to its "reality" than what we can provide ourselves. Yet the AIDS epidemic . . . is simultaneously an epidemic of a transmissible lethal disease and an epidemic of meanings or signification [the way in which language organizes rather than labels experience à la Saussure]. (Treichler 1988, p. 32)

A recurrent theme in AIDS activism is the need to separate these entangled social and scientific issues so that the infected individual can preserve his or her identity in the confusion. A profound sense of loss of selfhood was countered by activist assignment of the designation Person with AIDS (PWA). This label was viewed as an act of self-acclaimation, a naming to establish identity: "We do not see ourselves as victims. We will not be victimized" (Grover 1988). The activist rejects the victim label because it connotes a helplessness concomitant with loss of selfhood: "As a person with AIDS, I can attest to the sense of diminishment at seeing and hearing myself constantly referred to as an AIDS victim, an AIDS sufferer, an AIDS case – as anything but what I am, a person with AIDS. I am a person with a condition. I am not that condition" (Navarre 1988, p. 143). Even more to the point: "Victims suggest innocence. An innocence, by the inexorable logic that governs all relational terms, suggests guilt" (Sontag 1989, p. 11). Beyond seeking "self-empowerment" (Navarre 1988, p. 144), the PWA must counter the dehumanization that takes aim at "the most deep-seated sense of the self's fragility" (Gilman 1988, p. 271). Their fundamental political position is that "any separation of not-self ("AIDS victims") from self (the "general population") is no longer possible" (Treichler 1988, p. 66). "Ultimately, we cannot distinguish self from not-self: for "plague is life" and each of us has the plague within us; "no one, no one on earth is free from it" (ibid., p. 69; quote from A. Camus, *The Plague* [New York: The Modern Library], p. 229). By this reckoning the immune self metaphor has dramatically enveloped the entire body politic. The destruction of the immune system is not restricted to those infected with HIV, but represents a plague to us all. We are each subject both to the dangers of what is becoming pandemic and responsible for those particular individuals tragically suffering its immediate consequences. The immune self, *the* object of HIV destruction, *the identity* under attack, has now been incorporated into its broadest cultural context: A microcosm of the individual has become a microcosm of an integrated society, where each, in fear and trembling, must accept responsibility. As already noted, if one considers the semantic oppositions that are used to describe the AIDS crisis, "self and not self" leads the list (ibid., pp. 63–4), and these serve only to insulate the beseiged.

Thus, the self metaphor in immunology is more than an interesting example of how language reflects conceptual changes in science, and it is more than a case study of academic semiotics: The self metaphor offers a lesson in the work of public language and of politics in the broadest sense. What is clearly illuminated by these discussions concerning the self in the AIDS experience is the dependence of identity on so many *extraneous* definitions. First, according to the medical model, the self is destroyed by a dysfunctional immune system. Identity is concomitantly submerged into the closets of taboo, and despite the protests of AIDS activists and the powerful arguments for human rights, the agenda has already been established: A plague is amongst us, and those afflicted must be shunted aside. The PWA, despite protestations, becomes a self markedly altered – distorted, if you will – by an imposed social redefinition of identity. In response to each of these forces the existentialist notion of an authentic self (established, given, and indubitable) becomes contingent, if not chimerical, as it is shown to be fundamentally subject to definition from without. Reading selfhood through this prism reaffirms the modern sociological paradigm that individuals are social not merely by virtue of being part of a larger group, but in the definition of their innermost identity. The vulnerability of the immune self in the context of the AIDS crisis simply underscores the central role the metaphor has assumed. It has fully emerged, from an obscure source of immune reactivity to a full-fledged metaphor of the mind, possessing cognitive functions as a network. In addition, the immune self has been more broadly linked to the core function of embodiment, where it has assumed a new resonance with holistic ecological integration. Finally, the immune self has been assigned to serve as a microcosm of broad cultural perceptions and thus functions as a metaphor of the body politic. As a result, the language of immunology has been widely adopted, and the science has triumphed in the sense of having wide social relevance. In this view the imperialistic agenda has been largely fulfilled.

Does the self then have useful meaning when used so expansively and with so many implications? To come back, then, to where we began and ponder once again the significance of metaphor, it obviously functions as more than just "imagery." Metaphor illustrates, fashions, and conveys cumbersome concepts through its ability to project meaning on many levels. We cannot deny the power of these allusions, but at the same time we are compelled to scrutinize the message carefully. As shown in the case of AIDS, the immune system has been assigned responsibility in helping to create an expanded version of the self. Just as immunology borrowed from other discourses to erect a language for its theoretical growth, the science has in turn contributed to the construction of a far-expanded cultural notion of selfhood. But at the same time one must conclude, considering the prevalence of self in psychological and philosophical discourse, that its meanings are hopelessly complicated and intertwined in diverse agendas, swept away in this torrent of introspection. Charles Taylor is a particularly astute observer of this modern self-consciousness: "In one sense, humans always have a sense of self; but we see ourselves as having or being a self, and that is something new" (Taylor 1991, p. 307). Rousseau bequeathed an elusive self that, as is described in the concluding chapters, is even more problematic than he left it. The modern self is in disarray, and we can hardly be satisfied that *self* in immunology is *explanatory.* Personal identity, whether derived from the languages of art, literature, psychology, or philosophy does not offer a stable structure upon which to erect the immune metaphor. Dismembered (whether reassembled or left in parts), the self in our postmodern world remains "doubtful." This is the central concern I address in the remaining portion of this essay, but preceding a frankly philosophical analysis, I believe a critical assessment of the self metaphor within the scientific framework of immunology is warranted. Therefore, before we engage in a more far-reaching discussion placing this issue in the context of modern philosophical discourse, let us step back once again to briefly examine immune selfhood as an operative model.

C. AN OUTLINE OF A CRITIQUE

So, at the end of the twentieth century, following the spectacular advances in our understanding of immune function at the molecular level, we are still left to ponder the question, What is the immune self? Immunology has elegantly succeeded in offering a mechanical explanation of generation of (antibody) diversity (G.O.D.), which is the basis for understanding how an indefinite number of antigens may be countered with a finite number of antibodies; this is no less than an understanding of the grammar of the immune language. What is not known is how the host decides that an antigen should be countered. What are the logical consequences of the self defined in terms of the other or, negatively, as not the other? The question (leaving aside natural immune mechanisms) is thus refined to highlight that in explaining G.O.D. or MHC–antigenic complexing, we have only addressed the next immediate question: How is diversity and immune challenge mechanistically explained? Solving that problem would in fact be an extraordinary accomplishment, no less than the culmination of a century of immunologic research. So why is such an ungracious, if not gratuitous, distinction made?

A fundamental ambiguity regarding the nature of the immune system remains. On the one hand, it is responsible for defending the host, that is, preserving identity, but on a more fundamental level it defines that organismal identity. The first function lends itself easily to metaphoric language of defense and to related mechanical modeling. This is the punctuated self model, and it is effective in the context of asking questions, for example, about how antibody diversity is generated, because the underlying issue of *how* identity is established is not germane to the particular research question. G.O.D. *begins* with an assumed self. But the second issue, the definition of the self *arising* out of immune encounters and problems of system regulation, is far more nebulous. The problem is to clarify how to view the immune system that is, as one that defines the self or defends it. Although these aspects are not mutually exclusive, the dual function has, and does, cause ambiguity and confusion in formulating the theoretical foundation of the discipline. What is the self? It certainly is not a static entity by any organic parameter, least of all by its ever-changing immune history, where each encounter with environmental challenge provokes the self to decide acceptance, incorporation, or rejection and finally "immune incorporation" (i.e., antigenic processing and immune response). With each such deci-

191

sion, the self is altered. The point, of course, is that to know mechanisms of generating antibody diversity is but a subsequent partial solution to a more profound question concerning the boundaries of the self and how it makes that "cognitive" leap to encounter the other. The molecularization of immunology has not answered that question. Here we are not concerned with a readily available genetic definition, but why a given antigen is processed for immune destruction rather than simply eaten (i.e., incorporated). How is the foreign perceived before or even independent of MHC linkage?

When one compares the sophistication of knowledge concerning acquired and natural immunity, there is little doubt that the focus of fascination has been with the "smart weapons." The discoveries of the highly discriminating specificity of immunoglobulin chemistry and the elegant shuffling of gene fragments to custom-produce antibody are true testaments to the power of molecular biology. Nevertheless, natural immune mechanisms play a paramount role in the entire spectrum of infectious disease, cancer surveillance, and tissue repair (Nelson 1989), but the mechanisms of identifying targets for immune destruction by antibody-independent means remain obscure. Critics have argued that our preoccupation with acquired immunity has been misplaced, for it in fact probably represents the more minor or secondary mode of defense. As noted in the discussion of d'Herelle's argument with Bordet (Chapter 3), serology is not necessarily a reflection of immunity (Marshall 1959). From the problems of serology the theme of specificity has culminated in the sophisticated understanding of MHC-mediated recognition. Here we need only emphasize that perception, cognition, and response are the properties of other immune components that do not exclusively use the MHC to distinguish self from nonself. For instance, natural killer lymphocytes, various phagocytes (macrophages, neutrophils), and complement are effective first-line mediators of host defense. These mechanisms, along with the proper perspective on MHC-restricted recognition as one, albeit important, means of immune recognition, underscore once again the hierarchical complexity and emergent character of the immune system.

The focus on process that "cognitive" theory brings to the study of immunity utilizes a scientific approach that differs from a purely reductive one. Thus, whether we are preoccupied with MHC- or non-MHC-restricted recognition, the full scope of cognition at the molecular level remains unresolved. Note that antigen presenting cells (APCs) do not appear to have the capacity to differentiate self from nonself (Winchester et al. 1984), but there are mechanisms by which phagocytes engulf organisms by relatively nonspecific recognition mechanisms involving carbohydrate binding by proteins with a high degree of specificity for such interactions. These "primitive" recognition proteins are called lectins, and the process is termed lectinophagocytosis (Ofek and Sharon 1988). Although unrestricted MHC recognition processes are undoubtedly important in phagocyte surveillance functions (Nelson 1989), this issue has not been carefully examined and cannot be discussed theoretically with the same molecular sophistication as that afforded to T-cell MHC-mediated recognition processes. Further, I have already discussed in several contexts the temporal-dependent nature of selfhood, that is, how successive immune encounters dynamically define immune identity during the organism's life history.

MHC is a necessary component for defining the immune self, but it alone is certainly not sufficient (although no other genetic signature can make any better claim).[3] I do not contend that the MHC assumes the dominant genetic basis of immune selfhood, and there is no denial that the genome encoding it serves as the foundation of that identity, but I do resist the reductionism so clearly expressed by the following claim:

> Cloning of the MHC genes revealed that . . . self identity is hard-wired, directly encoded in only a handful of genes that differ in their primary DNA sequence from one individual to another. In other words, whereas a newly born individual inherits the potential for the complete diversity of antibody and T-cell receptor specificities, he inherits only his unique identity at the MHC locus. (Watson et al. 1992, p. 305)

The full philosophical argument as to why this position is an inadequate and an untenable definition of immune selfhood is the subject of the remaining chapters (see especially pp. 222ff.), and suffice it here simply to assert that this MHC signature of selfhood is no more *the self* than the genetically programmed structure of our respective nervous systems. We each begin with a nervous system that assimilates and processes information and then directs our behavior and thoughts. The issue is the *determinant* role of the prefigured neuronal organization that arises from the genome. Obviously we are to a certain extent "programmed" by our genes, but it is the epistatic life experience that confers *who* we are. Our abilities vary (IQ, motivation, emotional states, and so on, are to some extent genetic), but to deny the enormous component of identity that emerges in the complex interactive systems that we call mind is to fall victim to the worst restrictions of genetic reductionism (Tauber and Sarkar 1993). I am arguing that a similarly expansive logic applies to immune function, freely acknowledging the cognitive metaphor!

An interesting problem in this regard, and perhaps a vivid illustration, is graft versus host disease (GVHD) in syngeneic (identical twin) or autologous bone marrow transplantation (Rappeport 1990). This syndrome involves the reaction of immunocompetent donor lymphocytes against recipient tissue (usually manifested in skin rashes, but often also including hepatic or gastrointestinal disease), and may be accounted for by minor histo-incompatibility in "matched" donor–recipient pairs. But how can the reaction of the patient's own lymphocytes against self constituents be explained? GVHD in this context may be due to imbalances of a reconstituted immune system, obscure effects from storage of the lymphocytes, new exposure to altered self antigens, or failure to eliminate T cells with receptors for self antigens. Nonimmune mechanisms may also play a role (ibid.), but the evidence is strong that autologous GVHD occurs through autoimmunity.[4] As in the general case of autoimmune disease, we must conclude that the genetic signature of self – the MHC or its amplified family – is but one pole of a complex multidimensional axis, where dynamic interactions dictate recognition and reaction.

Let us return to examining the clonal selection theory (CST), which as a program of theory and research, is the closest thing to a paradigm that immunology possesses. First, it accepts the discrimination between self and nonself as the central issue of immunology, a perspective derived from the historical development of the field. From Paul Ehrlich's first side-chain theory, which postulated antibody formation in terms of a selective mechanism not significantly different from the current model, the underlying presupposition has been that the self is given (i.e., defined, formed, constant) and in need of protection. As already noted, during this early period immunity was postulated (and generally accepted) as *establishing* organismal identity, and (immune) defensive functions became subsidary to this primary role. We will return to this thesis, but in any case, twentieth-century immunology has followed Ehrlich's lead. To avoid self destruction, the notion of immunity as defensive in nature requires "ignorance" of the self; the avoidance of self destruction was provided by Burnet's concept of embryologic purging of those clones that would recognize (and thereby initiate attack on) the self. From this vantage point the object of immune activity is not the self, but the other; consequently antibody production is viewed as the process by which the organism defines nonself. Perhaps paradoxically this positive definition of the other then defines the self – negatively. The following argument is a summary of one given previously (Chernyak and Tauber 1991), restated here in large measure to offer a short orientation from which the phenomenological discussion of the next chapter proceeds.

Ehrlich was forced to view immune activity directed against the host as "*horror autotoxicus,*" and Burnet postulated the incomplete antibody repertoire to explain the sanctuary of the self against immune attack. Blanks in the immune library represent the negative definition of self. Ironically, perhaps, the self cannot be defined in *immunological* terms, because immunological self-definition would invoke immune reaction, that is, *horror autotoxicus.* Is the self defined by its selfness? Not according to the theory of embryonic purging of self-reactive clones. From the modern perspective, affinity (i.e., reaction) to the other is the normal expression of the immunological response. The other is the universe represented by antigen-specific antibodies and lymphocyte receptors. What is this universe? That which is so represented. Thus, immune presentation becomes the definition of the other. The matter becomes increasingly confused when we consider the idiotypic nature of certain antibody specificities and the general role of either reactive or nonreactive autoantibodies in immune regulation. These phenomena raise questions about the very nature of the immune network.

As already discussed, an elegant critique of CST has been developed by Coutinho, Stewart, Varela, and Vaz, who counterpropose what is referred to as the autonomous network theory (ANT) (Coutinho et al. 1984; Varela et al. 1988; Stewart and Varela 1989; Coutinho and Stewart 1991). They view the immunological process primarily as a cognitive system, which functions as a self-referential process. From this vantage the immune system is closer to the Metchnikovian formulation, where immunity is concerned with self-establishment, producing images not of the other, but of self-creating activity. This activity is the organism's self, and what deserves to be saved *existentially* is also saved *essentially* by retention of its images (Chernyak and Tauber 1991). Self, then, is the self-retained and, as such, its image. The hypothesis treats the other (nonself) as a kind of constraint utilized by the cognitive immune system to perform its self-referential functions. Of course, the caveat remains that cognition in immunology, as has been discussed, is an extended metaphor from neural processes and philosophy of mind. To envision "knowledge" as a definition of cognition is obviously inappropriate, but at the same time, there are end points in the immune reaction where the foreign is perceived as other and the self responds. At this point any theory accounting for cognition in immunity is most tentative, however, and the term is used here in a weak sense: as the perception of information qua information. This formulation is based on a powerful sense of self autonomy, and represents an important conceptual shift. But is ANT sufficient to base a theory of *recognition* on? The more salient question remains: How is the self truly self-defined? If ANT gives a positive definition of self and if the other is to be known only as incorporated in the self, then how is this nonself truly foreign? In the autopoietic (Varela 1979) foundation of ANT's earliest formulation (Vaz and Varela 1978) there is no satisfactory discernment of the other, and the matter is never resolved. A "true" other can only be defined by an external observer, and the interface, the "cognitive" mediation, is left unspecified.

At present the cognitive immune paradigm must seek a constitutive mode of "knowing." The immune system must not only function to identify itself as opposed to the other, but also constantly define self from itself. On such a basis immunity is a *process* that always provides for an open system of self-definition, constantly producing self and other from itself. In this formulation *the primary role of the immune system is self-identification and thus self-definition*. The foreign is recognized only when the normal defining process breaks down. In such a dynamic, dialectical process the immune system is constantly differentiating self and thereby establishing its identity. An analogous dynamic process is Jean Piaget's (1952) description (by a phenomenological interpretation) of how infants learn about their world. Rudimentary reflexes function repetitively in their own assimilative survey of the world, and when reflex movements come across data that cannot be assimilated, an accommodation results. A novel reaction occurs, comprised of a reorganization of movements, or other senses, marshaled for information. To assimilate information, the child accommodates to its environment by organizing it and reacting to it. (Husserl used a similar descriptive schema, described in Chapter 6.). As Zaner notes more generally, for the self to emerge into explicit wakefulness, disturbances or crises must occur; "in this sense, self is in essence always a *problēmā* to itself" (Zaner 1975, p. 172). From this perspective, "the genealogy of self thus signifies the emergence of heterogeneity, complexity, enriched reflexivity, from relative homogeneity" (ibid.). Instead of defining the other in self/nonself negative terms (i.e., self of internal origin and other of external origin), self is defined by a self-reflective process governed by its own rules and thereby providing the definition of the other. Perhaps orthodox immune theory will approximate a similar formulation. In these terms, tolerance, then, allows peptides to function in an "open" system, that is, an ongoing primary process of self-differentiation, whereas rejection or attack begins when a peptide no longer allows self-definition; the system is then "closed," and there is a breakdown of the self-discerning process. Various general schemes for such a model have been proposed (Varela et al. 1988; Chernyak and Tauber 1991; Cohen 1994) and suffice it to simply observe that the *cognition* of the ANT or immune homunculus theories remains to be defined. In other words, the immune cognitive paradigm must address how the "otherness" of the other is discerned. That is, what constitutes its nontransparency? How is "meaning" derived? This theoretical issue represents a profoundly difficult concern yet to be adequately approached, but these theorists explicitly recognize the problem: MHC is but the beginning of immune recognition, serving only to orient the system.

The notion of defining the self "through" the other – or perhaps better, in the *context* of the other – is at best a nascent idea in the epistemology of immunity, but it has a rich literature in philosophy (Taylor 1987; Theunissen 1977). Arising from the phenomenological, existentialist concern with intersubjectivity, later structuralist and poststructuralist philosophers have "decentered" the subject in favor of systemic correlations for its definition. But well before such radical assault on the self, beginning with Hegel, and recurrent throughout the nineteenth century, the *I* and *other* relationship has been viewed as one of mutual dependence. Beyond Kierkegaard's view of the self defined by a complex self-relational calculus (1849, 1955, p. 146), Ludwig Feuerbach stated a theme that would be echoed in the twentieth century from the sociology of George Herbert Mead to the religious philosophy of Martin Buber:

> The single man in *isolation* possesses in himself the *essence* of man neither as a *moral* nor as a *thinking* being. The *essence* of man is contained only in the community, in the *unity of man with man* – a unity, however, that rests on the *reality* of the *distinction* between "I" and "You." (Feuerbach 1843, 1972, paragraph 59, p. 244)

Self-discovery from this general vantage requires passage through otherness, specifically interaction with a different, nonobjectifiable "Thou." There are two basic strategies for such self-definition: (1) Starting from the premise of the ego, access to intersubjectivity is sought by construing the other as an alter ego ("Other-I"); (2) the self is derived in some manner form an original encounter with a Thou (Dallmayr 1984). Much of twentieth-century philosophical thought may be encapsulated by these two general approaches, which examine various solutions of how alterity is established in the nonreciprocity of relation,

197

that is, how the self both perceives and is defined by its encounter with the other. It is an enormously ambitious project to relate this dimension of the selfhood problem further in the context of immunology, and I attempt only an initial sketch from a particular phenomenological approach in the next chapter. But I believe that in the sense that this essay is a prolegomena to a philosophy of biology, the various approaches to alterity must be explored as to how they might be applied to describe the nature of biological identity and individuality. From the bare outline offered here, it appears that this topic would find rich philosophical mulch in immunology. My purpose, however, in this introductory work has been simply to note that fundamental issues pertaining to immune function with respect to its cognitive activity have been modeled on the assumption of the immune self as given, albeit continually changing, and in need of defense. This is the punctual ideal, the one widespread in the public domain but hardly adequate to model the phenomena.

The other dormant theme not adequately explored here concerns the philosophical and scientific concern with self-organization. In this organismically ordered universe, the question of the *telos* again arises: "Far from there being just a coincidence of inwardness with some indifferent parts of extencity that simply happen to serve as the center regions of a phenomenal outwardness, the particular parts – organisms – are obviously *organized for* inwardness, for internal identity, for individuality" (Jonas 1982, pp. 89–90). End-directed processes appear in reductionist explanation, even as the mechanistic evolutionist seeks interpretation entirely in terms of "chance and necessity" (Monod 1972). In post-World War II biology, teleology has become teleonomy (Pittendrigh 1958), to rid biology of finalism.

> Apparent purposes are observed in adaptation and development and classical biology was always mixed up with teleological, i.e., finalist thinking. In order to get rid of it, the goal of research was stated as looking for mechanisms with teleonomy, the difference being that in one case (teleology) one deals with purposeful end-directed processes governed by an "intelligent, designing mind," while in teleonomy one is looking for non-purposeful end-directed processes. (Atlan 1989, p. 240)

Causality must answer to both chance and purpose. "In other words, a causal role in evolution is played by internal relations as well as the external relation of natural selection" (Birch, 1988, p. 75). This topic, bridging modern evolutionary debate, recent theoretical research in chaos, and complexity and self-organization theory, is obviously highly germane with respect to placing the particular issue of immune selfhood in its widest scientific context. This task is more ambitious (and daunting) than the one tackled in this essay; it remains the crucial next step in placing immunology in its proper organismic context. (See, e.g., Jantsch 1980; Salthe 1985; Yates 1987; Mittenthal and Baskin 1992; Kauffman 1993, Atlan 1994.)

In summary, "What is the immunologic self?" remains ambiguous by our current molecular criteria. I believe that on the elusive–punctual axis I have used to structure this discussion immune selfhood remains closer to the nebulous region. Part of the problem simply stems from posing the issue in the metaphoric terms of cognition, which immediately identifies the sought "entity" as the same elusive category yet to be firmly articulated in any of the domains that invoke it – psychology, neuroscience, cognitive science, or philosophy. The second aspect of the dilemma is the more concrete question of *how* to represent immune cognition in its various guises. In this sense the science is still immature. Fundamentally I am doubtful whether immunologists have successfully posed the relevant questions concerning the self and nonself encounter as a perceptive event, which is not to say the problem is unrecognized (Varela and Coutinho 1991; Cohen 1992a). This is not a question to be resolved at the molecular level; like other cognitive problems, it depends on principles of organizational, systematic, cybernetic, hierarchical, and even logical analysis. I do not want to be viewed as denigrating the extraordinary accomplishments of molecular biological application to immunology. But G.O.D. must not be confused for solutions that in fact only address the phenomena to which such molecular mechanisms are appropriately applied. Rather, I only want to note that the biological issues demand complementary approaches, where biochemical and genetic processes are incorporated into a comprehensive biological perspective.

In advocating this orientation, I have made no attempt to comprehensively consider current immunological theory (see, e.g., Schwartz [1993] for an excellent discussion of autoimmunity that ignores the cognitive paradigm) or its paradoxes (Hoffman, Levy, and Nepom 1986). Although my general conceptual bias is evident, I readily admit that a rigorous scientific dissection of theory has been subordinated to a critique of immunology's language. But I make no apology, for in this analysis the inexorable demands of understanding the self metaphor have directed attention to a deeper inquiry of immunology's governing agenda: In probing the *self*'s multifarious meanings, I have exposed the basic foundations of its construction. One might inquire of any immunologist who uses *self,* or for that matter, *memory, learning,* and *recognition,* how such metaphors direct his theory and research. In some cases the metaphors may be found to stimulate fecund scientific models, and in other instances the language only serves to distort the scientific inquiry, reflecting unsuccessful approximations and false assumptions. Metaphor and theory share an underlying structure, and thus their meanings are reciprocal and intimately linked. The wider context this philosophical inquiry offers our endeavor is to broaden the intellectual horizon of immunology's central question, providing, I hope, some insight for understanding the *self* metaphor, and the science it serves.

6

The self and the phenomenological attitude

A. THE EARLY PHENOMENOLOGISTS

There is an extraordinary parallelism in fin-de-siècle concepts of the self in psychology, art, philosophy, and our subject, immunology. Such a sweeping claim may not be judicious to make, and it could be contested by arguments concerning the artificial construction of a *Zeitgeist* or the inadmissibility of extending analogical cases from one discipline or activity to another. But attempting that claim largely structures the remainder of this essay. In this chapter the phenomenological critique is used to explore further the basis of the immune self concept as understood from this position. The last chapters deal with metaphysical issues that, I believe, underlie our understanding of selfhood. To reach well beyond the confines of immunology is obviously risky, but aware of these dangers, I am still committed to sketch what I perceive to be a widely pervasive understanding of how the world was viewed (and what the nature of the perceiver was) during the same formative period that witnessed the birth of modern immunology. New conceptions of the knowing subject in his or her personal cosmos were evoked by a startling reassessment of the self. This is the same mission identified for immunology, but in a different scientific domain.

My exercise is pointed toward a critique of immunology as a discipline that has as its basis the *self* metaphor. This fundamental concept has the same philosophical construction of personal identity as that formed from novel late nineteenth-century psychological concepts. We must explore the notion of an authentic self as formulated at that time and the means by which that entity was postulated to function in its empirical cognizance. The case of close correspondence cannot be *proven,* but it postulates a position from which we might critically assess what the immune self *is* behind its metaphorical mask. It is this belief that justifies examining the possible spillover from related disciplines.

There is no accident that radically new formulations appeared in both psychology and philosophy in the 1890–1900 decade as the problem of selfhood was scrutinized in new ways from each field. Perhaps my focus on the philosophical discussions at the expense of the psychological is best captured by Thomas Nagel's quip: "We are in a sense trying to climb outside of our own minds, an effort that some would regard as insane and that I regard as philosophically fundamental" (Nagel 1986, p. 11). But philosophy *was* indebted to a newly emerging psychology, which had emerged as a distinct discipline by 1890. Its experimental, albeit introspective, methods originated thirty years earlier and continued to spread to many other modes of inquiry. The self-consciousness of the method and, more important, its consequences are exemplified by Ernst Mach's drawing of himself in his *Analysis of Sensations* (1886, 1914). We share, from his vantage, the view of a room, bracketed by the contour of his nose, mustache, and brow. The proportions of his lounging body, the room's chair, window, bookcase are singularly perspectival: "[T]he self includes – or, more precisely is what it sees" (Ryan 1991, p. 9).

This school of philosophically inclined psychologists, with Franz Brentano as the leading exponent, converted the Cartesian *cogito* to an identity defined by what the "I" perceived. More generally the familiar distinction between subject and object was no longer tenable, for the self became subsumed by consciousness, and defining consciousness became a central problem. The key texts dealing with this issue are Brentano's *Psychology from an Empiricist Point of View* (1874), Mach's *Analysis* (1886, 1914), and William James's *Principles of Psychology* (1890, 1983). These phenomenological psychologists began with the premise that the mind did not see the object "as is," but by integrating related perceptions. A total experience was thus constructed from imperfect and piecemeal data. A correcting mind formed the conscious image, for perception was based on an "interactive relationship between subject and object: the object was, in effect, partially 'created' by the act of seeing it" (Ryan 1991, p. 11). And Brentano went further: The object did not exist except with reference to the act of seeing, and conversely perception existed only in reference to its object. This relationship was called "intentional," and it served as the origin of twentieth-century phenomenology as expounded by Husserl.

Phenomenology means literally the description of appearances; the word *appearances,* like *phenomena,* need not carry any suggestion of unreality. It is interesting that the phenomenological critique recalls an earlier philosophical issue that in many ways deals with the same problem but formulates it differently: "In the end, after an honest effort, we will probably find ourselves agreeing with the philosopher [Kant] who asserts that no idea is fully congruent with experience, although he admits that idea and experience can and must be analogous" (Goethe 1818, 1988, p. 33). There are obviously complex Romantic antecedants to this view, which I have detailed elsewhere, as exemplified by Goethe's epistemological project (Tauber 1993) or from another perspective in Karl Marx's philosophy. Beyond the economic and political analyses, Marx enunciates a Romantic theme highly germane to what Nietzsche would reiterate in his own attempt to unify a self divorced from nature and its most essential being (discussed in the next chapter).

Marx and later Nietzsche identify the integration of some self identity with the body, with the organic, as both crucial and telling in their apparent separation. Two themes that reside in Marx's polemic are later developed as foci of twentieth-century phenomenological philosophy. The first concerns the primacy of defining a person by his or her activity. The credo of existential phenomenology was to become "Do not look for that elusive entity we call the self in some abstraction, but find it declared by the process of living." Marx regarded the core issue of alienation to be the fact that one's work in the industrialized domain was divorced from one's true essence, because produced objects had no direct relevance to one: One's labor had become an isolated commodity. The manifestation of that alienation is in one's self-defining activity: "[T]he personal physical and mental energy of the worker, his personal life (for what is life but activity?), as an activity which is directed against himself, independent of him and not belonging to him. This is self-alienation" (Marx 1844, 1963, p. 126). But beyond the primacy of activity, which Marx assigns to our animal nature ("The animal is one with its life activity. It does not distinguish the activity from itself" [ibid., p. 127]), there is our conscious life activity. This second component, consciousness as it relates to definable selfhood, becomes philosophy's preoccupation by the end of the century. Marx views alienated labor as fundamentally dehumanizing precisely because it undermines a person's consciousness as autonomous.

I cite these rather unexpected sources only to emphasize the deep preoccupation nineteenth-century philosophy had with the knowing self. The problem of describing what actually happens, in terms of human consciousness in connection with the "objective" world, would ultimately lead to a philosophy centered on these issues. There is a profound incommensurability between the objective, centerless physical world and the subjective, perspectival world of the self. The argument concerning the applicability of "objective" criteria to characterizing the self as mind must address the following questions: Does the mind itself have an objective character? What is its relation to those physical aspects of reality whose objective status is less doubtful? How does one understand, in a centerless world, a personal perspective? The claim that mentality exists outside objectivist analysis has strong proponents despite the dominance of logical empiricists and their allied cognitive scientists. The phenomenologists address these issues in their own characteristic fashion, but note here that the issue of the mind as subject to "objective" analysis is eschewed by James and Husserl. "Objectivity of whatever kind is not the test of reality. It is just one way of understanding reality" (Nagel 1986, p. 26) is the credo of attempts to verify the legitimacy of viewing mind as a nonobjective entity. At the base of these discussions is the effort to analyze the tension between regarding the self as characterized by its own perspective of organizing its world (i.e., what gives the self-locating philosophical thought its peculiar content) and the other aspect of the self's character, that is, its being perspectiveless, because the self constructs a centerless conception of the world by casting all perspectives into the content of that world. So what is phenomenology, how does it situate itself in these philosophical matters, and finally, why is such a strategy a useful means to critique immunology? I will present the case, first, in its historical context.

The first psychological phenomenologist was Friedrich Edward Beneke. In 1824 he published *Contributions to a Purely Psychological Psychopathology*, where, as an antiidealist (and antimaterialist), he argued that psychological phenomena should be viewed by themselves – studied under their own autonomic method and organized into a totally autonomous psychology. Probably the first attempt at psychological self-observation, it represents the first attempt at ego psychology, which was developed in the mid–nineteenth century by Wilhelm Griesinger (Harms 1962, p. 684). Especially influential in the German-speaking world, Griesinger began to chart ego function in the structure of both the normal and abnormal psyche. Freud was particularly indebted: His personal copy of Griesinger's famous text *Pathology and Therapy of Psychic Disease* was obviously studied carefully, and it is viewed as the main source of Freud's ego psychiatry (ibid.). But the phenomenological movement proper, generally regarded as being born with Brentano's work and especially associated with his student Husserl (later to be developed in distinct ways by Heidegger and Sartre), emphasized the description of human experience as directed toward objects. Brentano's examination of first-person knowledge relied on the medieval notion of intentionality. Like Mach and James, he was a trained scientist who believed that mental events have physical causes; he was thus committed to neurophysiology. After making the formal positivist bows, Brentano divided human experience into either mental or physical phenomena and identified "descriptive psychology" as enumerating, describing, and classifying mental phenomena, thus unifying the inner and external world by intentionality:

> . . . mental phenomena are acts which refer to objects whether those objects have both actual and intentional existence or intentional existence only; they are acts which either bestow a kind of existence on an object which has no actual existence, or add a second mode of existence to an object which does actually exist already. (Ibid., p. 28)

Mental phenomena are the exclusive objects of inner perception, and therefore they alone are directly perceived and yield self-evident knowledge. He argued that intentionality is common to all mental phenomena and peculiar to them. Thoughts have objects (even if unreal), and these constitute intentional objects. Intentionality is defined as an internal "aim" or "reference" to conscious content, whether real or imaginary; that is, in every case there is an *object* of thought. The intentional object is that which is present to consciousness. For instance, as an intentional object an apparition is a ghost, whose material source may be a fluttering sheet. What is at issue is the conscious interpretation, for the very existence of the intentional object depends on the mental state that refers or is directed to it. Thus, subjective experience, whether or not it is ratified by others, is "actual" in terms of its personal reality and must be dealt with like any other intentional object of thought. Consciousness is thus unified, but at the same time Brentano left the self somewhat unstable and lacking firm boundaries, as we will see in James's analysis. The elusive self-conscious self was left "suspended, so to speak, in its intentional relation with the object. This notion, in which the traditional dualism of subject and object begins to be broken down, was perhaps his most lasting contribution to philosophy and psychology" (Ryan 1991, p. 11).

Ernst Mach stood in opposition to Brentano's implicitly holistic conception of consciousness both in how it viewed intentional objects and the nature of its function and structure (an attack on the Kantian position that consciousness necessarily entailed self-consciousness). A physicist and mathematician by training, Mach was perhaps drawn to psychology through his interests in acoustics and optics. More pro-

205

foundly, both as scientific practitioner and philosopher, he was committed to an ambitious and highly influential approach to science that is characterized as positivist. (He was a seminal forerunner of the school of logical positivists known as the Vienna Circle [Frank 1941, 1961]). In his desire to rid science of any "contaminating" metaphysics and to unify the various disciplines under common methodologies and precepts, Mach was all the more powerfully drawn to psychology by his doctrine (and hope) that all scientific propositions might be formulated in terms of perceptions. Mach believed "genetic psychology" would take a place among the natural sciences, addressing the same sorts of questions about mental phenomena that other sciences pose about physical phenomena (McAlister 1982, p. 13). He reasoned that an objective understanding of the knowing perceiver would fulfill the ultimate goal of establishing the unification of science, from psychology to physiology to chemistry to physics (Cohen 1970). The nature of perception was thus crucial to his philosophy of science, which was based on strict empiricist doctrine: "Mach never maintained that our world consisted of complexes of perceptions, but that every scientific proposition was a statement about complexes of perceptions. Whether it be a proposition of physics, biology, or psychology, it can only be proved or refuted by comparison with observation" (Frank 1941, 1961, p. 90).

Metaphysics would have to be eliminated, because it was viewed as "contradictory to the economical function of science" (ibid., p. 91). The philosophical antecedents for this position are complex, but one important source was Rudolf Lotze (1817–81), who argued in the preceding generation that "Philosophy proper or Metaphysics gives no new knowledge but is merely an attempt to reconcile opinions gained by very different processes of thought and observation; by the rigid methods of science on the one side and the demand of practical life on the other" (Merz 1896, vol. 4, 1965, p. 749). Lotze, however, never progressed to Mach's ambition for a unitary basis of knowledge, acknowledging instead three distinct realms of observation, namely, fact, universal law, and values, by which meaning was derived. Thus, science, art, and value were distinct, but the positivist hierarchy ordered their relations: "Knowledge of existence depends upon knowledge of fact acquired through observation and experimentation. Consequently, the empirical sciences are the proper investigators of existence" (Gotesky 1967). In *The Science of Mechanics* (1883, 1893) Mach forcefully argues that there is no a priori knowledge, that all science is based on sensory experience, and thus all basic mechanical concepts must ultimately be derived from, and related to, perception. This antimetaphysical, empiricist sensationalism was ultimately structured by his reductionist philosophy. Mach's atomism harkens to the extreme position of Cartesian reductionism, where the world consists of incoherently organized, free-floating particles. Like the eighteenth-century associationists, Mach believed objects were but groupings of sense impressions congealing almost randomly, in unstable relations of proximity and linkage. In this regard his drawing may be recalled, where the self is not a conventional person, but simply the reduced entity of a particular field of vision (or more generally, perception). But if vision deals with an ad hoc world without clear and unified definition, then the self is similarly but "a collection of variously grouped perceptual elements, a 'mass of sensations, loosely bundled together'" (Mach 1886, 1914, p. 24; Ryan 1991, p. 12). Thus, Mach essentially returned to a Humean position propounding a self less stable and distinct than Brentano's, or as Mach himself dryly noted, "[T]he ego is not a definite, unalterably, sharply bounded unity" (Mach 1886, 1914, p. 19).

The problem was not only to define consciousness, but also to understand its unity and its connection with the means by which the world assumed meaning. As Merz concluded his magisterial review of nineteenth-century thought, he mused that the power of reductionism was limited, largely viewed with suspicion (except by its ardent proponents!), whereas a comprehensive unification of the world in and through consciousness was regarded as the key priority:

> The analytic process is irreversible. The point at which we start to synthesise or put together is purely arbitrary, fixed by our knowledge or rather our ignorance, and the product of such synthesis is accordingly artificial, not natural: the world of things, images of thought or practical constructions, is accordingly artificial; these are neither natural nor artistic. Now if this process of looking at things as a whole and not in their isolation has become the order of the day it surely must recommend itself also in dealing with the totality of things as revealed to us through consciousness. The unity which we are in search of in philosophical thought certainly exists as the felt continuity of consciousness, of what Kant termed the "unity of apperception." (Merz 1896, 1965, vol. 4, p. 776–7)

When William James entered the fray, he took the position that consciousness was to be viewed as a unified whole (not a composite of parts) that arose from the sense of continuity that holds the self intact regardless of gaps in consciousness. He derived this principle from Brentano's notion of inner perception, the reality of our sense of implicit coherence. And that notion in turn harkens back to Schopenhauer's view of the subject connecting the objects of his or her world through his or her own relation to them (discussed in Chapter 7). The phenomenological challenge, that is, the issue of thought's intentionality, is developed by regarding the object, even when it is not directly linked to consciousness, as being an object of potential perception or action. Even outside the boundaries of immediate perception, the "hat in the cloakroom" (namely, what is not *seen,* but is known to exist) thus is allowed to exist. Similarly the self, whose coherence was undermined by the problem of intentionality (the self dissolving in the unclear boundaries between subject and object) and by the problem of elementarism (the self dissolving as a transitory bundle of associations), was pragmatically reassessed: the self exists as a functioning, convenient, practical entity that we attach to our ever-changing sense impressions. It was an arduously argued position. So, in reviewing James's critical, if not radical, empiricism, we briefly consider the far-reaching effects of these psychological consternations.

B. WILLIAM JAMES

William James offered an epistemological facet of the self-defining self. The discussion here focuses on his radical empiricism, his creation of a psychology that identifies consciousness as a "selecting agency" (James 1890, 1983, p. 142), "the very hull on which our mental ship is built" (ibid., p. 640), forming the "nucleus of our inner self" (ibid., p. 423). This is the central motif of *The Principles of Psychology* (his *magnum opus,* begun in 1878 and published in 1890) and constitutes the basis of viewing the mind as an active, selective self, whose volition is dictated by attention: "[E]ach of us literally *chooses,* by his ways of attending to things, what sort of a universe he shall appear to himself to inhabit" (ibid., p. 401). The selective attending of consciousness is simply the extension of the cognitive expression of an active organism in its environment, and it is manifest from the simplest responses to the most complex interactions. Human action – volition, thought, reorganization, and manipulation of the environment – can be understood in light of the Darwinian position that individual variation is the basis for adaptive change and adjustment. Given that context, James asserts an ideal of freedom of action and the unique value of the individual (Seigfried, 1984). The world must then be organized from "one great blooming, buzzing confusion" (James, 1890, 1983, p. 402) by selective focus in two respects: attention and interests. Attention is regarded as voluntary action, not merely a passive response to external stimuli. In the same year that Metchnikoff published his first phagocytosis studies (1883), James wrote "On Some Omissions of Introspective Psychology" (the basis of Chapter 9 of *Principles*), where he attacked the simplistic empiricism of the Humean sense data epistemology and argued that simple sensations did not exist; there were rather the "results of discriminative attention, pushed to a very high degree" (ibid., p. 219). Moreover, the Lockean combination of "simple ideas" has no true permanence; only because of active sorting and combining are we able to experience an object by processing sensations as "stepping-stones to pass over the recognition for the realities whose presence they reveal" (ibid., p. 225).

During the same period painters, for instance, Monet (impressionistically) and Seurat (pointillistly), were demonstrating the extraordinary complexity of visual perception, where a largesse of highly variegated color was coalesced and processed as an image with certain uniformity and finiteness; absolute quantities or qualities were replaced with synthesized ratios and conglomerates. In large measure the radical orientation of that period rested upon a *self-consciousness* of such processing. Goethe had already been concerned in the 1790s with the construction of scientific facts in the context of accepted theory:

> It is easy to see the risk we run when we try to connect a single bit of evidence with an idea already formed. . . . Such efforts generally give rise to theories and systems which are a tribute to their author's intelligence [but may] harm the very progress of the human mind they had earlier assisted. (Goethe 1792, 1988, p. 14–15)

And John Ruskin by midcentury was acknowledging the role of attention in perception and aesthetics:

> . . . the mind vacant of knowledge and destitute of sensibility, and the external object becomes little more to us than it is to birds or insects. . . . On the other hand, let the reasoning powers be shrewd in excess, the knowledge vast, or sensibility intense . . . the object will suggest so much that it shall be soon itself forgotten, or become at the utmost, merely a kind of key note to the course of purposeful thought. (Ruskin 1901, p. 270)

But James was to carry what Goethe called the old truth – ("[W]e really have eyes and ears for what we know" (*Conversations with Eckermann*, 1824) – and what Ruskin called sight to formalized philosophy. The experiencing mind, that is, one adaptive and shifting along the dimension of time, in which experience remolds the "experiencer" and objects of attention and inquiry are never viewed twice in the same context or from the same perspective, demands an organizational, dynamic psychology. An atomistic view of an unchanging world simply cannot support James's psychology. Thus, beyond the active component of his system, the parallel to Metchnikoff resides in the holistic nature of thinking, where the Jamesian self acts in its full phenomenological context.

James's most consistent, and probably most lasting, contribution was his insistence on a phenomenal totality (Seigfried 1990, p. 88). "Thought suffused with the consciousness of all that dim context" (James 1890, 1983, p. 227) refers to all that structures our experience and that can only be analyzed reflectively. Our selection (and ordering) of experience is present in every perception, and in fact we "ignore most of the things before us" (ibid., p. 273). Thinking thus involves choice, and what interests us are those objects of practical or aesthetic importance. He seems to have concluded that the need for selectivity of consciousness is an ever-changing flux of reality won by arbitrarily arresting experience. Conceptualization requires isolating and distinguishing some aspects, excluding others, and ordering reality not by hard and fast divisions, but by personal processing. Experience, then, enters not as bare or raw data, but according to the needs of the one experiencing. Experience is thus manipulated to serve our needs, and the self becomes an inviolate "me," a self-defining, unique individual:

> Even the trodden worm . . . contrasts his own suffering self with the whole remaining universe, though we have no clear conception either of himself or of what the universe may be. He is for me a mere part of the world; for him it is I who am the mere part. Each of us dichotomizes the kosmos in a different place. (Ibid., p. 278)

James distinguishes the self from consciousness.[1] In the tenth chapter of *Principles,* "The Consciousness of Self," he observes that the world may obviously be divided between a self and a nonself, but upon immediately recognizing that the perceptive events are mediated by embodied reactions, for example, movement of the head, sensations perceived by sight or sound, we are forced to contemplate the visceral origin of this bodily knowledge.[2] James becomes preoccupied with bodily sensation as mediating consciousness:

> . . . the 'Self of selves,' when carefully examined, is found to consist mainly of the collection of these peculiar motions in the head or between the head and throat. I do not for a moment say that this is all it consists of . . . but I feel quite sure that these cephalic motions are the portions of my innermost activity of which I am most distinctly aware. . . . [It is also likely] that our entire feeling of spiritual activity . . . is really a feeling of

bodily activities whose exact nature is by most men overlooked. (Ibid., p. 288)

Consciousness ironically becomes another "object" of consciousness. This "self" is not consciousness itself. As an object, consciousness somehow resides separate – closer and more intimate to our true self, but always distinct. Consciousness is the stream of thoughts that, when contemplated as an object, must then be displaced by another core sense of selfness. The self cannot be purely experienced, but is represented by the *"sum total of all that he CAN call his"* (ibid., p. 279) These, of course, include emotional, spiritual, and material elements of identity beyond bodily sensations, but the embodiment reactions serve as the foundation of consciousness and, more fundamentally, selfhood.

This argument clearly may be related to the similar position taken by Sartre, Merleau-Ponty, and other later European phenomenologists, who place the body at the core of experience and the origin of reality:

> Its [consciousness] appropriations are therefore less to *itself* than to the most intimately felt *part of its present object, the body. . . . These are the real nucleus of our personal identity. . . .* They [body, its adjustments, reaction] are the kernel to which the *represented* parts of the Self are assimilated, accreted, and knit on. (Ibid., p. 323)

Interestingly, however, in a footnote accompanying this passage, James steps back from identifying the embodied, the sense of body, as *the* self. (Here we see a separation from the crucial role of the body for Nietzsche – and for that matter, for Schopenhauer.) For James, the body serves as an approximation, or better an approach toward defining the self, its distinction from consciousness. The self must remain ultimately elusive:

> The sense of my bodily existence, however obscurely recognized as such, *may* then be the absolute original of my conscious selfhood, the fundamental perception that *I am.* All appropriations *may* be made *to* it, *by a thought* not at the moment immediately cognized by itself. Whether these are not only logical possibilities but actual facts is something not yet dramatically decided in the text. (Ibid., p. 324)

James must return to an attempt at least to define a means of knowing the self. Nevertheless, since consciousness is world-directed, it is defined in terms of its objects, including the recognition of its own body as its "self." It cannot *be* the self, but rather is a function or condition of all objectification and is necessarily preobjective. Consciousness is only *known* in retrospect: "[I]t is not one of the *things experienced* at the

211

moment; this knowing is not immediately *known*. It is only known in subsequent reflection" (ibid., p. 290). Consciousness, then, is process, the function of objectification, but it is irretrievably distinct from its ontological source, its selfness.[3]

This is not the forum to evaluate Jamesian consciousness further, beyond noting that he struggled throughout his career to recount how we share a common world, to reconcile the individual's experience as consistent with a universe experienced in common (Seigfried 1990, pp. 82–97). At certain points he assumed a primordial experience (*Some Problems of Philosophy,* 1911, 1987), at others, he took a resigned attitude to the amazing connectedness of nature, viewing the consistency of experience responsive to our subjective purposes as "a miracle not yet exhaustively cleared up by any philosophy" ("Reflex action and theism," in *Will to Believe,* 1897, 1979, p. 96). The self identifies its unique stream of consciousness, or more properly a segment or limited portion of that flow. James's task was "the reinstatement of unity into the experience of selfhood while retaining the Humean idea that the only verifiable *I* is a momentary section of the stream of consciousness. Whenever the dichotomy between *I* and *me* is experienced, the *I* can be identified as that part of the stream which momentarily contrasts itself with everything else. . . ." (Myers 1986, p. 349). Thus James used "I" as a noun of position, like "this" or "here," and "my" designates the kind of emphasis that will define the self:

> Now, *what is this self of all selves?*
> Probably all men . . . would call it the *active* element in all consciousness; saying that whatever qualities a man's feelings may possess, or whatever content his thought may include, there is a spiritual something in him which seems to *go out* to meet these qualities and contents, while they seem to *come in* to be received by it. It is what welcomes or rejects. It presides over the perception of sensations. . . . It is the home of interest . . . that within us to which pleasure and pain, the pleasant and the painful speak. It is the source of effort and attention, and the place from which appears to emanate the fiats of will. (James 1890, 1983, p. 285)

Not only is there an evident indeterminacy here, but James may also be guilty of presenting a tautology: What is the self? That which the self appropriates. This issue must be reexamined in any critical appraisal of the phenomenological position. Even early criticism of phenomenology focused on the need for some provision to be made for a common entity underlying or uniting the successive states that might identify a self. Without this commonality, what conferred continuity? But James resolutely refused to see consciousness as anything more than processes filtered through the brain (Myers 1986, pp. 354–5), and thus he saw no ego or substantial self being assigned to support it.

Essentially James defaulted in postulating how mind functioned, having "grown up in ways of which at present we can give no account" (James 1890, 1983, p. 1280). Ultimately the nature of the experienced self is also left an open question, remaining as a description of an inter-active process of that self with the world. His later doubts concerning the role of rationalism still leaves an active self to order and experience the world, for as he wrote in *Varieties of Religious Experience*, nature is "a vast *plenum* in which our attention draws capricious lines in innu-merable directions" (James 1902, 1987, p. 394).

James has been regarded variously as antiphenomenologist (Hare and Chakrabarti 1980, pp. 232–3), a precursor of Whiteheadian process philosophy (Lowe 1949; 1990, pp. 104, 226, 345), and a modern her-meneutist (Seigfried 1990, pp. 173–207). But his epistemology must be viewed, in my opinion, as prephenomenological. His philosophy force-fully and energetically engaged the environment, where the self actively selects and constructs its world from the bewildering complexity of the surrounding plenum. The self must thereby differentiate itself, and it does so only by active engagement. The Metchnikovian self is vividly reiterated in this construction. Thus, in addition to Brentano, whose role in the birth of twentieth-century phenomenology is well known, James may fairly be assigned an important source for Husserl, specifi-cally in their shared goal of establishing a method of radical empiricism and rejection of psychologism (Wilshire 1968; Edie 1987).

James anticipated phenomenology's approach to the dual meanings of "primacy of perception." In the first sense, it designates the inescap-able character of the perceptual world and our embodiment in it. We are always tethered in the place and time in which our thinking takes place; the perceptual world is the constant and pervasive realm of con-sciousness. But a second thesis of primacy of perception is perhaps even more dominant: The perceptual world not only accompanies "other orders of experience, but 'higher orders' of experience, and more particularly categorical thought" (Edie 1987, p. 10). Thus, phenome-nology argues that consciousness resides at different, interrelated or-ders of reality ultimately grounded in perception: "It is not a question of reducing human knowledge to sensation, but of assisting at the birth of this knowledge, to make it as sensible as the sensible, to recover the consciousness of rationality" (Merleau-Ponty 1964, p. 25). One form of this position is psychologicism, which is simply defined as the belief that the structures of reasoning are to be identified with psychological processes. James rejected atomism, the view that experience is made up of finite bits of sensuous data (color, hardness, etc.): "No doubt it is often *convenient* to formulate the mental facts in an atomistic sort of way, and to treat the higher states of consciousness as if they were all built out of unchanging simple ideas. . . . But . . . we must never forget that we are talking symbolically, and there is nothing in nature to answer to our words. *A permanently existing "idea" . . . is as mythological an entity as the Jack of Spades"* (James 1890, 1983, pp. 229–30).

Anticipating Husserl, as well as the Gestalt psychologists, James argued that consciousness is cumulative, never identically in the same state, and consists of "transcendent 'objects' of conscious acts and their meanings, which are fixed by consciousness and stabilized through temporal and synthetic acts which themselves have none of the qualities or characters of their objects" (Edie 1987 p. 11). James maintained (as Husserl did in *Logische Untersuchungen*), that especially in abstract thinking, for example, in aesthetics, ethics, and science, there are *"relations amongst the objects of our thoughts which can in no intelligible sense whatever be interpreted as reproductions of the order of outer experience"* (James 1890, 1983, p. 1235). Judgment must be invoked, and this faculty functions independently of associative perceptual experience. Experience that demands comparison is not subject to simple conditions of perception; categorical thought has an atemporal and aspatial quality that is "ideal." Thus, the "laws of logic" and "forms of judgment" are in a fundamental sense independent of psychological processes; that is, they escape psychologism. But then the question is why judgment effectively mediates our world:

> It is a very peculiar world, and plays right into logic's hands. . . . Evidently it [the construction of relations from diverse categories] is a pure outcome of our sense for apprehending serial increase; and, unlike the several propositions themselves which make up the series (and which may all be empirical), it has nothing to do with the time- and space-order in which the things have been experienced. (Ibid., pp. 1246–7)

The conclusion is that the world we perceive and the one we think are the same, for logic and perception are consistently intwined. Objects as constituted in the ever-changing stream of consciousness are an *achievement* of the mind, with its property of intentioning or meaning-analyzing the object of thought. James distinguished the object of thought both from the act of consciousness in which it is given and from the reference object toward which it is directed. In other words, James's objects of thought belong neither to the physical world nor to the stream of experience; nevertheless, they are the means by which *objectification* of both consciousness and the world occurs (Wilshire 1968). "Object of thought" is nothing less than the mind's functional apparatus, which in the chapter entitled "Conception" is defined as the "subject of discourse" between these two poles – the perceived world and the perceiving senses: "It [the subject of discourse now called conception] properly denotes neither the mental state nor what the mental state signifies, but the relation between the two, namely the *function* of the mental state in signifying just that particular thing" (James 1890, 1983, p. 436) As James Edie notes, this is very close to the phenomenological notion of the phenomenon, for "the notion of 'object' . . . is more fundamental than 'being' or 'thing' or 'reality' because modes of being are correlated to modes of experiencing" (Edie 1987, p. 34).

C. EDMUND HUSSERL

Edmund Husserl radically extended James's philosophy of the self and how the world is known. Husserl shifted away from the description of experience toward a description of the objects of experience, which he called phenomena. Simply stated, he regarded phenomena as things that "appear," as essences that the mind intuited. The task of phenomenology was to describe these essences. He shunned empiricism and viewed the problem as an a priori task. Early in his career Husserl assumed the Cartesian position that the immediate knowledge one has of a conscious state is the foundation for understanding the nature of the mental state in general. The problem, then, was to isolate the intrinsic nature of the mental from the extraneous. Intentionality assumes a pivotal position in assigning "meaning" or "reference" as essential to every mental act. Certainly intentionality in its broad meaning did not originate with the phenomenologists. Husserl's project largely went beyond psychologism, whose crude origin as introspection dates at least to the earliest Romantics. (Consider Rousseau's claim made in *Émile:* "This power of my mind which brings my sensations together and compares them may be called by any name; let it be called attention, meditation, reflection, or what you will; it is still true that it is in me and not in things, that it is I alone who produce it, though I only produce it when I receive an impression from things. Though I am compelled to feel or not to feel, I am free to examine more or less what I feel" [Rousseau 1762, 1987, p. 232].) As will be shown, during most of his career Husserl eschewed psychologism, but did return to a psychological phenomenology in *The Crisis of the European Sciences* (1954).

215

To understand the nature of mentality is to reveal the fundamental operation of meaning, whereby the world is made intelligible. The mental act is the datum, the fundamental basis of knowledge, and of course, this raises the problem of defining the world independently of the knower. Husserl is plunged into skepticism, but in antithesis to Descartes, he doubts the observing subject himself. The self exists only through intentional acts existing as the subject and never as the object of consciousness. In making the "gaze" the privileged vehicle of the individual's relation to the world, twentieth-century phenomenological philosophers in general have concluded that consciousness is mediated by the corporeal intermediacy of the senses; that is, meaning depends quite literally on how we see things (Merleau-Ponty 1945, 1962). The way we are in the world and our consciousness of ourselves are based on the experience of "seeing," which thus becomes synonymous with "being aware" intentionally (Husserl 1913, 1982, p. 75ff). Intentionality is taken beyond Brentano's psychological meaning, but it also assumes an ontological significance:

> We then understand ourselves, *not as subjectivity which finds itself in a world ready-made, as in simple psychological reflection, but as a subjectivity bearing within itself, and achieving, all of the possible operations to which this world owes its becoming.* In other words, we understand ourselves in this revelation of intentional implications, in the interrogation of the origin of the sedimentation of sense from intentional operations, *as transcendental subjectivity,* where, by 'transcendental', nothing more is to be understood than the theme, originally inaugurated by Descartes, of a regressive inquiry concerning the ultimate source of all cognitive formations, of a reflection by the knowing subject on himself and on his cognitive life, the life in which all scientific formations valid for him have been purposefully produced and are preserved as available results. (Husserl 1948, 1973, pp. 49–50; see also pp. 75ff)

This experiential reality, in opposition to a "scientific reality," which is experienced abstractly, is thus fundamentally subjective. Objects function less "as they are" than "as they mean," and objects only mean for someone (Kohák 1978). "To see implies seeing meaningfully," and in this alignment, the gaze functions "less as the faculty that receives images than that which establishes a relation" (Morrissey 1988, p. xx). Gaze is thus a metaphor for understanding.

The twentieth-century phenomenological movement built upon the positivist revolt against Romantic speculation. Whereas empiricism reduced experience to passive sense perception, the positivists recognized *experience* as the ultimate ground and meaning of knowledge. Husserl extended this position, regarding experience in "its full breadth as primordial awareness presenting all our being" (Kohák 1978, p. xi).[4] This program, with its intense attention to experience *qua* experience, essentially suspends the ego and the natural object. The interaction referred to as perception *of* the natural world by a perceiving subject, that is, $S \rightleftarrows O$, is "bracketed" by the focused scrutiny of the arrows. Although the subject is always bumping up against the world, in Husserl's relentless reconstruction, the subject is discerned (and defined) in its experience, that is, as lived. I will focus upon *Ideas I* (1913a, 1982), which is very much a product of that immediate pre–World War I decade that witnessed the declaration of the abstract in art (e.g., Cubism and Schoenberg's new atonality) and the submerged literary subject and self-conscious selection of reality (e.g., the publication of James Joyce's *Dubliners* [1914] and the first part of Marcel Proust's *Remembrance of Things Past* [*Swann's Way*, 1913]). A strong case may be made that the essential problematics of our postmodern world, which took form around this time, persist as the established boundaries of twentieth-century identities and the discourses revolving about them (Kern 1983; Pippin 1991). To this end, any discussion of the *self* should seriously consider the lessons that might be derived from these philosophical deliberations. In *Ideas* Husserl argues for a radically reformulated notion of the self, which will now be briefly outlined.

Husserl begins by analyzing the commonsense world, the universe of "facts," which, if allowed, quickly degenerates into an uncritical epistemology. The world in which we live is constructed by projected meanings and is naively realistic. Each particular, for example, an object such as a chair, is truly a contingency, assuming its meaning in our experience. A chair is ultimately a chair in *principle,* and its particular fashion has only the broadest conditions or boundaries placed upon it. So any given "chair" is a chair both in fact and more saliently as it fulfills the principles of "chairness." Thus, we recognize chairs in an otherwise bewildering array of possibilities by their general common properties assigned by our gestalt experience of "chairs," by abstracting to their essence and appropriately applying the abstraction to any given case. This second dimension of "in principle" perception, then, is of a higher order, where meaningful wholes are presented or perhaps as gestalts are constructed. *Seeing* is a *process* that is more than the ordinary (or primary) perception of objects, but must include perception of principles as well. Husserl regards the *eidos,* the principle, as representing a kind of object altogether different from commonsensical reality: *"The essence (Eidos) is a new sort of object"* (Husserl 1913a, 1982, p. 9).

217

"[F]acts actually present themselves, not as common sense interprets them, but as structured by an *eidos,* as making sense" (Kohák 1978, p. 18). In short, we experience objects as embodying principles. Thus, fact and essence are intrinsically linked. Phenomenology is a study of the principles rather than a study of the facts. Husserl argues that it is the "logic" of phenomena that makes the particulars intelligible and susceptible to study. His philosophy, then, is committed to scrutinizing the underlying eidos, *independent* of all factual knowledge. (Note that the converse is not asserted: "*[P]ure eidetic truths contain not the slightest assertion about matters of fact.* And thus not even the most insignificant matter-of-fact truth can be deduced from pure eidetic truths alone" [Husserl 1913a, 1982, p. 11].) To break through the habits of common sense is to regain an awareness of our actual lived experience.

Bracketing (or *epochē*) was Husserl's means of focusing on experience in its lived immediacy, deliberately setting aside all theoretical assumptions (including common sense) and describing experience as faithfully as possible. For example, consider a theatrical stage, where the object world is suspended, existing but without meaning until the drama develops and both character and objects assume meaning in the narrative. "The world as 'real' is bracketed; it remains as lived. It is no longer the *explanans,* explaining human acts. It becomes the *explandum,* deriving its meaning from them" (Kohák 1978, p. 28). The drama presents experience as the subjects live it rather than from the viewpoint of an observer. Husserl, in treating the world as a stage, models how we might break the ordinary perception of the world as objective. The world does not disappear, but assumes a new meaning. This "staging" of the world is what Husserl attempts in bracketing: Reality is the *experiencing* of the object. Bracketing is an exercise to recognize the world as experience, or as Husserl writes, as phenomenon.

Bracketing thus reveals phenomena as we in fact encounter them, as subjective experience. Common sense interprets reality as the object, and we experience it incidentally. For Husserl, lived experience is now the object of inquiry, and the world explains it. To understand my experience, common sense assumes I need not know that I am experiencing, but must discover what in the world is causing it. (This, of course, is the scientific attitude as well.) Husserl responds with two rebukes: First, the subject in fact *imposes* an interpretive reality, and second, experience is a *subject's* activity, which may be objectively ordered but remains firmly embedded as the subject's own. Bracketing suspends the thesis of common sense that a transcendent reality "explains" experience. It

218

is an operation of treating phenomena as we encounter them, as subjective experiences. Thus, in bracketing, any transcendent reality posited as objectively given rather than as experientially derived is denied *explanatory* value. The subject confers experience. The world remains as a datum, but now as a datum to be explained, not as one that explains experience. In short: "It is the intelligibility of experience that makes the world intelligible, not vice versa" (ibid., p. 39).

How is the subject to be defined in this phenomenological formulation? No doubt the centrality of the subject's *experience* is asserted. The world exists "out there," but it assumes reality in our experience, otherwise being irrelevant to *us* as experienced. We accept its objective status, but it is in experience that it assumes meaning, so the issue is *how* we experience it as meaningful; here meaning is the central concern. Meaning is conferred in the subject's experiencing, and thus the I that perceives and brackets cannot be suspended. But in Husserl's system there is no natural subject, one being among others; neither a body nor a *res cogitans,* soul, psyche, or ego. The implicit self is "simply the sheer I-hood of all experience" (Kohák 1978, p. 29).

Fundamentally the self remains undefined, for the postulation of a cogito or distinct ego is suspended. However, even though the self is elusive as a *construct,* it is ever present as the experiencing subject. Experience is, of course, from the first-person perspective, that is, the world as experience rather than as a transcendental object, but "the ego in every sense shall be left out of consideration"; it is enough to deal solely with "*the stream of mental processes*" (Husserl 1913, 1982, p. 68). Husserl, thus, built on the Jamesian notion of the stream of consciousness (ibid., p. 87), the "manifold of thoughts and feelings," where the world of experience has its own logic, its own internal coherence, which is not determined by the laws of physics, that is, efficient causality, but by motivation. The Jamesian task of looking for the unity of the stream of consciousness drove him to introspection, to gaining awareness of the stream, and then to articulating that awareness.

Husserl delved into a new strata. In the resolute focus of viewing reality as made up of verbs, rather than of nouns, the subject also became a predicate of action, never an object. Husserl deployed the intentionality of Brentano to refashion the Jamesian query toward defining the consciousness of the act. He remained close to James in viewing coherence as derived from the active synthesis of perspectival views[5] (ibid., p. 88). What James called cognitive, Husserl shifted to intentional. He later argued (1937) that this purposiveness is built into the nature of our being – that is, it is transcendental – and therefore that meaning is contingent upon life itself and only provided with order through consciousness.

The descriptive phenomenology of *Ideas* was modified and amplified later by Husserl in *The Crisis of European Sciences* (1954). This work essentially metamorphosed his project into a transcendental philosophy. *Ideas* did not differentiate a pure from a transcendental ego. The Cartesian ego is empty of essential components, and the issue that Husserl left for the *Crisis* was how a subject could be both an experiencing entity and an object in its world. In the Cartesian modality the experiencing ego cannot be taken for itself and at the same time be made into an object. When Husserl asserted a transcendental ego, he formulated a self that was defined solely in relation to its intentional object (Kockelmans 1977, pp. 277ff). The transcendental ego of the *Crisis* constitutes itself continuously as existing, grasping itself not only as a flowing life and stream of experiences, but also as an "I" – a self. The basis of a meaningful intention, a subject *connected* to its world, is through transcendental subjectivity. In this scheme world consciousness is at the same time consciousness of the self *in* a world. The orientation is conferred by intentionality, which posits the subjective–objective poles. Subjectivity is not the subject, but the *ordering* of reality in terms of value and meaning; correspondingly objectivity is not the subject, but, again, an ordering, now in terms of extension and causality.

Reality's structure is explicated by transcendental subjectivity, but that is not to say the world *is* that subjectivity (Erazim Kohák, personal communication). We always perceive the world both as a segment of the external horizon and our own internal border, each flowing on in the unity of our perceptual conscious life. The harmony of that total perception is sustained through corrections, in what Husserl regarded as communalized perceptions (Husserl 1954, pp. 161ff). We share a common world, defined by others' perceptions as well as our own. Husserl critiqued those philosophies that took for granted that each "thing" appeared different to each person. Shared inventories lead to transcendental subjectivity, and in this broadened context personal intention does not belong solely to the isolated subject, but rests upon an integrated "intersubjectivity." These inventories are, of course, multilayered and overlapping from subject to subject, and it is this coherence that leads to unity of synthesis, an objective world (Husserl 1954, pp. 167–70). In short, the intersubjective constitution of the world is formed from elementary intentionalities. And it is the individual intentionality that represents the only actual – and genuine – way of explaining and of making the world intelligible.

Never regarded as a static state, Husserl's phenomenology constitutes an organism as always seeking, probing, observing, assimilating. The order of consciousness begins from the essence, that is, the eidetic, and the ideal, and the particular is interpretively derived from it (Husserl 1913a, 1982, p. 89). For instance, a table is seen as a trapezoid, even as we know that it is a rectangle and, more saliently, we know it is a table. Husserl thus argues for an eidetic science to study experience directly; in this endeavor the subject as a self emerges only in its acts. Thus, the agenda of Husserl's phenomenology is established. As lived, reality is the *experiencing* of an object, event, or emotion. In contrast from the usual, unreflective perspective (the natural attitude) reality is *the object,* and experience is only incidental to it. What Husserl calls the "thesis of the natural standpoint" is inverted in phenomenology by examining lived experience to determine how the world explains it. The subject is not defined via the traditional self–other configuration, but characterized only by the subject's activity. In this sense phenomenology suspends judgment on the subject *per se.* After the bracketing (the *epoché*) Husserl was left with a phenomenological residue that he called "the region of pure consciousness" – what we have referred to as lived experience:

> . . . consciousness has, in itself, a being of its own which in its own absolute sense, is not touched by the phenomenological exclusion [bracketing]. It therefore remains as the "phenomenological residuum," as a region of being which is of essential necessity quite unique and which can indeed become the field of a science of a novel kind: phenomenology. (Ibid., pp. 65–6)

This residue is rather perplexing and represents in early Husserl an undeveloped stage of his philosophy.[6] Husserl's "pure" or "transcendental" consciousness that remains after the reduction – the bracketing – is an attempt to name an as yet poorly defined ego that fluctuates among several levels. Two characteristics of the subject arising from this process are of particular interest here. First, the subject is freed from the limitations of the natural attitude, a release that has already been described as a basic concern of phenomenology. ("Does this pure ego have content? No, in the sense that it not a container; yes, in the sense that it is an aiming at something [i.e., intention]" [Lyotard 1986, p. 47].) And second, as a result of the reduction, the very nature of the self is exposed: "[T]he subject which is hidden from itself as part

221

of the world discovers itself as the foundation of the world" (Ricoeur, 1967, p. 26). Here we see the transcendental idealism of *Ideas I,* where intuition (in either its sensuous or its eidetic forms) "legitimates the sense of the world. . . . Transcendental idealism is of such a nature that intuition is not repudiated but founded" (ibid., p. 25). Any intuition functions by its intention. (For a detailed account of Husserl's concept of intuition and intentionality, see Levinas 1963, 1973; and various essays in Dreyfus 1982.)

The *epochē* thus assumes double significance: On the one hand, it isolates consciousness as phenomenological residue and thus allows an eidetic analysis of consciousness, and on the other, it allows consciousness to emerge as "absolute radicality," on which

> all transcendence is founded . . . [S]uch is the true significance of putting in parenthesis [bracketing]: it turns the gaze of consciousness back on itself, changes the direction of this gaze, and, in suspending the world, lifts the veil that separates the ego from its own truth. This suspension shows that the ego remains what it is – that is, interlaced with the world – and that its concrete content remains [in] flux. . . . (Lyotard 1986, p. 51)

Although the natural standpoint of our common sense may be Cartesian, our ordinary lived experience is not. For instance, reading this page cannot be reduced simply to thinking (a subjective event), but neither can it be viewed as a "thing" somehow divorced from process. In our awareness it is neither: The act includes both the object *and* the consciousness of it. Perhaps the best graphic depiction is Escher's etching of two hands drawing each other, or more simply the act of grasping our own hands: Which is subject? Which is object? Obviously the act is defined by viewing the process as one: "[I]t is an act, irreducible to either *res cogitans* or *res extensa*" (Kohák 1978, p. 50). Thus, identity emerges. I am a reader because I do the reading. To be human, to constitute a self is not a thing, an entity, or an object, but rather a mode or way of being, a way of functioning and living.[7]

Husserl can be interpreted as effecting an anti-Cartesian revolution. He has suspended realism, the reference to an entity-world, and what remains is everything in consciousness. The mirror of nature is shattered. More saliently it is the active faculty of consciousness that defines the object, which is always an entity *in* experience, the object *of* consciousness. And it is this discerned essence of the object that bestows its meaning:

> *We are not speaking of a relation between some* psychological occurance – called a mental process – and another real factual existence – called an object – nor of a *psychological connection* taking place in *objective* actual-

The self and the phenomenological attitude

ity between one and the other. Rather we are speaking of mental processes purely with respect to their essence, or of *pure essences* and of that which is "a priori" *included* in the essences *with unconditional necessity.* (Husserl 1913a, 1982, p. 73)

Thus, according to Husserl's argument, we live not in an objective world, but rather in one we constitute ourselves – in a broad sense, in a self-created world. Our intentions, as well as our facts, are value laden and inescapably perspectival. There are vivid examples from language and mythology that clearly illustrate (before our current fascination with the interpretative turns of hermeneutics) the relativity of meanings lived by peoples of different cultures. But staying solely in the confines of our own culture, consider the fierce disagreements about artistic intent or thematic meaning surrounding any great dramatic or narrative work (for instance, *Hamlet*). Our interactions are value laden at diverse levels, and we must conclude that as we live, we live meaningfully. In this view there are no subjects or objects independent of experience, for objects are subject constituted, and subjects are in turn defined by their activities. But ultimately the constitutive role of consciousness is an "irreducible experience." Obviously I am not the creator of my world, but in another sense I am the world I bump against. The object and the subject thus emerge in their encounter. Finally, when Husserl writes, "The shift of standpoint is a matter of the fullness of our freedom" (Der universelle Zweifelsversuch gehört in das Reich unserer vollkommenen Freiheit) (Husserl 1913b, 1950, p. 64), there is both an explicit epistemological argument and an underlying ethos. If Hume is correct in arguing that consciousness is the mirror of nature, then any shift would be impossible, but Husserl argues that consciousness is *self*-determined, free and active. What is seen is not determined solely by what is there, but also by what is projected. But this position also assumes an ethos, for if cognition shapes the world, it cannot do so in a neutral fashion, and if perspectival change is construed as a manifestation of human freedom, then such perspectivism might naturally serve as the means of achieving a human ideal. Thus, it is not surprising that we find both Jewish (e.g., Martin Buber and Emmanuel Levinas) and Christian theologians (e.g., Paul Tillich and Gabriel Marcel) building on such a phenomenological foundation.

223

D. REPERCUSSIONS

Returning to the persistent attempt to link the conceptual development of immunology to its relevant philosophical context, we must again address the potential consequences of a self that might still be viewed as problematic even with the phenomenological provisos. If we are not to fall back on a transcendental principle of unification (as Husserl did in later work), then the challenge remains as to what self is beyond its "self definition" in process. Certain general observations might be made concerning immune theory in relation to Husserlian phenomenology. This application is based upon both the metaphorical allusion to a cognitive self that employs perspective modalities to differentiate the self from the other and the more obvious and direct phenomena of immune recognition. This discussion is focused on the latter aspect. When Husserl expounds that we perceive our experience from both factual and eidetic viewpoints, he wishes to place the particular datum within the broadest principle of cognition. This principle focuses on the particular and bestows its meaning. For example, looking at a cathedral, we may view the arches and flying buttresses as well-articulated masonry, as a configuration of their geometric relationship, or more abstractly as a force diagram, calculating weight distribution and support stresses. We may also marvel at the aesthetic gestalt and resultant emotive power of the structure. When we turn to focus on the meaning of our perceptions, we dissect all these aspects of seeing. The complexity of the actual experience is thus exposed. The "facts" of reality are a meaningless aggregation of fragments that become intelligible only as structured by an eidos. In the immune encounter the scope of *to see* is only partially understood. Our best model presently is the MHC complexed with a peptide as a "perceptive" event. But from the Husserlian account this description is quite inadequate, for the eidos is derived from the system at large, where "cognition" resides.

There are two Husserlian dictates of interest here. First is the question of eidos, the issue that we must in principle understand what we are seeing in order to see it. Strange markings may be totally unintelligible, unrecognizable even as a form of writing, much less a discernible message, unless we know the particular language. To extend this analogy to MHC-restricted recognition, the science must establish the language by which such cognition occurs. It must involve reactions well beyond, and additional to, a lymphocyte "seeing" an inscription. What does the lymphocyte know? The foreign character of that peptide, its meaning, must be derived from the universe of its context. An overriding eidos must "adjudicate" its inclusion or exclusion from the universe of reactive lymphocytes. This is the basis of the essentially phenomenological critique found in Chapter 5. The notion of self is not used here, for selfhood as we understand it from a disengaged distance, emerges in the process of recognition. And this, the second key point of Husserlian phenomenology, cautions the inquiry to remain focused on the experiencing subject, for precisely by focusing on the act of perception, on the act of recognition, on the act of reaction, is the subject defined. In this sense there is no self residing as some preexisting and specified ontological entity. The immune self arises from immune activity. To think otherwise is to place ourselves outside, looking in. The self then becomes another object, just as the world is its object. But as Husserl demonstrates, selfness is not definable from the subject's vantage. How does the subject define itself? Its boundaries are only described in its bumping into "reality," in its meeting with the other. The immune self becomes defined through experience: in immunology's terms, in its immune encounters.

It is not my intent to further stretch Husserl's phenomenology of consciousness to a critique of immunology, but his inquiry has certain interesting repercussions for immune theory. First, consider that the metaphor of self immediately conjures a cognitive subject and implies certain borrowed assumptions of such an identity. The mimicking of neural network theory in immune network theory is but one example, but more subtly the immune self suggests certain properties of behavior borrowed from our mentalistic notions. Not only may these be scientifically unfounded, but the very model may distort and impede our theoretical construction of how the immune system works. With that caveat, we might nevertheless benefit from the critical dissection of what the self is from the perspective of a phenomenological critique. Specifically, what is the governing *eidos* of the self, and how is it defined by its activity?

My orientation in this discussion eschews the problem of mind–body duality and seeks the moorings of that duality in *process,* using a phenomenological orientation toward understanding the activity that confers unity and cohesiveness to establish identity. If immunology appropriates the term *self,* it must be cognizant of the philosophical baggage that accompanies that term. In the modern criticism of this issue, we have come to appreciate that from our subjective experience, every awareness (*noesis*) has a determinable orientation toward something (*noema*) perceived through corporeal schema. This is the cognitive ex-

225

pression of the immune model. Subjective awareness, then, is an organized product of an orientation, and it is the reactive, perceiving organism that is the orientational source with reference to the objects. Thus, as it perceives, it organizes, and from this seemingly obvious observation, beginning with Edmund Husserl and extending to Maurice Merleau-Ponty and other twentieth-century phenomenologists, the body is viewed as a subject, or as the self, but in an anti-Cartesian fashion: The organization of the world phenomenologically cannot define the self, but the composite array of behavior points to that nebulous construction. "Objectivity," even as presented in a particular or occasionally shifting perspective, still has an "unvarying region at the center of the variable perspective, distinguished as 'belonging' to itself as its 'body,' and behold, this phenomenal region happens in each case to coincide with some material system . . . we call organisms" (Jonas 1982, p. 89). Phenomenological philosophy attempts to achieve a coherent view of the world by concentrating on unfragmented experience, rather than on the division that ensues from a mental–physical orientation.

The mind–body problem is viewed as a result of a methodological or cultural by-product that makes reality appear dichotomous. The dualism becomes misplaced, for it renders or reifies the dual aspects of our reality (the natural world and our inner consciousness) into ontological "substances," whereas both aspects are oneself, not welded together, but existing as one embodied person. Critics have argued that this position is subjectivistic and idealistic (Farber 1967); however, phenomenologists understand the subjective mode as being firmly anchored in objective reality. The experience obtained through our senses is in fact the natural world, which includes the subject's body and all the rest of reality "in itself." But objectifying thought recognizes only two alternative modes of being, the *in itself* of spatial objects and the *for itself* of consciousness. Merleau-Ponty argued that this strict dichotomy of object and subject is at fault: The self appears at the nexus of objectivity and subjectivity. His ontological perspective remains ambiguous, but it appears to argue that the *relation,* or ongoing interactive *process,* between them generates the self. Merleau-Ponty attacked Cartesian doubt by "taking a blind plunge into 'doing'":

> It is true neither that my existence is in full possession of itself, nor that it is entirely estranged from itself, because it is action or doing, and because action is, by definition, the violent transition from what I have to what I aim to have, from what I am to what I intend to be. (Merleau-Ponty 1945, 1962, p. 382)

226

The self and the phenomenological attitude

When the phenomenologist deals with the *cogito,* he clearly establishes the equivalence of *I think* and *I am:*

> It is not the "I am" which is pre-eminently contained in the "I think," not my existence which is brought down to the consciousness which I have of it, but conversely the "I think," which is re-integrated into the transcending process of the "I am," and consciousness into existence. (Ibid., p. 383)

The implications of this attitude are far reaching. Self becomes defined in process:

> . . . *"self" is fundamentally a situated or positioned reflexivity oriented towards the world which is itself displayed, arranged, or organized strictly in reference to this reflexively oriented habitus.* The self is literally no-thing; nor is it identical with any of the delineated constituents. Neither noesis, organism, nor noema, *self is precisely the peculiarly complex reflexivity itself: that by virtue of which any reflexive referencing* ("itself," "myself," etc.) *is at all possible.* In that sense, self turns out to be the *eidos* of human life. (Zaner 1975, p. 168)

This phenomenological description has a cogent resonance with our thinking about immunology, since both share essentially similar problematics. (That phenomenological psychology arose at the same time that Metchnikoff engaged the issue of self in immunology is a tantalizing hint of a shared Zeitgeist, discussed in Chapter 8.) The self serves as an integrating entity. By its very "nature" self connotes an inclusiveness of being (perceiving, knowing, feeling, remembering); it entails the whole of our inner presence and mediates that core identity with respect to both the body and the environment. The self in its world is ever-changing and thus contingent, but the prevailing theme of its historical development is its persistence and representation as a holistic construction, forever appearing a manifestation of one faculty or another, but nevertheless remaining somehow coherent and intact. We know from conditions – schizophrenia and other diseases – that dissolve, disrupt, or fragment this integral sense of selfhood that a profound sundering and undermining occurs. Perhaps the existence of self becomes most evident through its dissolution. We recognize the artificial, if not destructive, analysis of Foucault (see Chapter 8) or Mach as rendering the fragile self into something else – unable to recompose itself once subjected to fragmentation. The self only exists intact, as a whole. At the same time, we understand *self* as only a metaphor: There is no such *thing* as a self. If anything, the self exists as *process.* It is the *doing* entity. We must then ask, to what purpose has immunology employed the self? Clearly not as an *entity:* There is only an operational definition of selfhood; the genetic signature conferred by MHC is but a fulcrum for distinguishing between self and nonself, and it must participate only as one component of an intricate array of recognition mechanisms and interlaced processing elements to effect an ultimate acceptance or rejection. Selfness thus arises in immune processing. And, of course, we know that MHC-mediated immunity is but one, perhaps the most intricate, of a variety of immune recognition mechanisms. What is the nature of the poorly characterized preimmune (or natural!) systems in this fundamental regard of final ingestion?

The self emerged in immunology to address a problem that remains unresolved, namely, how does immune activity emanate, and what does it protect? We know that in extreme cases, such as attempts at transplantation or in autoimmune disease, the self has been violated, but these are exceptions to the normal state. Toward what is the unstimulated economy of immune activity, that is, its background "noise" level, directed (Benner et al. 1982)? This low-grade, baseline behavior of "unstimulated" lymphocytes and APCs, processing dying tissue and generating large quantities of antibodies in a constantly replenished pool, represents a dynamic "core" activity that remains poised for action in what has been heretofore the central concern of immunology: the evoked immune response. The immune self more properly resides as the "residual" or quiescent state of this immune network; however, it is this poorly characterized condition of immune function that must be more carefully deciphered to understand better the character of immunity. Methodologically challenging, this examination acknowledges that immunity, beyond its provoked exhibitions, resides normally and prevalently in the tending of its endogenous functions – in the defining of self. If immunology is now groping inward to decipher the ontologic nature of the network, then we may readily accept the use of the term *self.* After all, the metaphor is but a tool, and as such it helps to pose the question and structure our inquiry. The language heralds a new challenge.

With this summary of the basic tenets of Husserlian phenomenology and its application to the critique of the immune self, I have endeavored to situate my arguments more firmly in the relevant philosophical discourse. This perspective is, of course, dictated by my general attitude regarding the construction of current theory and what I regard as a potentially useful restructuring. The discussion could more deeply address the philosophical underpinings of phenomenology and deal with the potent criticisms of its detractors, but I will close instead with an attempt to trace the faint outlines of immunology's metaphysical foundations from a different philosophical orientation. I regard this effort as a preliminary exercise, one requiring much further investigation, but I believe this essay has allowed us the ability already to scan the intellectual horizon and spot where immunology appears as part of its larger landscape. Finding those coordinates is a worthy endeavor not only because it offers an added dimension to our central theme of defining the meaning of the immune self, but also because this case study seems to exemplify how firmly science is embedded in its philosophical culture. The power and dominion of a scientific metaphor must be derived from its extracurricular sources, and thus to complete any analysis of an effective metaphor, such as the self, we must seek the deeper sources of its strength. To do so in this case, we return to Metchnikoff.

7

The self as organism: A philosophical consideration

A. THE SHARED VISION OF METCHNIKOFF AND NIETZSCHE

In a profound sense the success of immunology has rested upon the ascendency of its epistemology, and in this view it has been highly successful. The appropriation of the self must be regarded as an important element of that project. But beneath this triumph we can discern a metaphysical issue that we must further explore to truly discern the theoretical underpinnings of this science. The self metaphor depends on its metaphysical moorings to articulate its meaning; thus, despite ample Wittgensteinian warnings to the contrary, I remain committed to exploring these foundations. If epistemology enunciates *how* we know what we know, metaphysics (in a Whiteheadian [1925] tradition) proposes *what* there is for us to know. This is the underlying Metchnikovian challenge. Thus far, the inquiry has been oriented around the self, constructed by metaphor and analyzed in epistemological terms. There remain in fact underlying issues that require attention, and it is to these that we now direct our inquiry. Not surprisingly, considering the logic of my argument, we again turn to Metchnikoff.

The fundamental basis of Metchnikoff's thinking consists in an understanding of the indeterminateness of organismal integrity, maintained and perpetuated through an ongoing process of self-definition. The self could no longer be clearly delineated as a given entity. The "boundaries" of the organism are constantly being reestablished under the assault of temporal change and environmental challenge. A comparable pictorial image would be the cubist or fauvist vision of the object fusing into its contextual surroundings and thus blurred in its identity. But the challenge of the nature of identity remained and was resisted by an ethos that averred the organism's active striving for self-definition. This scientific perspective thus assumes a most intriguing philosophical position, one central to the concerns of the period. Indeed, if Metchnikoff had wished to discuss the full implications of his theory, he would have found sympathetic listeners among contemporaneous philosophers and artists. What appears to be a revolutionary concept in biology, arising from Darwinian problematics, in fact was found throughout the world of humanities as a response to the pessimism and nihilism of the period. It is in this broadened context of a fin-de-siècle Zeitgeist that our modern conception of selfhood may be beheld with remarkable clarity, frozen for a moment to offer an anchor to which we remain tethered (Tauber 1992b).

Metchnikoff's key insight resides in his concept of a disharmonious organism that requires a subsystem to take control of harmonizing disparate, competing centers of activity. This vision reflected his own psychology. During early adulthood he was afflicted with several severe depressive episodes, marked by a debilitating pessimism and punctuated by at least two suicide attempts (O. Metchnikoff 1924). One of the most remarkable aspects of the genesis of the phagocytosis theory is that it emerged during the psychological resolution of his personal difficulties (Tauber and Chernyak 1991). In many respects Metchnikoff's psychohistory provides fascinating insight into his ideas concerning the supposed objective world of embryology. Metchnikoff's scenario is not unique: It is echoed by the case of Friedrich Nietzsche. Both faced similar personal nihilistic challenges, and both responded intellectually in a similar fashion. Nietzsche was born in 1844; Metchnikoff, in 1845. Both suffered from episodic incapacitating psychosomatic depressive illness: Nietzsche was afflicted most severely in 1876–9, and Metchnikoff attempted suicide in 1873 and 1880. Both confronted their pessimism with a philosophical response in antithesis to Schopenhauer, who exerted profound influence on both during their early twenties. Beyond these interesting parallels, there is strong evidence that their respective philosophies closely echoed their psychological states (O. Metchnikoff 1924; Hayman 1980; Pletsch 1991).

There are similar points of intersection between James and Metchnikoff, although James would no doubt resent a comparison to Nietzsche, whose philosophy he regarded "as the sick shriekings of a dying rat" (James 1902, 1987, p. 47). (James was in fact referring to both Nietzsche and Schopenhauer.) James was only three years older than Metchnikoff, suffered similar psychosomatic illnesses, and again was most severely affected in his late twenties. The importance of these biographical facts is the strong evidence that their respective philosophies closely

231

followed their psychological state (O. Metchnikoff 1924; Barzun 1973; Myers 1986). It is possible, of course, that their respective trials and intellectual resolution of emotional predicaments simply reflect a broader cultural ethos and crisis. Whether they shared a similar Zeitgeist is entertained elsewhere (Tauber 1992b). The literature of such relationships is immense, but superficially we easily discern an abundance of cultural heroes representing in various images a similar psychohistorical pattern.

Metchnikoff's formulation of selfhood is most clearly echoed in Nietzsche's parallel construction. *Thus Spoke Zarathustra* (1885, 1959, composed between 1883 and 1885) and *The Will to Power* (1904, 1967, edited selections from his last notebooks, 1883–8) were written during the axial period in which Nietzsche most clearly elucidated his reaction to nihilism, by asserting the primacy of the will and self-construction (Corngold 1985). The coincidences of age and experience only underscore the similar visions of the self that these works embody. In the process of linking Metchnikoff's concepts to Nietzsche's vision, I risk not only overextending analogies, but simplifying the nuances of thought of two highly complex thinkers; however, I believe the comparison is warranted, as Nietzsche's mature position corresponds closely to the Metchnikovian vision of immunity. In discovering a common orientation, we enhance our understanding of the underlying metaphysical structure of self as they grasped it, elaborated and bequeathed to us.

Although the affirmation of the centrality of action and the activity of willing can be traced back to Kant and although the self's function in creating its object (e.g., its moral object for Johann Fichte and its aesthetic for Friedrich Schelling) preoccupied Romantic philosophers, Nietzsche is the consummate metaphysician of the active self, and more profoundly of the self striving for self-definition in a constant process of overcoming. The elusiveness of Nietzsche's entire thought may be unified as a philosophy of individuality. Despite the absence of general formulas, Nietzsche posits a striving hero struggling with active and self-conscious growth; what emerges is a modern Prometheus. The noble goal of perfectionism is crucial, but more fundamental is the struggle itself, for the actuating force is the will to struggle: "The wretched spiritual game of goals and intentions and motives is only a foreground – even though weak eyes may take them for the matter itself" (Nietzsche 1904, 1967, p. 518).[1] The context of the struggle is, of course, the war on nihilism; in the process a self-created moral universe is sought – never finally established, but always forming.

The self as organism

The self is to be actualized in its struggle, in the self-definition of its world as the manifestation of will to power. That energy is directed against nihilism and against convention; it opposes a false world with decadent values that negates the self as creator. The energy source of will and the sovereign arbiter of value is the self. We must delve into two different dimensions of Nietzsche's thinking to arrive at a composite understanding, but first, note that the issue of nihilism is ambiguous, in that for Nietzsche there are both active and passive forms, resulting in either an increased or a reduced "power of spirit" (ibid., p. 17). A "divine way of thinking" (ibid., p. 15) denies God but arises from "a reverence for the Self, a love of this life, and a desire to be creative" (Thiele 1990, p. 88). Spiritual anarchy results from the rejection of this vision of the self. Throughout this essay *nihilism* is used in the pejorative (i.e., passive) sense, although recognizing that Nietzsche employed the term with these double meanings in different contexts.

Let us begin with Nietzsche's epistemology. A fundamental skepticism that pervades his thought, leading to his well-known critique of morality, originates with how he defines subject–object relations (ibid.). Nietzsche uses the image of Man standing with his back toward a supposed reality and the mirror of his mind before him; the perceiver stands in the way of an unobstructed total view of the world. The thing-in-itself is, then, only an abstraction, an imagined existent in a world where the perceiver is absent, and approximated at a point beyond the multiple perceptions attained as the subject steps around it to obtain mirrored views from various angles. One can know the world only as one measures, perceives, and interprets it (ibid., p. 29). This perspectivism is similar to James's radical epistemology (discussed in Chapter 6), with all of its limitations of defining a common cosmos and connections between perceivers. But each philosophy allows the individual to become the ultimate interpreter of his or her world. Nietzsche's perspectivism is profound, reaching into the mind – "The human intellect cannot avoid seeing itself in its own perspectives and *only* in these" (Nietzsche 1882, 1974, p. 336) – and its object, the world: "There is absolutely no escape, no backway or bypath into the real world! We sit within our net, we spiders, and whatever we catch in it, we can catch nothing at all except that which allows itself to be caught in precisely our net" (Nietzsche 1881, 1982, p. 73).

Nietzsche's epistemology thus leads us to the second dimension of our concern, the metaphysical. Again using the image of the mirror, he argues "one loves one's desires and not what is desired" (Nietzsche 1886, 1966, p. 93). The true object is the self: "In the end, one experiences only oneself" (Nietzsche 1885, 1959, p. 264), but it is a multidimensional self. Perspectivism yields a dynamic, pluralistic, responsive and ever-changing self: Multilayered, the self can be uncovered as an onion is peeled, but it can never be defined or completely shown in its totality (Thiele 1990, p. 215). The self, like its object, is elusive and always developing; as Zarathustra proclaims: "Become what you are." It is in this context that Zarathustra's famous pronouncement "God is dead" takes on special significance for defining the self. If we follow a dialogical formula of Man's (as subject) relation to the deity (as object), self-definition is found in the encounter; if radically self-directed encounter defines the subject's other, that is, when the object-in-itself, or externally imposed, is dead, then only the self is left to define its other. Such a postulate must underlie Nietzsche's very metaphysics of will to power, in order to overcome the nihilism that results from values obtained from a source external the self proper. The self is ultimately left inviolate: "No matter how far a man may extend himself with his knowledge, no matter how objectively he may come to view himself, in the end it can yield to him nothing but his own biography" (Nietzsche 1878, 1986, p. 182).

Nietzsche, like Metchnikoff, viewed Man as subject to conflicting needs, instincts, and drives. The spectrum of human action results from this internal competition, and thus perspectivism is not only a means of interacting with the world, but a reflection of man's inner conflicts: "It is our needs that interpret the world; our drives and their For and Against. Every drive is a kind of lust to rule; each one has its perspective that it would like to compel all the other drives to accept as a norm" (Nietzsche 1904, 1967, p. 267). Nietzsche, again like Metchnikoff, views Man as disharmonious (unlike the biologist, however, Nietzsche viewed this condition as peculiar to humans!):

> ... a single individual contains within him a vast confusion of contradictory valuations and consequently of contradictory drives. This is the expression of the diseased condition in man, in contrast to the animals in which all existing instincts answer to quite definite tasks. (Ibid., p. 149)

Not only is the dissonant (or more benignly, pluralistic) state uniquely human, but the aim of its harmonization serves as the object of human will: "To become master of the chaos one is; to compel one's chaos to become form: to become logical, simple, unambiguous, mathematics, *law* – that is the grand ambition here" (ibid., p. 444). Heirs of Darwin, both Nietzsche and Metchnikoff pictured humans as suffering from an almost inveterate internal struggle whose resolution required a special mechanism. This is not to suggest that either Metchnikoff or Nietzsche was a Darwinian, as the word is commonly understood. Metchnikoff, of course, had a highly sophisticated understanding of Darwin's scientific argument, but for most of his career he was highly critical of it. (For his relation to Darwinian theory, see Tauber and Chernyak 1991, especially Chapter 4.) Nietzsche's (mis)understanding of Darwin's theory is examined by many commentators (see, e.g., Mostert 1979; Smith 1986; 1987), but in the context of this discussion perhaps it is most important to note how Nietzsche distanced himself from Darwin's interpreters, who limited the lessons of struggle: "The whole of English Darwinism breathes something like the musty air of English overpopulation. . . . The struggle for existence is only an *exception,* a temporary restriction always revolves around superiority, around growth and expansion, around power – in accordance with the will to power which is the will of life" (Nietzsche 1887, 1967, p. 292). Whether the will to power, arising from the self alone and subject to no other force or influence, the postmodernist credo (Steiner 1989), can serve that function remains an open question. To what extent this position was a consequence of particular understandings of biology deserves more careful scrutiny, a topic which I will consider in the third section of this chapter, but first allow me to situate Nietzsche's philosophy in the historical context of post-Kantian philosophy. Doing so permits us a deeper understanding of Nietzsche's and of Metchnikoff's projects.

B. ANTECEDENTS OF THE NIETZSCHIAN SELF IN KANT AND SCHOPENHAUER

To situate Nietzsche's philosophy of the self properly, a brief survey of the epistemological positions to which he responded is warranted. I make no pretense to identify comprehensively the antecedent issues that formed the crucible for his ideas, but there is clearly a progression that can be traced from Immanuel Kant's transcendental subject to Arthur Schopenhauer's attempt to place that subject *in* nature's context. Nietzsche can be read as having expounded Schopenhauer's formulation as much as he ultimately rejected it. But to understand these precursors is more fully to appreciate why Nietzsche regarded the biological self not only as crucial to his philosophy, but in fact as the foundation of the entire edifice. So let us begin by sketching Kant's contribution.

Kant took the "pure subject" of Descartes and Locke – and the somewhat ambiguous posture of Rousseau – and showed that the self was capable of knowing itself only because it could also know the world as an object. For this discussion, we may simply summarize Kant's epic fusion of rationalism and empiricism in what he termed the Transcendental Deduction. Experience *qua* experience is structured according to a priori categories of understanding. These included the principles that every event has a cause and that objects have substance and exist in space and time. As a priori principles, these cannot be established empirically. Knowledge arises from the synthesis of concept and experience, and as it is "transcendental," that is, nonobservable as a process, it must be presupposed. Synthetic a priori knowledge is possible because we can establish that experience as it must conform to the categories of understanding. Kant's understanding of the synthetic a priori depends on the synthesis of concept and experience. We can have such knowledge of reality only as "phenomena" – as objects of empirical enquiry. Phenomena are discoverable; they enter into relationship with us. This was a crucial refutation to those rationalists who attempted to describe reality as apprehended by the pure intellect, and it represents the foundation in establishing the reality of the phenomenal world – the world of appearance. To apprehend this reality, a subject is required.

In direct response to Hume, the transcendental argument moves from the nature of experience back to understanding the subject of experience; that is, it arrives at a view of what we must be like in order to have experience as we do. Thus, the observing subject is defined by inference: The self is observed as the external world is, even as ultimate reality remains elusive. The noumenon, whether subject or object, is an existent, though in itself an unknowable, reality that reason must postulate. Transcending experience and all rational knowledge, reason must assume the existence of noumena as the source for all science and philosophy. Because the shape of reality is partly formed by the mediation of the observer, the things in themselves are insurmountably a translation or an interpretation. The self thus emerges, since in the idea of a thought, in every mental content, is embedded the notion of a subject that has an immediate and intuitive unity. Kant refers to this unity as the Transcendental Unity of Apperception.[2] *Apperception* in this sense means "self-consciousness," and *transcendental* indicates that the unity of the self is known as the presupposition of all (self)-knowledge. In Kant's scheme, the self is an "original unity" whose mental state are not adjuncts but properties. Identity is corollary to the first-person presupposition that the self exists as an object.

The transcendental unity of apperception is of crucial importance to Kantian idealism; moreover, it has far-reaching implications as the foundation or starting point of all later philosophical treatments of the self. This position is no less than the basis of the modernity project to define personal autonomy and self-identity (reconsidered in Chapter 8). At issue is the unity of the self, not only as an object, but more crucially as a subject, not something that simply can be known, but as a willing, feeling, knowing agent. Since much of post-Kantian philosophy, and especially the contemporary postmodernist reaction, is directed at such a "knowing subject," we would do well to describe it in some detail. Kant uses the principle of transcendental apperception in justifying synthetic a priori knowledge and the unity of thinking and of self-consciousness in general. It thus serves as the supreme or original principle of all human experience:

> No knowledge can take place in us, no conjunction or unity of one kind of knowledge with another, without that unity of consciousness which precedes all data of intuition, and without reference to which no representation of objects is possible. This pure, original, and unchangeable consciousness I shall call *transcendental apperception.* (Kant 1781, 1966, [A107], p. 105)

There are three claims (or features) for this faculty: identity, unity, and self-consciousness. The subject must be identical through time, for without such identity, the ability to recall and maintain continuity would fail. The basis of unity refers to the requirement of an active subject to unify its experience. This ability in turn rests upon the third feature, that of self-consciousness, which refers to the capacity to reflect on its own unity and identity: "That is, as the word 'apperception' already indicates, Kant is arguing that self-consciousness is . . . a condition of experience" (Pippin 1989, p. 19). Thus, Kant argues for an active and self-aware self, and this is pivotal: The self must reflect upon itself to *attain* its autonomy: "Being *in* a subjective state . . . does not count as having an experience *of* and so being aware of that state unless I apply a certain determinant concept . . . and judge that I am in such a state, something I must *do* and be able to know that I am doing" (ibid.).

Kant also referred to apperception as a logical condition: It must be logically possible to ascribe representations to myself, that is, to become conscious. The objects of empirical consciousness are not the objects of the self or the mind; thus there is a clear distinction between the world as object and the mind's ability to attend to its own representing activity as a distinct capacity. This is an important distinction between the Kantian self and the Cartesian. The Cartesian view is that all consciousness, including all Kantian experience, is a species of self-

The immune self

consciousness (ibid., p. 20). According to Kant, on the other hand, self-consciousness is a condition of experience, because the unity necessary to appreciate an object by an identical subject depends on the implicit ability to be self-aware. A subject that is "perceiving, imagining, remembering, and so on is an inseparable component of *what it is* to perceive, imagine, remember, and so on" (ibid., p. 21).

The problem and nature of the unified experiencing subject (an issue already discussed in some detail in Chapter 6) is pinpointed by a closer examination of transcendental apperception:

> The consciousness of oneself . . . is . . . empirical only, and always transient. There can be no fixed or permanent self in that stream of internal phenomena. What is necessarily to be represented as numerically identical with itself, cannot be thought as such by means of empirical data only. It must be a condition which precedes all experience, and in fact renders it possible, for thus only could such a transcendental supposition acquire validity. (Kant 1781, 1966, [A107], p. 105)

Paton (1951, pp. 102–5) illustrates Kantian self-identity criteria by supposing we hear a clock striking twelve times. We hear each stroke (synthesis of apprehension in intuition), we hold the memory (recall) of earlier strokes (synthesis of reproduction in imagination), and finally we must be conscious of counting to grasp the unity of the process (the synthesis of recognition in the concept). Thus, to perceive and understand the clock striking the hour, there must be (1) one consciousness and (2) a concept under which the various appearances are united. Under these conditions the string of subjective sensations becomes an object that in this sense is dependent on the unity of the self, as the unity of the self is dependent on it or is manifested in regard to it. Thus, the self emerges, and the objective world follows, for every judgment involves a synthesis or unification of representations in consciousness, whereby the representations are conceptualized so as to be referred or related to an object[3]:

> The empirical consciousness which accompanies various representations, is itself various and disunited, and without reference to the identity of the subject. Such a relation takes place, not by my simply accompanying every relation with consciousness, but by my adding one to the other and being conscious of that act of adding, that is, of that synthesis. Only because I am able to connect the manifold of given representations in one consciousness, is it possible for me to represent to myself the identity of the consciousness in these representations, that is, only under the supposition of some synthetical unity of apperception does the analytical unity of apperception become possible. (Kant 1787, 1966, [B133], p. 78)

238

There is no confusion between the necessary identity of the self as knowing and the quite different question of the identity of the self as something known. The issue here is to describe what Paton calls the self-identity of the subject-self, as opposed to the object-self, that is, the transcendental as opposed to the empirical self. The former represents the necessary condition of any object's being known. There must be one self which is integrated in its different apprehensions if there is to be cognition of any temporal reality and of an ordered world. Even if the self sets itself beside other objects in an objective world, "[N]one the less it is still the subject of all objects whatsoever, including that self upon which it seeks to reflect. The subject self is always the centre of its whole world however much it may be thought of as passing from one apprehension to another" (Paton 1951, p. 104).[4]

The general epistemological problem raised by the *Critique of Pure Reason* was the impossibility of knowing anything about the thing-in-itself, since it bears no relation to the phenomenal world, which is the sole object of understanding. But the world *is* phenomenal and therefore fully objective, real and discernible to a knowing subject, because everything beyond phenomenality is an empty phantasm. However, for one case Kant allows immediacy of knowledge, namely, that of knowledge of ourselves. Such knowledge is uniquely practical and is available in the immediacy of our bodies and as an exercise of our autonomy. The direct knowledge of our body is of a radically different order from any other mode or object of apprehension. By allowing knowledge of the transcendental self as *practical* knowledge, Kant's critical successors remarked, he explores the reaches of subjectivity, rather than establishing the foundations of true objectivity. When he refers to "our" experience or mind, the question is whether the "our" is to be regarded in a general sense (i.e., of human beings conceived impartially) or in the specific idealist sense in which "our" refers to the abstract subject (the *I* that is engaged in the intellectual construction of the world). "This ambiguity is crucial, since, depending on its interpretation, we seem drawn either towards an impersonal metaphysics, or towards a highly solipsistic epistemology" (Scruton 1982, p. 161).

This philosophical ambiguity introduced by Kant plagued nineteenth-century philosophy. One of the earliest critiques, and the single most important response to his epistemologically based concept of the self, was offered by Arthur Schopenhauer:

> . . . a consciousness of one's own self and a consciousness of other things, are in truth given to us immediately, and the two are given in such a fundamentally different way that no other difference compares with

this. About *himself* everyone knows directly, about everything else only very indirectly. This is the fact and the problem. (Schopenhauer 1819, 1969, vol. 2, p. 192)

Schopenhauer recognized that Kant had not established the basis of this inner sense. In his magnum opus, *The World as Will and Representation* (1819), Schopenhauer attempts to incorporate both an idealist conception of the subject as the nonworldly origin of objective experience and an embodied conception of the subject as an organic part of nature. There is an intractable tension in such a construction, where the self arises within the world and yet somehow peers *at* nature from an external and detached perspective. The attempt to describe the nature of the self and its relation to the world along these two disparate axes, namely, of a separate self that is still *in* its world, prefigures much of the post-Kantian discussion of what constitutes identity and selfhood. But Schopenhauer is also particularly important for this discussion because of his profound influence on both Metchnikoff and Nietzsche, whose intersecting ideas about the self are brought under careful scrutiny later in this chapter.

First, let me briefly summarize *The World as Will*, which begins with Kant's transcendental idealism. The world of which we have knowledge is a world of appearance or, to use Schopenhauer's concept, "representation" (*things*-in-themselves are unknowable). Although there *is* an empirical world composed of spatiotemporal objects, that world is "exhausted" in what the subject perceives as its object:

The world is my representation. . . . [N]o truth is more certain, more independent of all others, and less in need of proof than this, namely that everything that exists for knowledge, and hence the whole of this world, is only object in relation to the subject, perception of the perceiver, in a word, representation. . . . Everything that in any way belongs and can belong to the world is inevitably associated with this being-conditioned by the subject, and it exists only for the subject. The world is representation. (Ibid., vol. 1, p. 3)

In other words, all objects presuppose a subject, and it is the subject that confers connectedness and identity on the world. Thus, a great deal depends on the nature of the subject. This knowing subject is not conceived as an individual, but rather as the necessary correlate of all objects; as such, it is never an object itself: "That which knows all things and is known by none is the subject. It is accordingly the supporter of the world, the universal condition of all that appears, of all objects, and is always presupposed; for whatever exists, exists only for the subject" (ibid., p. 5). Importantly, it follows that the subject of representations

240

is itself unknowable, "like an eye that cannot see itself" (Janaway 1989, p. 6). Schopenhauer has thus established a perplexing double riddle: one, of the self as unknowable to itself, and two, of the transcendental world as known only as the self's representation. It is noteworthy that for Schopenhauer knowledge of the phenomenal world of appearance is necessarily illusory. Going beyond Kant, Schopenhauer embraces the posture of Eastern philosophy: Reality is hidden behind the veil of *Mâyâ*, or "the veil of deception" that causes us "to see a world of which one cannot say either that it is or that it is not" (Schopenhauer 1819, 1969, vol. 1, p. 8). The question whether knowledge of the world *in itself* is possible rests at the heart of the matter, and thus the profound problem of post-Kantian selfhood is laid open. Schopenhauer attempts to solve his double paradox by asserting that in fact the subject and the object are both part of a universal will.

It is not necessary to explore in detail the metaphysical basis of Schopenhauer's notion of will, other than to note that it underlies the world in its entirety and manifests itself in the multiplicities of all things we know in experience. Schopenhauer's "will" may be summarized as "the force exemplified in the constitution and motion of everything in the universe from the cosmic wheeling of galaxies to the perpetual whirl of subatomic particles" (Magee 1983, pp. 142). Bryan Magee goes on to argue that the word *will* was chosen because Schopenhauer believed this was the nearest we, as experiencing subjects, can come to a direct apprehension of its meaning as an inner sense of the mundane drive of life. Schopenhauer's central doctrine is "quite simply that [the will] is noumenally at one with the force that drives everything else in the universe" (ibid.). As George Simmel observed this is essentially a metaphysical concept. "Physics looks for causes, but the will is never a cause. The relation of will to phenomena is never similar to that of cause to what is caused" (1907, 1991, p. 42). "We know how to find reasons for every single physical movement, but we cannot provide any reasons for movement and causality as such. Schopenhauer understands the infinity and the incompleteness of will, which are consequences of its being the metaphysical one beyond which there is no being" (ibid., p. 44).

> [W]ithout representation, I am not knowing subject but mere, blind will; in just the same way, without me as subject of knowledge, the thing known is not object, but mere will, blind impulse. In itself, that is to say outside the representation, this will is one and the same with mine; only in the world as representation . . . are we separated out as known and knowing individual. (Schopenhauer 1819, 1969, vol. 1, p. 180)

As Simmel adroitly puts it: "Schopenhauer has reduced man to the denominator shared by the totality of the world" (1907, 1991, p. 34).

Schopenhauer thus incorporates Kant's practical knowledge into his philosophy as knowledge of the will. The will is what constitutes the thing-in-itself, and all appearances are manifestations of will, knowable to the intellect through the application of Kantian categories. But in contrast to knowledge of appearances, the will is known "immediately," and not by the intellect, but rather *through* our action. The will *in* us is both more powerful and more primary than the intellect: "The intellect is the secondary phenomenon, the organism the primary, that is, the immediate phenomenal appearance of the will; the will is metaphysical, the intellect physical; the intellect, like its objects, is mere phenomenon, the will alone is thing-in-itself" (Schopenhauer 1819, 1969, vol. 2, p. 201). Interestingly Schopenhauer's argument shares a similar structure to that of the argument of Descartes, who invokes "thinking" to posit selfhood. Knowledge of our own willed actions, for Schopenhauer, is uniquely immediate: "[W]e know what we will because our action itself expresses our will. Willing . . . is acting" (Janaway 1989, p. 7). With the denial of an inner or purely mental act of willing as preceding action, Cartesian dualism is foreclosed and subjects of will are thus embodied. For Schopenhauer the immediacy of practical knowledge is integral to the sense of our essential being. This insight into how action situates and reveals the subject provides access to the underlying thing-in-itself as existing *in* ourselves rather than being external and forever beyond our grasp. As embodied will, our actions define, or better, expose, that will, leading ultimately to its full recognition:

> For the whole process [consciousness] is the *self-knowledge of the will;* it starts and returns to the will, and constitutes what Kant called the *phenomenon* as opposed to the thing in itself. Therefore what *becomes known,* what *becomes representation,* is the *will;* and this representation is what we call the *body.* (Schopenhauer 1819, 1969, vol. 2, p. 259)

The entire world is but an empirical objecthood that manifests the same essence with our very being, the unconscious striving force that Schopenhauer called will (ibid.). The intellect, as a natural manifestation of the will, becomes conscious of itself. This reflective triumph is the result of the human position at the pinnacle of organic development; it is nothing less than the capacity to appreciate one's own nature, one's fundamental being as will.

Schopenhauer's famous pessimism arises from what he perceived to be our inescapable torment from powerful and unfulfilled desires.

Seeing us as subject to unknown forces within us, he might well be credited as a major forerunner of the Freudian unconscious: "Consciousness is the mere surface of our mind, and of this as of the globe, we do not know the interior, only the crust" (ibid., vol. 2, p. 136). The insight that consciousness is comparable to the surface of the earth, the inside of which is unknown to us, is a lucid testament to Schopenhauer's sensitivity to the dynamic complexity of the psyche. The Schopenhauerian will is the direct precursor of Freud's id (Ellenberger 1970). But Freud would attempt to control and rationalize the unconscious, to objectify its function and thus attempt to liberate us from its powerful influence, if not dominance. In a strong sense Schopenhauer was a Romantic in his engaged stance to renew contact with those mysterious forces of our instinctual nature, and this was a key element of Nietzsche's early admiration (Nietzsche 1874, 1983). In any case, subject to such insatiable instinctual drives, Man exhibits his inhumane nature as a characteristic of the unharnessed will. Art is conceived as a means to escape the ruthless will, and thus aesthetics plays an important role in Schopenhauer's philosophy. (See Simmel 1907, 1991, pp. 75–104; Magee 1983, pp. 164–88; Hamyln 1980, pp. 110ff; Janaway 1989, pp. 275–9.) Again, Nietzsche was to follow Schopenhauer's lead as viewing art as a means for self-transfiguration (Young 1992).

The careful explication of the nature of the will and Man's relation to it are secondary, for our purposes, with respect to how Schopenhauer transformed Kant's epistemology, both extending the strengths and deciphering the weaknesses of that powerful philosophy. The key issue is to situate the subject. Kant had carefully discerned the distinction between appearance and the thing-in-itself. Thus, knowledge was limited by the a priori conditions of appearance, leaving the perceiving subject as a transcendental principle to unify representations. Schopenhauer asserted, first, that subject and object were necessary correlates, and more radically, the modes of organization of objects stemmed *from* the subject, rather than existing in themselves. Second, he situated the subject in the biological domain, where both body and mind were manifestations of blind purposiveness, the primacy of the will. This second element allowed Schopenhauer a teleological corollary, which he built into an elaborate explanation of human action that I have not discussed. What is directly germane is to note that the Kantian split between pure subject and mere object is found lacking in the Schopenhaurian context of the self-awareness of the will, of purpose, of subjecthood. The willing subject is another manifestation of the substratum of will, but unique in recognizing its own being.

The knowledge of the will cannot be separated from that of the body, so the experience of the body is the crucial window to knowing the will. The body may not be viewed as simply just another object, a "mere representation of the knowing subject. . . . The body occurs in consciousness in quite another way, *toto genere* different, that is denoted by the word *will*" (Schopenhauer 1819, 1969, vol. 1, p. 103), which not only distinguishes the body from every other object, but allows the subject to know it "*in* itself" (ibid.). We should note that there is a deep issue at stake between Kant and Schopenhauer. Where Kant had established the axiom of the phenomenality of the world (i.e., all that can be known are phenomena, not the thing-in-itself), Schopenhauer came to an opposing position: The world as appearance is unreal, and reality must be sought beyond it: "Kant senses reality as a category which produces experience, whereas Schopenhauer, who thirsts for the metaphysical absolute, senses reality in opposition to experience" (Simmel 1907, 1991, p. 20). That reality is the will, and the effort to know the will is to distill consciousness, that is, to know ourselves, namely, the body.

Schopenhauer could not rest content with leaving the matter at this stage of development, and in fact *The World as Will* does not establish a consistent and final philosophy of the self as a single relation that pertains between self and the world. The work ends with a penultimate position, where the very notion of distinct individuals is questioned. The unique subject, the knowing nidus of conscious experience, the source of imposed will on the surrounding world, is disclaimed and concluded to be illusory. The subject of representation and will adopts the highest viewpoint from which the self, as a distinct entity, is absent. By ceasing to strive for itself, or for the possession of an external reality, the Schopenhauerian self loses its boundaries as separate being and merges into a mystical state of nothingness. Schopenhauer thus closes the metaphysical circle of identity. Having begun by separating the self from reality, he concludes by allowing it once again to fuse with an undifferentiated will. But the undefined self now appears to be a metaphysical epitome, having willed its final integration.

The Schopenhauerian self is both pure subject of knowledge and subject of will. The self is neither a spatiotemporal individual nor an immaterial substance; "rather, it is analogous to an 'extensionless point,' to become a viewpoint to know the world, yet distinct from the content of what is known" (Janaway 1989, p. 296).

> . . . the world as representation . . . has two essential necessary, and inseparable halves. The one half is the *object*, whose forms are space and time. . . . But the other half, the subject, does not lie in space and time,

for it is whole and undivided in every representing being. . . . Therefore these halves are inseparable even in thought, for each of the two has meaning and existence only through and for the other. . . . They limit each other immediately; where the object begins, the subject ceases. (Schopenhauer 1819, 1969, vol. 1, p. 5)

The subject is thus essentially detached from its world, and it can only know that world as it comprehends reality's organization. This Kantian position is modified, however. Schopenhauer believed that the subject could become a passive mirror of the world, overcoming the subject-dependent forms of organization to become a pure subject of knowing (*Erkennen*).[5] Again, there is no empirical self to be identified, and in fact as a mirror we can finally lose any sense of being an individual within the world. The sense of selfhood remains embedded in the primary awareness of the self as a striving being, a self-aware willing entity. Spanning the continuum from rational thinking to unconscious action, the willing organism is physiologically based and teleologically organized. The will is thus embodied and expressed as the need to propagate and prolong life.

Schopenhauer's twofold construction of the self is inconsistent. The embodied, striving subject remains cut off from the epistemological conception of a nonempirical subject concerned with pure knowing. In the latter case the "extensionless point" cannot explain how phenomena are known. According to the theory of will, knowledge of the world stems from physiological structure and function; because such ability serves the primacy of the will as embodied survivorship, the explanatory power of this body-based self over the epistemological abstract formulation had an enormous influence on later nineteenth-century thought, especially on Nietzsche's philosophy, as discussed in the next section. Anticipating Nietzsche, Schopenhauer argued that empirical knowledge arises from and is dictated by the needs of the organism. He thus attempted to extend Kant's unity of apperception to include bodily striving. Schopenhauer discerned the inconsistency between his dual characterization of the self. On the one hand, the knowing subject is dependent upon forms of representation of the mind that are the result of organismic organization (ordinary empirical knowledge), and on the other hand, the pure epistemological subject mirrors the essence of nature, perceiving in an expanded fashion a reality independent of the forms representative of mind, that is, beyond normal categories of perception. In essence, we transcend ordinary experience of survival-dependent behavior. It is this window into the mystical world as a whole that tenuously links the dual construction of Schopenhauer's notion of the self. The Schopenhauerian synthesis is to fuse the willing subject and the knowing subject together into the unity of consciousness, the one "I," the self. "Schopenhauer's metaphysics are rooted in the idea of the subject as possessor of phenomenon and thing-in-itself, as citizen of two worlds" (Simmel 1907, 1991, p. 22). But an irredeemable schism remains, and it prevents a genuine individuality. Conceptual understanding keeps us removed from our individual essence, and Schopenhauer leaves us with the will alone, basically uncharacterized:

> Not only the consciousness of other things, i.e. the apprehension of the external world, but also *self-consciousness* . . . contains a knower and a known, otherwise it would not be *consciousness*. For *consciousness* consists in knowing, but knowing requires a knower and a known. Therefore self-consciousness could not exist if there were not in it a known opposed to the knower and different therefrom. Thus, just as there can be no object without a subject, so there can be no subject without an object, in other words, no knower without something different from this that is known. Therefore, a consciousness that was through and through pure intelligence would be impossible. . . . Therefore in self-consciousness the known, consequently the will, must be the first and original thing; the knower on the other hand, must be only the secondary thing, that which has been added, the mirror. (Schopenhauer 1819, 1969, vol. 2, p. 202)

Hume recognized that if we search within for that perceiving subject we so readily assign as our *self*, we find ourselves encountering the various objects of consciousness – passing thoughts, images, feelings etc. – but never any entity separate from these contents. Consciousness is always *of* something; it always has an object, and the perceiving subject can never be encountered as such. "What is the 'I' that contemplates 'my consciousness of x,' and to what does 'my' in the latter phrase refer? Whatever it is, it is systematically elusive – we never grasp it" (Magee 1983, p. 107). The perceiver *is* that which is perceived, or as Hume said, the self is but a bundle of perceptions. Schopenhauer's analysis, although radically different, concludes in close agreement to Hume (see Chapter 4): The perceiving self is nowhere to be found in the world of experience. The metaphysical self of Schopenhauer exists as the sustainer of the world but cannot itself enter the world, so in agreement with Hume, "subjects and objects are able to exist at all only as correlates of each other, [and] more to the point . . . their *structures* are correlative" (ibid.).

Kantian thought demanded the unity of the subject. Perception, knowledge, objectivity reside in a consciousness; they belong to an au-

The self as organism

thor of actions and purposes. Schopenhauer extended the issue by recognizing that a self formulated solely in philosophical terms is inadequate: The psychological source of identity must be explored. He attempted to expose the origin, development, and manifestation of drives, which were all relegated to a primary will:

> ... it is *the will* alone that is permanent and unchangeable in consciousness. It is the will that holds all ideas and representations together as means to its ends. ... Fundamentally it is the will that is spoken of whenever "I" occurs in judgement. Therefore the will is the time and ultimate point of unity of consciousness, and the bond of all its functions and acts. It does not, however, itself belong to the intellect but is only its root, origin, and controller. (Schopenhauer 1819, 1969, vol. 2, p. 140)

Thus, underneath the epistemological issues of how knowledge, perception, and objectivity are possible, lies the metaphysical substratum of the will. Schopenhauer radically transforms the Kantian formula of understanding the world: What is not phenomenal must be dominated by our metaphysical being, our being-in-itself:

> The turn to the subject and the centering of all world-categories in the ego – is the primary bearer of the Kantian distinction, both sides of which are given in ourselves, in the only existents that we do not know merely from the outside. ... We are both spectators and actors, phenomenon and radical cause of phenomenon. (Simmel 1907, 1991, p. 23)

From this position the psyche was to be viewed as a unified whole. There is the conscious ego, which strives to make sense of its world and at the same time expresses the will, which stems from unconscious drives. Schopenhauer readily acknowledged the interplay of the different aspects of rational consciousness as a product of unconscious drives, of the will of striving. In short, he placed the self firmly in the nature of organic creatures, whose actions and knowledge emanate from a primal will. The entire framework of the Kantian conception of the subject of knowledge is called into doubt, as already mentioned, because Kant's transcendental ego did not provide an explanation of how that which knows is at the same time a part of the knowable world of objects. Kantian epistemology becomes explicable, according to Schopenhauer, only as a secondary, epiphenomenal outgrowth of organism as will; it is in willing, not knowing, that our essence lies (Janaway 1989, p. 350). Simply put, Kant's pure a priori "I" contains no account of its instantiation by empirical beings. Schopenhauer attempted to place the subject in its world, and this fecund theme was profoundly extended by Nietzsche.

Later Nietzsche would reinvoke this primary element, this "will to life," in a transfigured role as "will to power." He built upon Schopenhauer's refutation of rationality as the deep-seated and basic essence of Man. Each regarded reason as essentially an accident, at best a tool in the hands of the will, and each dethroned rationality, replacing it with the primal will (albeit formulated in different ways). Fundamentally both philosophers shared the view that the will is not posited against, but resides outside, rationality, and perhaps most saliently outside its contradiction (viz. the irrational). The influence of Schopenhauer is plain and explicit in Nietzsche's writing; the importance of the will per se will be explored in the next section, where we consider Nietzsche's biologicism. However, it may be noted here in passing that Nietzsche quickly diverges from Schopenhauer's vision of the will, with crucial implications for a definition of the self. The first difference lies in the role of perspective. Nietzsche rejected the possibility of the pure knowing subject and saw the multiplicities of perspectives as integral to the *nature* of knowing: "There is *only* perspective seeing, *only* a perspective 'knowing;' . . . the *more* eyes, different eyes, we can use to observe one thing, the more complete . . . our objectivity" (Nietzsche 1887, 1967, p. 119). Never a mere mirror of the world as in Schopenhauer's formulation, rather, the subject always achieves a partial reflection of the world limited by the perspectival nature of its own drives and interpretations. Recognizing the singularity and partiality of our particular vision frees us from its limitation. From this position Nietzsche attacks Schopenhauer's basic foundation of the pure self, where the world is the undistorted object of the subject's awareness.

For Schopenhauer the self is a pure nonobject abstracted from all specific objects, states, or activities, whereas Nietzsche argued that this was only a singular interpretation serving particular ends (Janaway 1989, p. 351). The very notion of a Cartesian "I think" as the explicit conduit of Schopenhauer's subject (i.e., as a reflective will) is attacked:

"Thinking," as epistemologists conceive it, simply does not occur: it is a quite arbitrary fiction, arrived at by selecting the element from the process and eliminating all the rest, an artificial arrangement for the purpose of intelligibility.

The "spirit" [self], something that thinks: . . . is a second derivative of that false introspection which believes in "thinking": first an act is imagined which simply does not occur, "thinking," and secondly a subject-substratum in which every act of thinking, and nothing else, has its origin: that is to say, both the deed and the doer are fictitious. (Nietzsche 1904, 1967, p. 264)

The self as organism

And as an even more direct challenge to Schopenhauer, Nietzsche argued: "One must not look for phenomenalism in the wrong place. . . . [N]othing is so much deception as this inner world which we observe with the famous 'inner sense'" (ibid.).

Nietzsche discarded the "I" that thinks (not to mention "thinking"). This position derives from his very different concept of the will, which he viewed as a multiplicity of forces in competition: Thinking is only the occasional manifestation of such activity, and it is coalesced by value-laden needs and drives. But most profoundly Nietzsche contended that there is no need to posit a thinking subject, nor even a subject of willing:

> The assumption of one single subject is perhaps unnecessary; perhaps it is just as permissable to assume a multiplicity of subjects, whose interaction and struggle is the basis of our thought and our consciousness in general? . . . *My hypotheses:* The subject as multiplicity. (Ibid., p. 270)

Instead of belonging to a single, unified agent that thinks, wills, and acts, the mind is viewed by Nietzsche as a perpetual state of becoming, and the self is viewed as a continuous process of becoming: "There are only interpretations, but there is no 'I' that is the subject of the interpretation. The 'I' is a fiction whose author is an aggregate of manifestations of the will to power" (Janaway 1989, p. 355). The body not only provides the unifying element, the body *is* the self:

> Body am I entirely, and nothing else; and soul is only a word for something about the body. . . . "I" you say, and are proud of the word. But greater is that in which you do not wish to have faith – your body and its great reason: that does not say "I," but does "I." (Nietzsche 1885, 1959, p. 146)

C. NIETZSCHE'S BIOLOGICISM

Nietzsche's philosophy is thoroughly permeated by, even tethered to, a biological self-consciousness. We need not fully explore his personal rich psychohistory, with his various illnesses and chronic suffering, to appreciate the enormous preoccupation Nietzsche had with the body. He clearly had an obsessive concern with disease, both as metaphor and as object, and that concern accentuated the biological as a central pillar of his philosophy. For Nietzsche health is attained through struggle, a struggle that is a test of strength to overcome destructive forces. The health of the organism is thus measured not in the avoidance of sickness (both a case of happenstance and for the most part unattain-

249

able in real terms), but by the extent to which sickness is tolerated and ultimately overcome (Nietzsche 1881, 1982, p. 122; 1882, 1974, pp. 91–2, 176–7). In the struggle to master those destructive forces, albeit after profound suffering, the organism grows stronger – "What does not destroy me, makes me stronger" (Nietzsche 1888b, 1959, p. 467).

Suffering and vulnerability are thus transfigured into strength and vitality. That process, extending from the most primitive organism (ironically borrowing the amoeboid phagocyte that Metchnikoff would make famous) to modern Man's search for morality, is rooted in the primordial will to power. The will to power, Nietzsche's existential fulcrum and organizing principle, is developed in multifarious and often astounding ways. The will, which stems from the organic realm, suffuses his argument with a biological ethos that appears in the most unexpected places and in turn gives seed for later critics to argue what is philosophy, poetics, or psychology. As this application raises immediate controversy, the extent to which Nietzsche utilizes biology to provide the vantage point from which to critique nihilism and erect a new philosophy requires careful assessment.

In his scathing skepticism Nietzsche must stand on philosophically firm ground from which to launch his battle. His starting point is thus Man as an undisguised organic creature. Searching for Man's essence as reflected in his various endeavors (aesthetics, ethics, logic, epistemology), Nietzsche finds the basis of each activity to be biological. The areas that Nietzsche regards as ascendant human behavior, for example, creative work, art and aesthetic experience, are but manifestations of an unencumbered will to power. When he perceives a debilitating morality, he sees it as an example of the will's weakening or corruption (as will be shown, by inferior competing forces). But in either case, organic function, whether viewed from an evolutionary or a physiological perspective, leads to a new vision of Man, governed by laws uncontaminated by anthropocentric or moral distortions, which not only disguise and distort Man's true nature, but hamper the achievement of life's purpose: developing, overcoming, and becoming through the will to power. Nietzsche argued that our true moral values are to be discovered by understanding and accepting our basic and undisguised biological nature: "Fundamental innovations: In place of 'moral values,' purely naturalistic values. Naturalization of morality" (Nietzsche 1904, 1967, p. 255). Not only would this strategy invoke a healthier psychology, but, more crucial to Nietzsche's purpose, it would lead to an antinihilistic ethic: "Moral values are illusory values compared with physiological values" (ibid., p. 210).

The self as organism

Nietzsche began his reconstruction of values, what he would call the transvaluation (*Umwertung*) of all values, by first acknowledging and then building upon Man's animal nature. In celebrating evolutionary origins as the legitimate foundation of his critique, Nietzsche viewed the body as the only valid arbiter of Man's essential being. Our intellectualisms, rationality, morality have but distorted Man's true nature and diseased it. Therefore, Nietzsche maintained, we must first recognize our primary bodily essence:

> The animal functions are, as a matter of principle, a million times more important than all our beautiful moods and heights of consciousness. (Ibid., p. 355)

> Is there a more dangerous aberration than contempt for the body? (Ibid., p. 525)

> The strength and power of the senses – this is the essential thing in a well-constituted and complete man: the splendid "animal" must be given first – what could any "humanization" matter otherwise! (Ibid., p. 538)

Nietzsche's biological views are crucial to understanding the will to power, which is no less than the underlying driving principle of the primary organic force. Manifest in various derivative forms as the basis of human behavior, the instincts are sublimated in both creative and destructive activities, so that all human endeavors can be traced to them. Nietzsche argued that inferior drives compete and ultimately corrupt the primary force, the will to power, which in fact is the sole source of all of life's ascendency and perfection (we will consider this contention later). His mission, then, is to discover both a means to rescue Man from the nihilistic abyss that a perverse morality had bequeathed him and to provide an ethic consistent with a profound cognizance of Man's animal nature, that is, creating willing to power: "In man *creature* and *creator* are united: in man there is . . . clay, dirt, nonsense, chaos; but in man there is also creator . . . spectator divinity, and seventh day" (Nietzsche 1886, 1966, p. 154).

Nietzsche's discussion was first established in the coordinates of an organismal dimension and then raised along the second axis of the eternal recurrence, whose purest expression is the *Übermensch*. These two coordinates are inexorably interwoven together, with the eternal recurrence arising as the ethical affirmation of the biological bedrock of the will to power. Numerous commentators have dealt with various aspects of Nietzsche's biological conceptions relevant to our theme, including evolution and Darwinism (e.g., Fouillée 1901; Mostert 1979; Smith

251

1986, 1987), Nietzsche's concept of forces (e.g., Deleuze 1983, pp. 39–72; Moles 1990), the role of the body in Nietzsche's philosophy (e.g., Foucault 1984a; Warren 1988; Blondel 1991; Kōgaku 1991), and more specific issues regarding the unconscious (e.g., Jung 1934–9, 1988; Ellenberger 1970; Golumb 1989). I cannot here offer a review of this literature, for my initiative resides in offering a focused examination of Nietzsche's vision of Man as a biological subject and the implications of that vision for the subject–object relation.

If one simply characterizes the classical object as given, that is, independent of a subject, and the modern (i.e., Kantian) object as known only by the attributes bestowed by a knowing subject, then the postmodern situation confounds the subject–object relationship as a wavering exchange. Nietzsche's philosophy is perhaps a primary source of that elusiveness: "'Subject', 'object', 'attribute' – these distinctions are fabricated. . . ." (Nietzsche 1904, 1967, p. 294). But in an attempt to defy these obstructions, we will endeavor to discover how Nietzsche's biological argument transcends its metaphorical constraints, and how a true psychology ultimately emerges. In this context Nietzsche denies the body the status of an object, and firmly establishes it as the basis of our selfhood: "Essential; to start from the *body* and employ it as guide. It is the much richer phenomenon, which allows of clearer observations. Belief in the body is better established than belief in the spirit" (ibid., p. 289). He eschews the disjointed Cartesian mind–body dualism through a discourse that fuses the subjective "I" to its biological, that is, instinctual or emotional, identity.

As already discussed, Nietzsche clearly understood that the self required definition from multiple perspectives, and that the complexity even of a biological model was a forbidding challenge in any attempt to resolve the problem of multiplicity of identity. But precisely the perspectivism of that problem provided Nietzsche with the orientation to erect his philosophy. Of the several models he might have considered (outlined in Chapter 2), he ultimately rejected the cooperative and collective schemes as insufficient for his purpose and adopted the striving, evolving model. He was ignorant of the fact that a suitable theory, very similar in its intellectual structure to his own thinking, was at hand by the mid-1880s. Thus, only as metaphor would he engage any scientific vision to legitimatize his conclusions. He would use health as the metaphysical vehicle for his program of psychological and moral healing, and for him health was defined exclusively within the organic context: "Psychological healing must be put back on to a *physiological* basis" (ibid., pp. 134–5). However, it is not primarily in this domain that we find Nietzsche's preoccupation with the organic. He sought an even more profound stratum into which to sink his philosophic piles. Nietzsche rejected the collective biological perspective. He would not be beguiled by any relationship other than competitive striving:

> The greater the impulse towards unity, the more firmly may one conclude that weakness is present; the greater the impulse towards variety, differentiation, inner decay, the more force is present. (Ibid., p. 346)

> All events, all motion, all becoming, as a determination of degrees and relations of force, as a *struggle.* (Ibid., p. 299)

The concept of the self is heavily laden with cultural influences, and biological theory in this regard is also subject to such underlying constraints (e.g., Tauber 1991a; Longino 1990; Mendelsohn, Weingart, and Whitley 1977). Within nature's teeming struggle, Nietzsche could still invoke a collective ethos: "We are more than the individuals: We are the whole chain as well, with the tasks of all the futures of that chain" (Nietzsche 1904, 1967, p. 366). Nietzsche's cognizance of biological unity, irrespective of strife and competition, is but a corollary to the profound sense of evolutionary connection that pervaded his thought. Although acknowledging the cooperative model, where identity is largely assumed to stem from the collective (Roland 1988; Markus 1991; Gilbert 1992), he chose the competitive motif. Nietzsche's biology is thus based on a motif of struggle (e.g., competition of conflicting instincts within the individual). (We need not speculate how he understood Darwin's thought, which he apparently knew only through interpreters and popularizers and not in the particulars of his comprehension of evolutionary mechanics [Mostert 1979; Smith 1986, 1987].) Despite his pessimism regarding natural selection as applied to humans (Nietzsche 1904, 1967, pp. 216–18, and 361–5) – not an uncommon response even among sophisticated students of evolutionary biology of the time (Tauber and Chernyak 1991, pp. 68–100) – and his conclusion that evolutionary progress was an anthropocentric delusion (Nietzsche 1887, 1967, pp. 76–9; 1888a, 1959, p. 571; 1904, 1967a, p. 363) (a distinctly appropriate interpretation at a time when Man was viewed as the crown of the evolutionary tree), Nietzsche based his metaphysics of the will to power on the primacy of biological struggle. If all human endeavor originated in humankind's biological heritage, then the character of that endeavor was one of ceaseless striving, struggle, and competition. The imperative also directed that drive to an ethical end:

> What is good? Everything that heightens the feeling of power in man, the will to power, power itself.

The immune self

What is bad? Everything that is born of weakness.
What is happiness? The feeling that power is *growing,* that resistance
is overcome. (Nietzsche 1888a, 1959, p. 570)

Nietzsche begins with the proposition that Man is disjointed from his organic being: "Through the long succession of millenia, man has not known himself physiologically: he does not know himself even today" (Nietzsche 1904, 1967, p. 132). This leads him to his critique of Christianity, as the source of false values and nihilism. The remedy is first to give people back the courage to recognize and value their natural drives. And what are these natural drives? As would be expected from an evolutionary orientation, they are the heritage of multiple instincts in competition: "Every drive is a kind of lust to rule; each one has its perspective that it would like to compel all the other drives to accept as a norm" (ibid., p. 267). Thus, not surprisingly Man is viewed as out of balance, dissonant, and subject to the onslaughts of the internal struggle of conflicting instincts:

> A single individual contains within him a vast confusion of contradictory valuations and consequently of contradictory drives. This is the expression of the diseased condition in man, in contrast to the animals in which all existing instincts answer to quite definite tasks. (Ibid., p. 149)

The famous critique of nihilistic values, which are seen as deriving from the imposition of unnatural (civilizing) constraints on Man's true biological being, must yield its primacy to the more fundamental issue of first defining Man's true biological character. Man begins with a hierarchy of instinctual conflicts, so how is balance to be achieved? In fact, that balance is an impossible ideal. Nietzsche does not seek to harmonize conflicting instincts. Rather, stronger forces constrain weaker ones within an authoritarian or even oligarchic structure (e.g., Nietzsche 1887, 1967, pp. 73–6). Nietzsche is no democrat (Nietzsche 1886, 1966, p. 201; 1887, 1967, pp. 134–6; 1888a, 1959, pp. 643–7); he did not espouse harmonization or consolidation or majority rule. He applauded conflict as the natural and inevitable expression of the will to power (Moles 1990). He saw amalgamation and concentration of diverse forces as a sign of weakness, whereas he held the tendency to separate and distance the ruling force from the inferior ones to be an expression of strength: "[T]he strong are as naturally inclined to *separate* as the weak are to *congregate*" (Nietzsche 1887, 1967, p. 136). Thus, the lower forces accommodate to one another, and in sacrificing their independent existence, they assume specialized, subordinate functions. Self-preservation (Nietzsche 1886, 1966, p. 21; 1904, 1967, p. 345) or hunger (1904, 1967, p. 345), and sexual drives (ibid., p. 347) are none of them primary; each is derived from, and expressive of, the will to power. The stronger forces have the task of maintaining the system's integrity and therefore must limit the subordination of weaker functions in order to sustain them: "[E]very oligarchy conceals the lust for *tyranny;* every oligarchy constantly trembles with the tension each member feels in maintaining control over this lust" (Nietzsche 1887, 1967, p. 136). A balanced tension is thus struck between the weaker and stronger: "[H]e who cannot obey himself is commanded" (Nietzsche 1885, 1959, p. 226). The commanded obeys outside forces, and the commander obeys itself (Moles 1990, p. 138). The ascendency of power, achieved in constant striving, forms the basis of Nietzsche's ethics. Since the distinguishing quality of living force is its ability to control itself (which is the expression of spirituality; that is, spirit is self-mastery), the higher such control, the greater the spiritual attainment. The individual then attains his identity – his freedom – in assuming responsibility for it. "For what is freedom? That one has the will to assume responsibility for oneself" (Nietzsche 1888b, 1959, p. 542).

Where no stasis is allowed, the process of struggling forces is dynamic. For Nietzsche, the strong are not guaranteed always to prevail. Since the very nature of the will is to strive, forever to seek greater power, there are ongoing and ever-changing relations, such that dominated spirits may regenerate and triumph over commanding forces: "The species do *not* grow in perfection: the weak prevail over the strong again and again" (ibid., p. 523). This process, which Nietzsche saw throughout organic nature, is part of what he termed self-overcoming (*Selbstüberwindung*) ("And life itself confided this secret to me: 'Behold,' it said, 'I am *that which must always overcome itself*'" (Nietzsche, 1885, 1959, p. 225]), and is personified in the Übermensch.[6] The process of self-overcoming emerges from the struggle of the dominated: "Along stealthy paths the weaker steals into the castle and into the very heart of the more powerful – and there steals power" (ibid., p. 227). To the extent this dynamic is integral to life, whether of the individual, organism, species, civilization, or morality, then as a law it applies to all selves: "All great things bring about their own destruction through an act of self-overcoming; thus the law of life will have it, the law of the necessity of 'self-overcoming' in the nature of life" (Nietzsche 1887, 1967, p. 161). And the corollary to ceaseless struggle is suffering, which Nietzsche celebrated: "The discipline of suffering, of *great* suffering –

do you not know that only *this* discipline has created all enhancements of man so far? That tension of the soul in unhappiness which cultivates its strength. . . ." (Nietzsche 1886, 1966, p. 154).

But from whence does struggle arise? Here the fundamental basis of Nietzsche's biological concept is revealed, namely, the active organism. The will to power defines activity, where all life forces seek opposition and resistance and, in the process of overcoming, find advancement. This vector defines an active, as opposed to a reactive, organism (Deleuze 1983; Moles 1990, pp. 135–47). Rather than adapt, the organism attempts to assimilate or force its environment and others to adapt to it: "Against the doctrine of the influence of the milieu and external causes: the force within is infinitely superior; much that looks like external influence is merely its adaptation from within" (Nietzsche 1904, 1967, p. 47). The prominent activity is, then, self-assertion, self-aggrandizement – the quest to overcome. Such self-definition in tension and opposition to the other exists equally within the organism itself. Its constituents (whether individual organs or functions) must constantly be redefined within the dynamic context of the ever-changing individual. The self is thus subject to vigilant scrutiny and molding: "[W]ith every real growth in the whole, the 'meaning' of the individual organs also changes" (Nietzsche 1887, 1967, p. 78). In fact, Nietzsche views internal selection as essential to strengthening the organism and determining its survival. If the organism effectively deals with the dying (Nietzsche 1882, 1974; 1885, 1959, p. 100), effete, damaged, or atavistic, it becomes that much more empowered and competitive. Active selection determines its strength and health, and the enhanced ability to compete with others thereby follows. Thus, adaptation is viewed as "an activity of the second rank, a more reactivity; indeed, life itself has been defined as a more and more efficient inner adaptation to external conditions" (Nietzsche 1887, 1967, p. 79). Ultimately, then, the self decides what to appropriate for itself from the other, whether the other is an object of the external world or a constituent of itself:

> The spirit's power to appropriate the foreign stands revealed in its inclination to assimilate the new to the old, to simplify the manifold, and to overload or repulse whatever is totally contradictory – just as it involuntarily emphasizes certain features and lines in what is foreign, as every piece of the "external world," retouching and falsifying the whole to suit itself. Its intent in all this is to incorporate new "experience," to file new things in old files – growth, the feeling of increased power. (Nietzsche 1886, 1966, p. 160)

The self as organism

The subject, then, is a dynamic construct. In constant encounter with its other, both internal and external, the subject becomes renewed in its ever-changing adjustment in response to inferior forces seeking their own self-aggrandizement and in the face of the environment, which is the external challenge to the organism. Metchnikoff's autonomous phagocyte may parallel what some critics also discern as an ambiguous posture in Nietzsche. As David Levin notes:

> . . . he comes within a breath of recognizing something like a corporeal intentionality, i.e. a 'functioning intentionality' of the body which is anterior to acts of 'judgment' and makes them possible: 'Before judgment occurs,' he writes, 'there is a cognitive activity that does not enter consciousness,' but which operates through the living body [1904, 1967, p. 289]. Nietzsche even hints . . . at a cognitive functioning of the body which is able to unify its field of being and constitutes a personal identity without the intervention of a transcendental ego [ibid., pp. 270–1; 281; 294–5]. (Levin 1985, p. 35)

But such hints are not explicitly developed and remain only tantalizingly implicated, as they are in Metchnikoff's thought. In this regard both assume a certain poetic license.

"The sphere of a subject constantly growing or decreasing, the center of the system constantly shifting" (Nietzsche 1904, 1967, p. 270); this model of the self serves to emphasize that aspect of the individual that must be regarded as the "same" from conception to death, that is, an individual temporally integrated but in fact forever altering in the calculus of its development. The crucial insight, then, is to recognize that the initial subject is inexorably changed in its encounter with the other. The startling philosophical consequence is that the subject–object duality is blurred, and thus the clear (Cartesian) divide is irreparably undermined. Always subject to change, this version of the self is a most profound constant in Nietzsche's thought; perhaps it is even the constant upon which his philosophy is built. From this position a new vision of the self emerges in Nietzsche's philosophy, but before considering this implication, we must note that this conception of the subject echoes the vision of Metchnikoff. If the scientist had wished to elaborate upon the full implications of his theory, he would perhaps have found a sympathetic listener in Nietzsche, whose will to power resonates as the deeper philosophical foundation of the phagocytosis hypothesis.

There is no evidence of which I am aware that Nietzsche knew of Metchnikoff, but by 1907, and probably well before then, considering Nietzsche's influence in fin-de-siècle France, Metchnikoff knew the philosopher's work. But it is doubtful whether Nietzsche exerted any direct influence on the biologist. First, Metchnikoff had established his own mature philosophical (and scientific) positions well before Nietzsche became widely discussed. And second, Metchnikoff made his own assessment: "A German critique has reproached me for my ignorance of Nietzsche's works. I have read several of them, but the mixture of genius and madness in them makes them difficult to use. In this connection Moebius' volume, *Über das Pathologische bei Nietzsche* (Weisbaden 1902) is of interest" (Metchnikoff 1907, p. 230). And the expected next question must be posed: If Metchnikoff's phagocytosis theory was so important in establishing the modern concept of the organism in the context of immunity and, more generally, as a dialectical creature with self-defining properties and unstable boundaries, then what are the broader implications of Nietzsche's philosophy for our metaphysical understanding of ourselves as biological entities? (Note, "dialectical" in this sense is not Hegelian, signifying two forces creating a third; Nietzsche's self is composed of multiple forces in free competition achieving temporary order of dominance.)

According to Nietzsche, the will cannot be reduced to notions such as "life force," vitalism, evolutionary development, or materialistic process, for it forms the core essence of what we perceive as life and, more fundamentally, of our very consciousness. "Being – we have no idea of it apart from the idea of 'living'. – How can anything dead 'be'?" (Nietzsche 1904, 1967, p. 312). Nietzsche's ontology, then, begins with the organic, for biological nature is not merely an influencing force, but the driving force of all human endeavors[7]:

> All "purposes," "aims," "meaning" are only modes of expression and metamorphoses of one will that is inherent in all events: the will to power. To have purposes, aims, intentions, willing in general, is the same thing as willing to be stronger, willing to grow – and, in addition, willing the means to this. The most universal and basic instinct in all doing and willing has for precisely this reason remained the least known and most hidden, because *in praxi* we always follow its commandments, because we *are* this commandment. (Ibid., p. 356)

> [T]he body . . . will have to be an incarnate will to power, it will strive to grow, spread, seize, become predominant – not from any morality or unmorality but because it is *living* and because life simply *is* will to power. (Nietzsche 1886, 1966, p. 203)

The will, then, resides at the very core of our being, forming the foundation of Nietzsche's metaphysics and extending well beyond his concern

The self as organism

for understanding biological drives or even the self per se. In fact, for Nietzsche the self had become the will. The basic question, then, was how to distinguish Man as a special case of the will, for there is abundant evidence that Nietzsche assigned the will to power as the general underlying basis of the manifestation of force and cause. And life in its basic essence is the will to power: "That all driving force is will to power, that there is no other physical, dynamic or psychic force except this" (Nietzsche 1904, 1967, p. 366). And in the deepest strata of Nietzsche's thought, we ultimately discover the core issue: "[T]he innermost essence of being is will to power" (ibid., p. 369).

Nietzsche persistently seeks to declare the vital immediacy of being, the need for the full acceptance of the here and now and of the matrix of experience. The layers of morality and rationality that hide or encumber that primary encounter are stripped away as he attempts to plummet into the foundations of our experience. He thus distinguishes what would appear to be a guiding principle of biological processes, that of development and evolution (what he calls becoming), from being. Becoming, which is never complete or finished, but rather is always tentative and subject to change, is contrasted with being: Dionysus is cast against Apollo. That manifesto of Nietzsche's early work remains a recurrent theme of his philosophy and is most succinctly stated as follows: "To impose upon becoming the character of being – that is the supreme will to power" (ibid., p. 330). Thus, the will is engaged in the ontological project. As Zarathustra says, "Become what you are" (Nietzsche 1885, 1959, p. 351), or as Heidegger wrote, "All being is for Nietzsche a becoming" (1961, 1979, p. 7). The basic active drive will not allow self-satisfaction, but it does allow rapture and self-fulfillment. The means of rapture are clearly aesthetic. Consistent with the primacy of the will ("art reminds us of states of animal vigor" [Nietzsche 1904, 1967, p. 422]), art serves as the conduit to the sensuous, the Dionysian. (This theme dates to Nietzsche's earliest writings (1872, 1956, p. 93) and remains a constant refrain throughout his work (Heidegger 1961, 1979; Young 1992).) Nietzsche thrusts art to the core of Man's endeavor, the essential, purposeful, life-sustaining and life-enhancing task. The basic instinct of the artistic is not necessarily directed toward "art"; rather, it is focused on "life" (Nietzsche 1888b, 1959, p. 529). (I further develop this theme elsewhere [Tauber 1995a].) Art as essentially creative "becoming" captures Nietzsche's rejection of a final state; in the process of growth we must recognize the value of experience in itself, rather than as a vehicle toward a final destination:

259

The immune self

Becoming must be explained without recourse to final intentions; becoming must appear justified at every moment. . . . [T]he present must absolutely not be justified by reference to a future, nor the past by reference to the present. To this end it is necessary to deny a total consciousness of becoming a "God," to avoid bringing all events under the aegis of a being who feels and knows but does not *will*. (Nietzsche 1904, 1967, p. 377)

Zarathustra thus proclaims, "God is dead" from this fundamental metaphysical vantage point, and from this position the prophet's attack on modern Man is launched: Man is degenerate, disoriented, and diseased.

Nietzsche's sensitivity to the biological, then, resides at several levels: The organic is the true basis of Man; to understand his psychology is to comprehend his instinctual drives and behavior. These, in turn, are but manifestations of the will to power. It is this will from which Man must understand his ontogeny (birth of being), and the metaphysics of this will secures the acceptance of the present being as not only the real basis of Man's existence, but the source of its true moral dimension. Thus, evolution as a biological force is subordinated to the will (ibid., p. 368), and the will remains the mysterious nexus of life itself: "What has been the relation of the total organic process to the rest of nature? – That is where its fundamental will stands revealed" (ibid., p. 368). From this position Nietzsche's psychology emerges. Its imperative was to show that we are not separate from our body: We *are* the body, the instincts, the unconscious. It is not enough to separate a hierarchy of selves, for example, an id, an ego, and a superego, each controlling a different aspect of behavior and all somehow consolidated in the individual. The individual *is* his primordial will, expressed by different affects, passions, behaviors, intellects, but each a manifestation of a single individualizing force, groping for power: selfhood. And that selfhood is found, as Zarathustra acclaims, in the truth of the body:

> I want to speak to the despisers of the body. . . .
> Behind your thoughts and feelings . . . there stands a mighty ruler, an unknown sage – whose name is self. In your body he dwells; he is your body. (Nietzsche 1885, 1959, p. 146)

The *self*, neither the I nor the ego, is the center of the person, of individuality. The I is an invented projection, and Nietzsche as psychologist discerns that if the ego attempts to control the self, the consequences are dire.[8] This view obviously anticipates Freud – that is, the self is composed of dissonant factions, and the ego (along with all conscious thought) is but a deceptive veneer of instincts in conflict. But unlike Freud, Nietzsche found healing in the reassertion of the core self, in

260

ever-evolving self-actualization. Nietzsche celebrated the self, a self that is fundamentally and profoundly active, dialectical with its experience, and ever-changing and growing. The spirit thus evolves, accentuating its elusive nature. The problem of seeking *identity* in our evolutionary episteme remains.

D. THE EPISTEME OF EVOLUTION

The very conceptual foundation of our understanding of organismal identity and integrity resides in a construction that echoes Nietzsche's essential principles governing the self. Even though he arrives at his formulation independently of Metchnikoff's science and Metchnikoff in turn does not rely on Nietzsche's philosophy, the two thinkers essentially shared a common vision: The self is not a given, established object, but lives as a dynamic and dialectical entity, evolving in time – developmentally and experientially. Its boundaries are ever changing, and thus the self is constantly in the process of redefinition. These principles are implicit in many areas of modern biology besides immunology (embryology, neurosciences, ecology, medicine), but the full implications of such thinking have yet to be fully explored. Powerful reductionist paradigms seek to obviate the need to incorporate such dialectical logic, but they do so only to their own detriment and impoverishment (Tauber 1991a; Tauber and Sarkar 1992). Innovative scientific interest in these issues is appearing with increasing force. I would argue that the science inevitably must respond to the question that has been recognized as a philosophical problem essentially since Heraclitus: "One cannot step twice into the same river, nor can one grasp any mortal substance in a stable condition, but it scatters and again gathers; it forms and dissolves, and approaches and departs" (Kahn 1979, p. 53). What is the essence of a changing entity? This was the question Aristotle posed as the central question of metaphysics, and was reiterated by Hume as the basic problem of the *self*. And despite our pervasive anti-metaphysical attitudes, we are still left with this implicit, nagging issue. It may be a problem of grammar (à la Wittgenstein), but modern reductionist science must expand to accommodate an answer.

In many respects, *On the Origin of Species* was the culmination of an evolutionary paradigm that had been developing momentum over the preceding century (Eisley 1961; Mayr 1982). The Chain of Being (Lovejoy 1936) constructed a hierarchical order. With the evolutionists this

261

linear order assumed two new caveats: the introduction of a vast scale of time and the mutability of forms. "A universe not made but being made continuously" (Eisley 1961, p. 9) was a concept whose seeds might be discerned in the naturalist writings of Gilbert White (1789, 1985, pp. 82–3), who recognized ancestral prototypes of domesticated birds; moreover, doubts of the fixity of species were implicated by John Ray's earlier fossil studies and the classification schema of Carolus Linnaeus (Eisley 1961, pp. 16, 23, 25). The growing sophistication of taxonomy, paleontology, and comparative anatomy led increasingly to the intimation that the transformation of life forms was possible, an idea that incidentally coincided with emerging theories of cosmic evolution. By the mid–seventeenth century Comte de Buffon had already identified "every significant ingredient which was to be incorporated into Darwin's synthesis of 1859" (ibid., p. 39), namely, the struggle for existence, the variation of forms within species, and a grand scale of time that allowed extinction. Missing was a mechanism of change, which was not supplied until Darwin postulated how heritable biological variation and selective struggle might generate organic divergence and, over time, new varieties and eventually species. We need not trace this history, but simply note that the notions of struggle, evolution, and progress were omnipresent after 1800 (Bury 1932; Zirkle 1941; Eisley 1961; Nisbet 1980; Mayr 1982; Desmond and Moore 1991).

With Darwin a materialistic basis was established for explaining the emergence of species and their transformation. Notwithstanding that few truly understood the argument of natural selection and fewer still accepted it, clearly the preceding conception of natural order had been shaken to its foundation. Around the turn of our century (e.g., Theodor Boveri and E. Rádl) it was recognized that the crucial factor introduced by Darwin was to bring historical analysis at center stage of biological thinking:

> One could not possibly conceive of the true nature of an animal by any analysis, be it ever so profound, or by any comparison with other forms, however comprehensive, because there lies hidden in the organism traces of the past that only historical research is able to reveal. (Attributed to Rádl [1905] in Cassirer 1950, p. 171)

Most biologists of the last half of the nineteenth century lacked the profound historicism required to appreciate Darwinism (Cassirer 1950, pp. 160–75). In the Newtonian world the object was given; there was no ontological issue of the subject–object relationship. In the post-Darwinian world, evolution is an ongoing dynamic, where the organic object is under constant pressure to change and in fact is changing by natural selection, neutral evolution, or other undiscovered mechanisms. Indeed a parallel intellectual revolution has occurred in physics: The object in the world of Einstein and Bohr has become relativistic and indeterminant. No longer given, the object as a modern construct is defined as held fleetingly as subject in the present and as object in its past (Whitehead 1925). The object has become less an entity than a process, dynamically evolving and ever different. Being has been fundamentally transformed in a Heraclitean flux into a becoming. This was the fundamental challenge to which Metchnikoff and Nietzsche responded. My discussion, centered on the organism, clearly is guided by this process. By preserving its self-identity in the face of an ever-changing flux of matter, the organism must serve as the basic ontological sense of the self: the emergent self-identity of form in a world to which it must adjust, always in a milieu and polarized by otherness.

The crucial role of evolution as a philosophical construct dates clearly to Georg Wilhelm Friedrich Hegel, a generation before Darwin. It was to Hegel that Nietzsche responded in terms of the implications of the will to power (albeit the notion of struggle and the particulars of biological evolution are most easily seen as being framed by Darwin). There are, of course, vast differences between Nietzsche and Hegel. Hegel, as historian, was concerned with self-realization on the vastest possible scale, applying the creative principle through the realm of nature, psychology, the state, and the histories of art, religion, and philosophy. The philosophy of history commanded him in particular, and in that analysis he articulated the evolutionary process. Nietzsche's purpose, on the other hand, was centered on the individual and his or her self-actualization. It was this perspective that "led him to the conception of a vast plurality of individual wills to power and culminated in a monadological pluralism" (Kaufmann 1974, p. 243). Other major differences between the two philosophers were Nietzsche's dominant concern with an antinihilistic ethic and his consequent domineering negativism. This critical stance and the accompanying rhetoric(!) set him quite apart from Hegel (see Löwith 1964; Kaufmann 1974).

Considering that Nietzsche remains at the center of our interests, how might he best be situated in relation to Hegel and Schopenhauer? Despite the many critical remarks he makes in regard to them, Nietzsche does glean an essential element from each, speaking of "the astonishing stroke of *Hegel,* who . . . first introduced the decisive concept of 'development' into science" (Nietzsche 1882, 1974, p. 305) and of Schopenhauer's seminal recognition of "the problem of the *value of existence*" (ibid., p. 306); moreover, he may also be credited with first introducing atheism as a valid philosophical position in opposition to Hegel (ibid., p. 307). Thus, Nietzsche acknowledged his debt to Hegel and Schopenhauer and realized that, despite his rejection of their respective philosophies, Schopenhauer had first posed the relevant question "Has existence any meaning at all?" (ibid., p. 308), and Hegel had prepared the grounds for Nietzsche's response to the question of History as evolution. "The two great philosophical points of view (devised by Germans): a) that of *becoming, of development,* b) that according to the *value of existence.* . . . Everything becomes and recurs eternally – escape is impossible!" (Nietzsche 1904, 1967, p. 545). Thus, Nietzsche developed his value of theory within the context of these predecessors. But the eternal return was uniquely his own. Regarding Schopenhauer's relation to Hegel, Nietzsche wrote: "As a philosopher, Schopenhauer was the *first* admitted and inexorable atheist among us Germans. This was the background of his enmity against Hegel" (Nietzsche 1882, 1974, p. 307). Nietzsche regarded Schopenhauer's atheism as "the locus of his whole integrity. . . . In this severity, if anywhere, we are *good* Europeans and heirs of Europe's longest and most courageous self-overcoming" (ibid.). (For other comments regarding the complex relationships between Nietzsche, Hegel, and Schopenhauer, see Simmel 1907, 1991; Merz 1896, 1965; Löwith 1964; Ackermann 1990; and Blondel, 1991, pp. 60–2. Regarding Nietzsche's critique of Hegel pertaining to the self-actualization of the individual in History, see Warren 1988, pp. 85–7.) Let me now summarize Hegel's most profound contribution to Nietzsche's philosophy.

We need not turn to Hegel's scientific understanding of biology (it was ironically naive even by the standards of his time[9]) to argue his cardinal contribution of placing the metaphysical problems of evolution squarely before us. Without going into detail, we can say that the Hegelian dialectic essentially attempts to rediscover the continuity underlying the artificially isolated aspects that traditional logic creates of dynamic reality. The struggle between the thesis and antithesis is resolved by an ascent to a higher plane from which it can be comprehended and reconciled. The ascent is the process of synthesis, which generates a new concept that in turn spawns its own negation, and so on, until the whole of reality is exposed. The historicity of the dialectic pushes to a metaphysics that begins at "being" and concludes with the great Whole, which he calls *Geist,* the cosmic spirit. The precedent to all concepts are "being" and its negation, "nothing." Their dialectical opposition is resolved in passage to a new concept of "becoming." The world is now perceived as "becoming" rather than "being." The dialectical process yields a higher truth, determinant being, which is that more familiar and less abstract form of existence. Evolution in Hegel's comprehensive philosophical system thus assumes the pivotal operation in understanding the very nature of our world: "Hegel calls becoming the first concrete concept; this means that the two antithetical concepts, Being and Non-Being, are mere abstractions" (Milič Čapek 1984, p. 115).[10]

Perhaps the clearest synopsis of the Hegelian insight was offered by Heraclitus twenty-three hundred years earlier in two fragments. The first is the famous observation that one does not step into the same river twice; that is, new waters (time) must always wash over us. This aphorism emphasizes the temporality of existence and its *changing* character. But Hegel's dialectic is more accurately situated in another Heraclitean fragment:

> "We step and do not step into the same river; we are and we are not".[11]
> The first part of this fragment defies any consistent visualization, since it consists of two conflicting and incompatible images. The clue to this contradiction is in the second part, where Heraclitus moves up to a higher level of generality in saying: "We are and we are not". It would be difficult to anticipate more fully Hegel's view that becoming synthesizes being and non-being. (Ibid., p. 116)

In turning to the metaphysical orientation of *becoming,* as opposed to being, Hegel radically reoriented philosophy: Nietzsche, as did virtually every nineteenth-century philosopher, replied, and in so doing, found his characteristic response to the Darwinian revolution. With his usual brashness, Nietzsche proclaimed:

> . . . the astonishing stroke of *Hegel,* who struck right through all our logical habits and bad habits when he dared to teach that species concepts [*die Artbegriffe*] develop out of each other. With this proposition the minds of Europe were preformed for the last great scientific movement, Darwinism – for without Hegel there could have been no Darwin. (Nietzsche 1882, 1974, p. 305)

Whereas Schopenhauer bequeathed Nietzsche with the will as a conduit to the body, the immediacy of experience and the central role of the instincts, it was to Hegel that Nietzsche reacted in establishing his own vision of evolution as a metaphysical foundation for the philosophy of the eternal recurrence. Nietzsche's philosophy assumes its most profound biological orientation in the eternal recurrence, which is fundamentally organic in its implicit connotations of renewal, regeneration, return. There are those who understand eternal recurrence as a cosmological principle; the reasons for rejecting this interpretation are

The immune self

amply argued elsewhere (Nehamas 1985, pp. 142–67). Nietzsche uses eternal recurrence not as a theory of the world, but of the self. The interpretation of the eternal recurrence must reside in a consistent reading of Nietzsche's concepts of the will to power and its corollary, becoming as true being: "Let us think this thought in its most terrible form: existence as it is, without meaning or aim, yet recurring inevitably without any finale of nothingness: the 'eternal recurrence'" (Nietzsche 1904, 1967, p. 35). In a profound sense, Nietzsche envisions that the eternal recurrence is the fulfillment of living each moment, each act, each choice without the demurrals of past remorse or future judgment. We are enjoined to live as if each moment is to be relived, unchanged, into eternity. The eternal recurrence is the final destination of a deeply rooted evolutionary process, a calling that should become an ethic of our biological being, independent of any transcendent principle. With that perspective, each moment is not only immutable, but precious and forever accountable to ourselves. Nietzsche's recurrence does not refer "to a life precisely like this one, but to *this very life*" (Williams 1952, p. 100), imbuing the quality of eternity into every moment.

Nietzsche thus leads us to a supreme self-awareness of the ultimate and inescapable responsibility for our acts. The last element of the ethic, then, is to accept the irrevocability of every choice, allowing us to assume the mandate of responsibility for our life, a life to be lived again and again, eternally. It is in this construction that Nietzsche captured the essential feature of the self as an evolutionary entity and thus clearly articulated the radical reorientation ushered in by Darwinian theory. The Nietzschean self is never given, but always in flux; boundaries are indistinct and only become more firmly defined either by the self's encounter with an other or within the subject's embedded context. Consequently, the self is subject to individual and collective history and thus dependent on complex relations with its object, environment, and history. So, in the broadest intellectual formulation organismic identity, a concept some biologists regard as problematic and requiring careful scrutiny, is no less than the epistemologic issue of establishing the subject–object relationship. This was the complex dialectical issue presented by Hegel and later developed by Nietzsche.

The model that probably best describes Nietzsche's biological conception is $S:S_o$, the dialectical self encountering the other in an ever-dynamic decision of assimilation or rejection – the active self, the self that defines itself and at the same time is always becoming, that can only be viewed from the outside, bounded by change and fluctuation. The self changes through its history with its other: The subject molds and is in turn altered by its object. Heidegger eloquently describes the process in Nietzsche's aesthetic, a rich interpretation that attempts to fuse the seemingly disparate aspects of will to power and eternal recurrence as two facets of one and the same concept. In classical terminology the will to power is the essence (*essentia*) of all things, whereas the eternal recurrence of the same is its existence (*existentia*). In the language of transcendental – phenomenological (discussed in Chapter 6) – philosophy the will to power is the thing in itself (*noumenon*), and the eternal recurrence of the same is appearance (*phainomenon*). In Heidegger's terminology of "the ontic-ontological difference," which refers to the fundamental difference between Being and beings, the will to power in Nietzsche's metaphysics stands for Being, and the eternal recurrence of the same stands for the multiplicity of being (see Heidegger 1961, 1979, vol. 3, p. 168). As Heidegger joined the will to power and the eternal recurrence of the same into one and the same thought, he tried to "complete" Nietzsche's thinking and to end the Western project known as metaphysics (Behler 1991, p. 23), notwithstanding Kaufmann's denial that the will might be viewed as a metaphysical principle at all (1974, e.g., p. 204, 420).

The strong parallel between the thought of Nietzsche and that of Metchnikoff, with a shared vision of the self emerging from their respective philosophies, constitutes the common foundation where we must ultimately seek a fecund synthesis. Nietzsche's concept of the self – both epistemological and psychological – may be understood as a dynamic, dialectical organismal model. Nietzsche sought to expand our self-conception, and to reorient our psychology from a divorced subject–object orientation to a self *in* its world, ever conscious of its relationships, but freed from falsified restrictions of experience. To the extent that our self-image determines our orientation to the world, Nietzsche's concept must be credited with redefining our entire understanding of the world with which we must engage, in providing a self in full encounter with its object – whether in the internal or external cosmos. "The true world – attainable for the sage, the pious, the virtuous man; he lives in it, *he is it* (Nietzsche 1888b, 1959, p. 485). Metchnikoff's biological formulation of the self is echoed in Nietzsche's metaphysics, where self-affirmation is always the way to return to one's essence, origins, and, in the process, becoming. Zarathustra is the prophet proclaiming that all being is will to power, and this self-assertion is the basis of erecting the concept of the Übermensch, the champion of post-nihilism. To master oneself requires the greatest power of harnessing the chaos of life into a continual self-overcoming of life. The assertion

267

of will in its purest expression is Zarathustra's herald: The self is to *overcome itself again and again.* Nihilism resulted from the disparagement of the basic human value to maintain contact with our true being – unsublimated and direct. To assert Being or the basis of human existence with the rhapsodic joy of eternal recurrence was Nietzsche's anthem to Man: a calling to deny nihilism and assert his true self. But that self, like the Metchnikovian organismal construct, is ever-evolving and always becoming – never given as a final entity, defined and complete:

> Put briefly: perhaps the entire evolution of the spirit is a question of the body; it is the history of the development of a higher body that emerges into our sensibility. The organic is rising to yet higher levels. Our lust for knowledge of nature is a means through which the body desires to perfect itself. Or rather: hundreds of thousands of experiments are made to change the nourishment, the mode of living and of dwelling of the body; consciousness and evaluations in the body, all kinds of pleasure and displeasure, are signs of these changes and experiments. In the long run, it is not a question of man at all: he is to be overcome. (Nietzsche 1904, 1967, p. 358)

8

The search for identity

A. FROM NIETZSCHE TO FOUCAULT

Those who argue that we are now in a postmodernist age often cite Nietzsche's attack on the Enlightenment (a critique profoundly and disturbingly extended by Heidegger) as the origin of a new vision of the self. As already discussed, Nietzsche's mature position corresponds closely to the presentation of Metchnikoff's theory of immunity and the midpoint of James's writings of *Principles.* Thus *Spoke Zarathustra* (composed between 1883 and 1885) and *The Will to Power* (edited selections from his last notebooks, 1883–8) were written during this period. If Metchnikoff offered a biological formulation of the self, and James an epistemological definition, then Nietzsche is the metaphysician who proclaims self-affirmation and self-assertion as the process of returning to the self's essence and origins, as the process, that is, of becoming. Nietzsche makes no attempt to re-create a Kantian enlightenment ideal of critical self-consciousness based on assumed "values" or an assumed epistemological foundation (Behler 1991, pp. 20ff). There is no traditional appeal to nature or any transcendental subject as a source of value:

> Self-determination or Nietzschean "affirmation" could no longer be linked in any way to the cosmos, one's true self, real happiness, complete rational autonomy, or one's realization within the historical community. What the idea of a modern epoch had sown and what Kant had cultivated, Nietzsche would now reap. (Pippin, 1991, p. 120)

The assertion of will in its purest expression is Zarathustra's herald: "I am that *which must overcome itself again and again.*" That self, like the Metchnikovian organismal construct, is ever-evolving, always becoming – never given as an entity, defined and complete. Or perhaps more fundamentally, "for Nietzsche, the problem lies deeper: it is the having of selves at all that is first in question" (Strong 1988, p. 162). The imbroglio resides in the inability to rely solely on the self as a source of meaning and value, which results from a relativistic heumeneutic that is constantly subject to interpretation, full of cultural and historical bias. From this perspective the critical stance modernity assumed of history and Man generated a philosophy that essentially destroyed itself: Nietzsche showed that the modern faith in rationality was merely a contingent strategy in a struggle for power (Foucault's later anthem) and that thus modernity naively misunderstood itself. This is the herald of the postmodernist position: Modernity is the story of failed autonomy. Our view of Nietzsche hinges on how he is placed in this issue, situating him as either the first postmodernist or the last modernist (Koelb 1990; Behler 1991).

The question concerning the legitimacy and implications of the postmodern attack on the self is central to our project. From the persistent attempt to link the conceptual development of immunology to its relevant philosophical analysis, we must again address the potential consequences of a *self* that might still be viewed as problematic even with the phenomenological provisos I have previously outlined (Chapter 6). If we are not to fall back on a transcendental principle of unification (as Husserl did in later work; see Chapter 6), then the challenge remains open as to what *self is* beyond its "self-definition" in process. The philosophical assault on the self has proceeded from various quarters, but I have chosen to illustrate it from the standpoint of Nietzsche. In tracing how, by 1890, philosophy and biology had firmly merged their respective treatments of the active subject, called self in one discourse and organism in the other, I seek to extend the implications of the modern debate. In the review of early immunology we found philosophical resonance between Metchnikoff's views and Nietzsche's biologicism. Implicit in that discussion was the recognition that the parallels discerned between biologist and philosopher were tentative and suggestive. They share a general orientation arising from a common notion of an organismic-based biology and a conception of experience as self-aggrandizing and dialectical. The self by these fin-de-siècle criteria evoked dynamic and elusive boundaries. In an ever-self-conscious post-Darwinism and at times almost poetic evocation of an evolving subject, Metchnikoff and Nietzsche shared an antinihilistic ethos that both framed and directed their response. The narrative described how reductionist principles dominated immunology into the first half of the twentieth century and the original Metchnikovian problematic finally reappeared in the 1940's. With the emergence of the self as an explicit metaphor after World War II, we were readily justified in seeking an explication of that concept. Now let us reintroduce Nietzsche's construction in the context of a deconstructed self, in vogue today.

The philosophical exposition of the "doubtful self" is probably best articulated by Michel Foucault. He falls squarely in the deconstructive critical reaction to the Nietzschean self. Not only did Foucault himself acknowledge his indebtedness to Nietzsche, but others have called him "Nietzsche's closest successor" (Megill 1985, p. 30), based on the comparison both of their substantive views and the methodology of their respective critiques. When Foucault summarized his endeavor to create a history of the different modes by which in our culture human beings are made subjects, that is, the objectivization of the individual and his or her subjection to control (Foucault 1983a, p. 208), there is a strong resemblance to Nietzsche's orientation. But Foucault's most obvious intersection with Nietzsche centers on the genealogical analysis that probes the developmental process by which the self is defined and that specifically addresses the question of how Man has become an "object," rather than self-asserted "subjectedness." Nietzsche and Foucault share a profound skepticism toward the history of their own times, and their common purpose is to make history "a curative science" – since "historical sense has more in common with medicine than philosophy" (Foucault 1971, 1984a, p. 90). From this starting point they quickly diverge.

It is through the skeptical method shared with Nietzsche that Foucault hoped to achieve insight to rectify the modern condition. The genealogist in Foucault's formulation is a diagnostician concentrating on the relations of power, knowledge, and the body in modern society to determine what it is that conditions and institutionalizes each. In "Nietzsche, Genealogy, History" (1971, 1984a), Foucault distinguished such analysis from that of his earlier opus concerning the archaeology of knowledge. The explicit parallelism with Nietzsche in such discourse is strangely implicit, for at no point does he endorse Nietzsche's larger purpose or signify an indebtedness. They perhaps shared a method, not a project. This distancing in fact reflects an appropriate separation and deserves special scrutiny. A major portion of "Nietzsche" concerns the meaning of genealogy, which is made explicit in how Nietzsche employed terms such as *Ursprung,* (origin, extraction) *Entstehung* (origin, emergence), and *Herkunft* (origin, descent). Foucault then explains that instead of discovering true origins, the genealogist discovers "the secret that they [things] have no essence or that essence was fabricated in a piecemeal fashion from alien forms" (1971, 1984a), and thus history "teaches how to laugh at the solemnities of the origin" (ibid., p. 79). The genealogy of values and knowledge is committed to the endeavor to trace the development and emergence of socially constructed truths. Foucault's genealogist does not seek meaning in deep analysis. The character of eros or sex or illness is revealed when viewed from the right distance: In surface events, minor details, subtle change, the hidden is revealed. Interpretation becomes a never-ending task. As Foucault proclaimed earlier in "Nietzsche, Freud, Marx," "[T]here is nothing absolutely primary to interpret because, when all is said and done, underneath it all everything is already interpretation" (Foucault 1967, 1986). Interpretation thus arises from social constructions.

What is defined as objective masks subjective motivations, and this issue is closely connected with Foucault's concern with power as defined in *Discipline and Punish* (1975, 1977, pp. 27–8). To understand Foucaul-dian power-knowledge in its various disguises becomes the object of the genealogist's mission (1971, 1984a, pp. 86–8). All knowledge remains relative serving as an instrument or expression of power and can emerge in history only as a result of societal forces. Here is the key that separates Foucault from nineteenth-century attempts to define and capture the self. Although Nietzsche and Foucault each discover the self, they arrive at their respective disparate conclusions from an ethos derived from very different origins: Nietzsche regarded Man as self-creative, self-renewing, and ultimately self-responsible. This dynamic process of self-definition bespeaks a potential emanating from some origin and striving toward an ideal. Foucault's vision of the self as a construct, as a product of historical process hapless in the churning of societal forces, thus begins with a cultural entity far removed from the true origins Nietzsche discerned and the potential inherent in that description. Man *became* an object in the course of history, but his true identity defies such a restriction in Nietzsche's ethical cosmos. Foucault's kinship in fact goes beyond a similar diagnosis but includes an echoed cry of Nietzschean individualization: "Maybe the target nowadays is not to discover what we are, but to refuse what we are" (Foucault 1983a, p. 216).

Perhaps the clearest way to illustrate the dissolution of the Foucaul-dian self, is to summarize the core concept of power as understood by Nietzsche and Foucault, the derivative subject/object of power – the body – and finally the resultant position of their analyses. Nietzsche understood power as the subject of his analysis. The universe, with Man as the product of competing drives and instincts, was seen as being governed by the will to power, a concrete force, whose various manifestations created and formed life. Power *is*. As subject, not object, power – albeit through the body – expresses not only the primeval energetics of the organic, but creates all those expressions of human endeavor that eventuate in culture and its history. To understand Man is to recognize how the will to power is governed, controlled, perverted into the negative, nihilistic depiction of late nineteenth-century European civilization. Ironically, without the antecedent genealogy of power as construed by Nietzsche, Foucault joins the critical attack on culture at the level of dissecting how power controls.

Foucault analyzes power as a societal weapon to establish hegemony over the individual's body, action, and thought. In this scheme the body becomes an object, and power becomes the means of control. Power is divorced from its origins and appears as a radically formulated schema of analysis. And it is precisely this point that highlights the fundamental difference between Nietzsche and Foucault. The basis of Nietzsche's critique of modern culture, philosophy, and history was the notion that each conspired to contort the body, which is the true subject, into an object. Nietzsche railed against rationalism and sought to recognize the primacy of the body as the foundation of his transvaluation of values. When Foucault wrote, "[N]othing in man – *not even his body* – is sufficiently stable to serve as the basis for Self-recognition or for understanding. . . ." (1971, 1984a, p. 88; emphasis added), he revealed a fundamental misunderstanding of, and irreparable separation from, Nietzsche. Power for Foucault is an object for analysis, whereas for Nietzsche power is the true subject – the protagonist of his inquiry. For Foucault power is but a *means* to define the self, and therefore it cannot constitute the self's very basis, as in Nietzsche's understanding. Consequently their respective conceptualizations of power and of the body as the expression (Nietzsche) or object (Foucault) of power project divergent avenues of thought. Following a summary description of their divergence, we will briefly consider its implications.

As already discussed in Chapter 7, Nietzsche regarded power as dependent on – and tethered to – a biological self-consciousness. The critique of Man thus originates in Nietzsche's concepts of biologicism, that is, of Man as an organic creature. In celebrating evolutionary origins as the legitimate foundation of his critique, Nietzsche views the body as the only valid arbiter of Man's essential being. Our intellectualisms, rationality, and morality have distorted Man's true nature and diseased it. Thus, he writes, "our most sacred convictions, the unchanging elements in our supreme value, are judgements of our muscles"

(Nietzsche 1904, 1967, p. 173), and "belief in the body is more funda-mental than belief in the soul" (ibid., p. 271). Therefore, Nietzsche ar-gues, we must first recognize our primary bodily essence. And that es-sence is but the expression of the will to power.

Nietzsche's mission, then, is both to discover a means to rescue Man from the nihilistic abyss that a perverse morality has created and to provide an ethic that originates in a profound cognizance of Man's ani-mal nature, that is, a creature willing to power. Organic function, whether viewed from an evolutionary or a physiological perspective, leads to a new vision of Man as developing, overcoming, and becoming through the will to power. Nietzsche began with the biological as the foundation of his new ethics. From this organic foundation the true nature of Man must emerge. Our moral values are, then, to be discov-ered by examining and understanding our basic and undisguised bio-logical nature. Not only would this strategy invoke a healthier psychol-ogy, but, more crucial to Nietzsche's purpose, it leads to an antinihilistic ethic. Since the very nature of the will is to strive, forever to seek greater power, there are ongoing and ever-changing relations through which dominated spirits can reassess themselves again and triumph over com-manding forces. This process, which Nietzsche saw throughout organic nature and termed self-overcoming, attains its ultimate realization with the Übermensch. The process of self-overcoming emerges from the struggle of the dominated. To the extent that this dynamic is integral to the individual, whether organism, species, or society, the law applies to all selves: "All great things bring about their own destruction through an act of self-overcoming; thus the law of life will have it, the law of the necessity of 'self-overcoming' is the nature of life" (Nietzsche, 1887, 1967, p. 161). The subject, then, is a dialectical construct. In constant encounter with its other, both internal and external, the subject be-comes renewed in its ever-changing adjustment to inferior forces seek-ing their own self-aggrandizement and with the environment, which serves as the external challenge to the organism.

Foucault's vision stands in stark contrast: Instead of an autonomous, striving, self-defining entity, the body is objectified to become but an-other focus of power. It undergoes, if you will, a reductionist critique. Foucault examined the body in its various guises as the object of bio-power. In dissecting the biomedical logos in *Birth of the Clinic* (1963, 1973) he analyzes how the body assumed a new identity in medical science of postrevolutionary France. In the ascendency of pathoclinical correlations and the new clinical institutions of the new medicine, Fou-cault saw another exercise of power, whereby disease was reduced to novel authority and control. The body, so well defined under the scrutiny of medical objectivity, was discovered to be subject to the same expressions of knowledge/power operative in other discursive analyses, particularly in the analysis of punitive practices and sexuality. Foucault brilliantly linked the body's subtle and minute social practices with the large-scale organization of power, but he failed to recognize or accept that the body's corruption and sickness, its enslavement by work, its desires molded by cultural expectation were perversions of what Nietzsche regarded as a true essence. The body's pollution was a given for Foucault, the inevitable result of societal power in history. In Foucault the body is scrutinized as any other "discursive object" in the genealogical endeavor; although it is a crucial focus of power/knowledge, it still remains simply another object of inquiry:

> . . . . the body is also directly involved in a political field; power relations have an immediate hold upon it; they invest it, mark it, train it, torture it, force it to carry out tasks, to perform ceremonies, to emit signs . . . there may be a "knowledge" of the body that is not exactly the science of its functioning, and a mastery of its forces that is more than the ability to conquer them; this knowledge and this mastery constitute what might be called the political technology of the body. (Foucault 1975, 1977, pp. 25–6)

Nietzsche begins his critique with a firm metaphysical foundation: There is power, it is manifest as the will to power, the body is its expression, and an ideal arises from this conceptualization. Foucault does not offer a means of escape, neither an ideal of self-renewal nor an ethic are his concern.[1]

Foucault claimed to have isolated the mechanisms by which power operates ("meticulous rituals of power"), the manner in which power is localized ("political technologies of the body"), and the dynamics of how power works ("a microphysics of power"). His is a fecund sociological analysis. He clearly acknowledges that his conception of power becomes a phenomenology of behavior. Like Marx, he does not analyze power in terms of its "why," that is, in terms of causal explanation, but rather seeks to trace "how" it acts.[2] The "how" is identified through the means by which power is exercised and the consequences of such exerted power between individuals. The term *power* is used to designate relationships (Foucault 1983a): Foucault analyzes power relations, not power itself. In the context of Foucault's role as Nietzsche's successor, a glaring problem arises, as Foucault claims that power does not exist: "Something called Power, with or without a capital letter, which is assumed to exist universally in a concentrated or diffused form, does not

275

The immune self

exist. Power exists only when it is put into action" (ibid., p. 219). Foucault takes this position because of his distrust of metaphysical inquiry:

> To begin the analysis with a "how" is to suggest that *power as such does not exist.* . . . The little question, what happens? although flat and empirical, once it is scrutinized is seen to *avoid accusing a metaphysics or an ontology of power of being fraudalent;* rather it attempts a critical investigation into the thematics of power. (Ibid., p. 217, emphasis added).

Foucault, then, forfeits functioning as anything more than a critic. Although his timorous dealing with the *philosophical* issue left him untainted with metaphysics, it also left him bereft of a self.

Foucault does not seek a new form of the self; his relentless analysis would not endeavor to "replace one kind of subjectivation with another" (Scott 1990, p. 55). The very idea of the self is always in question and represents the radical extension of Nietzsche's essential ethic of overcoming the self. For Nietzsche the self evolves; for Foucault its identity (as authenticity) is always in doubt. There is no effort to reconstitute an idealized self, for, on the contrary, the self is always inescapably a construct. Forming the self; defining internal states of the mind, formations, and transformations of desire; and conformity to mores and cultural principles are our practices. How the self relates to itself defines the individual's identity, and this project has become a massive cultural preoccupation and obsession with self-mastery. But the fulcrum of Foucault's argument is this: Not only is the self elusive (and its boundaries indistinct), but there is danger in actually constituting a self, since it then becomes the object of power and of limiting definition. In *The Order of Things* Foucault's objects of inquiry are defined "not by ideal essences outside them, but by the historical forces buried within them" (Gutting 1989, p. 181). When modern Man is thus scrutinized, the Cartesian notion of the *cogito* is an insufficient accounting of human consciousness, which is also inextricably connected with an unthought dimension (Foucault 1966, 1970, pp. 322–8). This *cogito–unthought* duality expresses the fundamental characteristic of Man as both an experiencing subject and a never fully understood object of that experience. After rejecting the ontological quest as the way of resolving this fundamental imbroglio, Foucault recommends abandoning philosophy's quest for Man. This philosophical anthropology is a dead end, and in the "death of Man," heralded by and buried within Nietzsche's "death of God" (ibid., p. 342), the new opportunity that arises in the "void left by man's disappearance [is] . . . the unfolding of a space in which it is once more possible to think" (ibid.).

Foucault's argument is as thoroughly self-overcoming as Nietzsche's,

276

but it loses its metaphysical mooring. By rupturing a stable, final basis for representation to organize knowledge (viz. experience), there can be no quest for origins or essence of knowledge: "In his [Foucault's] thought no identity seeks itself, is imminent or near to itself, liberates itself by proper self-appropriation, or fulfills itself. His analysis of these characteristics of modern thought is without a teleology" (Scott 1990, p. 82). Nietzsche would not fall into this essentially nihilistic position. Allowing a hierarchy of selves, each controlling a different aspect of behavior and somehow consolidated in the individual, a dominant self-identity emerges. The individual *is* his or her primordial will, expressed by different affects, passions, behavior, intellects, but each a manifestation of a single individualizing force groping for power: selfhood. And Nietzsche grounded that selfhood, as Zarathustra acclaims, in the truth of the body:

> I want to speak to the despisers of the body . . . [B]ody am I entirely, and nothing else; and soul is only a word for something about the body. . . .
>
> "I", you say, and are proud of the word. But greater is . . . your body and its great reason: that does not say "I", but does "I." . . . Your self laughs at your ego and at its bold leaps. . . . Even in your folly and contempt, you despisers of the body, you serve your self. (Nietzsche 1885, 1959, p. 146–7)

That self is fundamentally and profoundly active, dialectical with its experience, and ever-changing and growing. The spirit thus evolves. Nietzsche's philosophy proclaims the vigorous assertion of will, freedom, and choice: Shackled with suffering, conflict, and struggle, the self occupies a world of constant strife, but it remains essentially hopeful, since the self is that which asserts its will. The biological model Nietzsche chose was dialectical and active, where the crucial requirements of the creative, invigorated living individual are fulfilled by this most organismal of portraits of the self. The self must define itself, by experiencing, assimilating, responding – asserting its will. There is the ever-present danger of narcissistic self-engulfment, but the model Nietzsche proposes is built on a self defined by an outwardly directed dynamic process. Always in flux, Nietzsche's core concept of the self is will, and in affirming the primacy of the will – and Man's reawakened awareness of it – Nietzsche affirmed Life. He demanded direct encounter with our true being, firmly rooted in the organic (emotional, instinctual) realm. His entire philosophy is thereby profoundly indebted to the biological concepts upon which this edifice is built. On the other hand, the narrow confines of Foucault's formulation allows only monolithic analyses and thereby fails to allow for power's expression as both domination *and* self-rule (Taylor 1986).

Although Foucault and Nietzsche share a critical attitude toward the Enlightenment and its ideals ("Reason – the despotic enlightenment" [Foucault 1966, 1989, p. 12] and employ similar irrational rhetorical techniques (Megill 1985), there is an irredeemable gulf between the critic and the philosopher. Nietzsche offered a philosophical foundation on which he built a defined telos, aspiring to a poetic ideal and governed by a commitment to a conception of the self firmly anchored in its organic identity. From that position a psychology and an ethic emerged. Foucault could not offer any such orientation: Man remained a poorly defined product of his history – derived from the development of culture, subjected to the phlogiston character of power (Paternek 1987, p. 119). All knowledge is reduced to disguised power; all selfhood is historical contingency. Where is the self? We are simply given a critical vision that leaves us without identity and in fathomless suspicion. In some profound Foucauldian sense, "we" ourselves are always in question (Rajchman 1991, p. 146). In contrast, Nietzsche affirms the individuality of power, the active quest of the self, and the ethos of Man's Will to Power as the source of his true essence. From action to description – a sorry fate for deconstructionism, which began with Nietzsche's attempt to demolish in order to build again. The extent to which Nietzsche was successful is another question (Tauber 1995b). And the answer to *that* question is to decide the fate of the self. The entire issue rests on a metaphysical construction, and it is in this venue in which I will conclude.

B. CONCLUSION

I went in search of myself.
– Heraclitus, XXVIII, in Kahn 1979, p. 41

We perceive Nietzsche's pervasive influence (or perhaps the resonance of his screams) throughout the culture of the period 1890–1910. His fundamental challenge was to define the self, for his indictments reflected the disjointed sense of selfhood that marked modern consciousness. Whether traced to the Romantics or to Kierkegaard's angst, there was in the fin-de-siècle a widespread concern both to integrate human experience in a bewildering reductionist-positivist context and to discover meaning in an ascending secularization. I have discussed this issue in the context of immunology's emergence as a scientific discipline, split between an emerging reductionist immunochemistry and an older descriptive biology. Beyond viewing the birth of immunology as another case of the reductionism–holism controversy in biology, we have considered it in a broadened cultural context of the search for identity in fin-de-siècle Europe. The problematics of that period are evident in radical reappraisals of the self in psychology, philosophy, and the arts. I have already attempted to sketch the controversies within immunology as in part arising from the clash between an increasingly dominant positivism and a resistant neo-Romanticism. This was the context of the predicament of many of Metchnikoff's own contemporaries, and it was in this crucible that the prevailing scientific arguments and ethos forming immunology were engaged. If we were content simply to describe the intellectual issues of Metchnikoff's endeavor in the terms in which he formulated them, the narrative might end here. But the very source of our study is to examine how his intellectual odyssey progressed to our own day.

Immunology, both in its theoretical thrust and in actual practice, has exhibited a fundamentally altered vector of inquiry since World War II, in that it has sought to define the *origins,* or perhaps better, the *basis,* of immune identity, as opposed to its more restricted concern with identifying mechanisms of immunochemical specificity. The case was made that a self-conscious introduction of the self metaphor appeared to capture the scientific probe of that expanded program. Burnet offered the notion of self hesitantly precisely because it was not a well-defined concept; yet at the same time it pointed to an elusive issue he recognized as the truly relevant nidus of immunology. This novel recognition of selfhood as a theme spawned a theory to account for tolerance, autoimmunity, and the various questions surrounding the definition of how organismic identity is attained. Since Burnet's tentative proposal, the self metaphor has grown dramatically; I showed how it has infiltrated the popular imagination, forming an important pillar of modern selfhood. From that position I charted the basic outline of a phenomenological approach to the identity problem and to defining the self *in process* and then suggested how such a critique might be applied to current immunological theory. Again, the discussion might naturally end there, but there is one final point to be raised, if only to broaden the perspective on this endeavor; also, I hope, it will deepen our understanding of the formative issues underlying the discussion.

The question of identity no longer resides as a topic of philosophy alone in our bookstores and libraries; it is spread over several domains, including linguistics, anthropology, cognitive science, psychology, sociology, religion, and literary criticism, as well as in a new area called cultural criticism, which draws upon all the aforementioned disciplines, to focus on "modernity" and its problems. The theme of selfhood is apparent in the titles of certain of cultural criticism's works, such as *Situating the Self, The Saturated Self, Constructions of the Self, Modernity and Self-identity, The Listening Self,* and *The Evolving Self.* These works, among many others, address the questions, What is modernism? What is postmodern culture? Is there a postmodern identity?[3]

The modernist view of the self that serves as the implicit contrast to postmodern notions throughout this discussion is the one offered by Kant, namely, the transcendental apperception of the ego. As discussed in Chapter 7, this was, to a large extent, an operational definition of selfhood: the inner, fundamental, and unchanging sense of a unity of our consciousness. From this orientation, the nineteenth century witnessed a significant philosophical turn, which we may refer to as postmodern. Beginning with Hegel, and most clearly articulated with Nietzsche, the sovereign subject relates to only that which it constructs or confronts. The realization of the self is determined in a complex duality of its own self-consciousness, which is intimately dependent on its relation to the other, whether God, nature, culture, history, or other selves. In other words, otherness becomes constitutive. The centrality of the self's relation to the other assumes its more radical postmodern orientation as an expression of poststructuralism. Structuralism understands meaning to be a function of the relations among the components of any cultural formation or our very consciousness. For instance, the pictures of the mind's world assume their meaning, value, and significance from their relationships, that is their "place" in a structure. The self is thus constructed. But the postmodernists broadly argue that any structure crumbles when we recognize that no part can assume participation outside its relation to other parts. In other words, there is no center, no organizing principle privileged over structure and thus able to dominate its structural domain. From this perspective, there is nothing "natural" about cultural structures (e.g., language, kinship systems, social and economic hierarchies, sexual norms, religious beliefs), no transcendental significance to limit "meanings," and only power explains the hegemony of one view over another. In this scenario, the phenomenological insistence on the self's dependence on the other has been radically challenged: Not only has the self's autonomy been rendered meaningless, *any* construction of the self is regarded as arbitrary.

Postmodernist critiques of selfhood do qualify as a radical challenge, if not a refutation of such an entity as formulated by classical modernist criteria. The common element of a postmodernist view of selfhood seems to be that the self, like the rest of the world, has no reference point and must be regarded as having melted away. When the subject is "decentered," no longer an origin or a source, it becomes only the contingent result or product of multiple social and psychological forces. On this view, the unity of the self was at best a deceptive construction. Its very authenticity has been fundamentally challenged. In short, from this perspective, the self should more appropriately be viewed as a contingency or an interpretive scheme. If, however, the self is a contingency, there is no unity by which the self may be organized to confront its world. The postmodern view of the self disallows Kant's modern subject to determine for itself, completely and unconditionally, what to accept as evidence about the nature of its world or its organization. Self-determination has been replaced by arbitrary choice, a construction based on the unsteady assumptions of cultural practice and historical chance. We have already encountered the origin of this decentered subject in Nietzsche's overcoming self.

What is at stake is not only *what* the self is, but whether a self even *exists*. Modernity posits the possibility of attaining individuality and self-determination. In fact this has represented its core mission. The progress toward freedom, whether conceived as mastered nature, political liberty, or individual autonomy, has rested on the assumption of the existence of a self whose essential nature is self-assertion and growing self-consciousness. Kant is viewed as the first thoroughgoing modernist; his critical reflection focused on how "reason can now completely determine for itself what is to count as nature itself" (Pippin 1991, p. 12). Irrespective of the limits of such an enterprise, which Kant construed as the thing-in-itself, remaining inaccessible because of our reliance on our reality-based a priori categories, the knowing subject is affirmed and, more important, legitimized. The essential point is one of autonomy. Kantian rationality would free us from passion, tradition, prejudice and liberate us in order to attain self-determination: "*Sapere aude!* [Dare to know!] 'Have courage to use your own reason!' – that is the motto of enlightenment" (Kant 1784, 1959, p. 85). Alternatively Descartes is often assigned the distinction of being the first modernist. After all, he argued that Man's very existence and knowledge were dependent on the act of thinking. His *method* of reconstructing the world of experience was to arrive at this core residue of selfhood, and his new attempt (which incidentally was to mathematize the world in what is referred to as Descartes' Dream) certainly resided squarely on the self-reliant rational and thinking subject.

However, Kant is the exemplary modernist, because the skepticism that was Descartes' starting point – and the unresolvable schism between the empirical world and the rational subject he engendered – remained a philosophical quagmire until Kant's epic forging empiricism and rationalism on the foundation of a knowing subject. Thus, although Descartes may be credited as one of the founding modernists, Kant truly freed the "encumbered" self. Kant liberated the subject to determine autonomously what to accept as evidence about the nature of things and ultimately what to regard as an appropriate evaluation of action. The Kantian position was no longer tied to the dogmatic guarantee that the order of nature and thought were identical; as Pippin notes, Kant's idealism that a "spontaneous subject determines for itself critically what is to count as an objective claim about nature . . . marks the emergence of a wholly modern philosophical project" (1991, p. 47). Specifically Kant rejected the rationalist view that things-in-themselves correspond to "conceptual truths" and the empiricist claim that sensory data alone could constitute the foundation of knowledge. Each is rejected under the watchful critical faculty of a self-determining judge and formulator of experience. Nature is legislated; that is, it is understood and ordered by the rational faculty. Conversely natural science assumes its authority as the successful product of that endeavor. And most germane to our topic, identity becomes self-grounded; that is, just as the world assumes its meaning from a legislating rationality, the self – the knowing source of that rationality – is the measure of experience. To be modern, for Kant, meant to be radically critical; the subject could rely only on itself in a self-determined process of examination.

This position depends on a substantive self: the unity, or pure ego, of consciousness (described in Chapter 7). Suffice it to review here that the knowing subject, preserving identity through the passage of time, possessing self-awareness, independent in a certain sense of experience, required a form of receptivity. The mind also required discriminating and synthesizing functions that would unify sensory data into conceptual wholes. Experience was thus unified by an autonomous self, and knowledge was forged under these epistemic conditions in terms of "phenomena" and not as things-in-themselves (that is, as things existing independently of these conditions). Nevertheless nonempirical knowledge concerning cognition remained possible in Kant's scheme. The foundations of James's and Husserl's phenomenological inquiries can be traced back to this position, but more important for this point in the discussion, the knowing nature of an autonomous self was affirmed and established.

Kant's transcendental idealism has been subject to voluminous critiques and has been attacked as "supported by nothing on earth and suspended from nothing in heaven" (Beck 1978, p. 30). The issues reside on how to "ground" reason adequately in material reality, in that the representations of the mind are posited as partially independent of any given sensations and the integrative functions that discriminate and synthesize experience seem to be undetermined. But as Kant endeavored to articulate these autonomous principles, he allowed reason to ground itself by reference only to its own requirements; the "transcendental skepticism" that resulted from this position (we do not know things-in-themselves, but only things that are subject to these requirements) has bequeathed the philosophic problem of modernity – the autonomous, knowing self – in its full, but unresolved expression (Pippin 1991, p. 60). The issue since Kant, tackled by Hegel and Fichte and from Schelling to Heidegger, has been – beyond positing an elusive transcendental assumption of reason's freedom – to provide a grounded account for the self-legislated constraints on experience or action. It has turned out that the "nature" of Man, the a priori categories, the metaphysical teleological assumption, the "postulates" of practical reason, simply could not be critically substantiated. (Schopenhauer's critique of Kant served as the basis of Nietzsche's later response, but the staunchest defense, built along a different strategy, may be found in Hegel's assertion of self-determination [see Taylor 1975, 1989; Pippin 1989, 1991].)

Modernism also came under assault from other, nonphilosophical sources. By mid-nineteenth century, the limits of human self-determination and self-understanding were increasingly being questioned. This disquiet arose from several quarters: the demystification of Nature's order in the Darwinian thesis, the antipositivist reaction to scientific mastery as an overinflated ideal, largely irrelevant to many areas of human concern, and finally, the vision that modernity was a spiritual desert where, in the sundering of the link of Man to deity or in the dehumanization decried by social critics such as Marx, Man became not free, but isolated and forlorn (Ollman 1976). The postmodernist reaction, most simply stated, is that self-independence is illusory, autonomy is contingent, and the self is a mere construction of convenience and convention.

We see diverse responses to this crisis: Henri Bergson extended Nietzsche's heady enthusiasm for self-assertion (even self-creation) and reacted against an uncritical positivism in the widely popular and influential *Creative Evolution* (1907), which would also serve as the basis of a new ethics (Bergson 1946, p. 252). And, of course, Freudian psychology was a complex response to integrate the personality, not in a transcendental domain, but in the realm of reality inhabited by a self-conscious ego. The arts also reflected the search for the ego and its reality. Impressionism arose from several origins, but a crucial factor was the influence of scientists and philosophers of the period who regarded the body and soul, self and world, matter and energy as "elements within a single vast flux. Indeed, perhaps the best way to understand empiricist monism is to think of an impressionist painting, with its various flecks of color overflowing the discrete boundaries of individual objects" (Ryan 1991, p. 17).

Seurat, heavily influenced by James Maxwell, Helmholtz, and especially the American physicist Ogden Rood, was motivated to develop the pointillism technique, which allowed bits of colors to be integrated by the observer rather than mixed on the painter's palette or canvas. Félix Fénéon, a critic who reviewed the eighth Impressionistic show in 1886 (where the famous *Sunday Afternoon on the Island of La Grande-Jatte* was exhibited), cited Rood's theory as the basis for creating maximum luminosity, and more saliently he presented science as a model of abstraction that had important authority for the new group (Ward 1986, p. 456).[4] It was the crucial recognition of the later Impressionists and more clearly of the post-Impressionists of the 1890s that objects have no fixed and continuous identity, but change constantly from moment to moment as they are perceived. This, of course, is the lesson taught by James and other empiricism-oriented psychologists and philosophers of the period. More significant, however, are the implications for the perceiving subject. The artist demands the viewer to integrate, assimilate, and finish the painting: "Impressions" are emphasized as a way of acknowledging how individual perceptions are constructed. But at the same time, perspective increasingly collapses; that is, as foreground and background merge, there arises a visual confusion that "makes it impossible to say . . . exactly where the observer of the scene is standing. . . . The eye is constantly being frustrated" (Ryan 1991, p. 18). Seurat's *Sunday Afternoon* visually exemplified this paradoxical relationship between extreme subjectivity and the eradication of any clear point of view. This lesson was taken to its logical and pictoral conclusion by the cubist experiment, which began with Pablo Picasso's *Les Demoiselles d'Avignon* of 1907 and continued in fervent artistic dialogue with Georges Braque until the beginning of World War I (Rubin

1989). Their "anonymous art" (extended to hiding their signature on the back of the paintings) was designed, as Picasso said, "to set up a new order" that was independent of the artist (ibid., p. 19). The forceful if not shocking demand on the viewer to sort out the objects, define their relationship, and choose a perspective from several points of view embodied a prophetic message from cubism for the rest of the twentieth century. The work of art, as a pure idea, disengaged from the artist and realizable by anyone, is encapsulated by Picasso's quip, "[Y]ou'll see, I'm going to hold on to the Guitar; but I shall sell its plan. Everyone will be able to make it himself" (ibid., p. 20). In the process, of course, there was a fair degree of consternation and confusion.

This confusion is also evident in the literary product of the period. The mind was viewed as hopelessly "subjective," or "individualized." With such subjectivity, the external (objective) world might easily be distorted, and writers explored this new view of reality with startling, novel techniques of style and perspective. "The new problem which empiricism seemed to raise was that subjectivity was not itself a stable entity. Developing a single, consistent point of view no longer appeared to be a fruitful way of representing reality" (Ryan 1991, p. 19). So, as William James, Ernst Mach, and Franz Brentano were questioning perception and identity, Gertrude Stein, James Joyce, Franz Kafka, and Henry James were asking how to represent what we see: A table is known to be a rectangle but is seen as a trapezoid, so how it is to be represented? And beyond the natural world, how are emotional responses to be depicted? From which perspective are personal experiences to be described? This philosophical restlessness challenged the attempt to impose coherence on experience, whether on the canvas, in the novel, or in their very lives. Where was the logic to link an increasingly fragmented reality? James's popularizations of his pragmatism, for example, "Does Consciousness Exist?" (1904, 1987) and "Is Radical Empiricism Solipsistic?" (1905, 1987), were commonsensical responses, but the anxious question drove a more radical inquiry best described in the descriptive phenomenology of Edmund Husserl, who bequeathed a fecund but still incomplete philosophy of the self.

There is a lingering and recalcitrant ambiguity residing at the heart of the self problem. We intuit the reality of the self, but it does not exist as an *entity*. How to define the self is a central philosophical issue, and the punctuated summary of its modern history has been given in order to circumscribe the boundaries within which Metchnikoff might have understood the problem and translated it into his own research. His scientific – and philosophical – response would serve as an important

285

conceptual element in the new field of immunology. He clearly understood that the self as such was not a given, but represented an ever-changing evolutionary entity, both in the grand Darwinian scheme and in the minidrama occurring over the life span of the individual, most dramatically witnessed in embryonic development. Metchnikoff saw the same general processes occurring later in the adult organism in a somewhat different format, which he called physiological inflammation and more specifically immunity. Thus, the immune reaction was for him but a special case of the developmental process. Underlying each of these activities was the implicit problem of how the individual might be discerned and ultimately defined. I have already discussed this question in its various guises, and now I wish to simply emphasize certain key characteristics.

The self is defined by its encounter, whether social or historical, with the other. Identity is etched not solely by a self-defining ego, but in response to its world, undoubtedly subject to external forces. By the nineteenth century the predominant philosophical attitude was to view one's actions as dependent only on oneself: "The new identity as self-defining subject was won by breaking free of the larger matrix of a cosmic order and its claims" (Taylor 1975, p. 560). This understanding followed from the autonomous knower of Locke, the Rousseaunian imperative of obeying only oneself, and the Romantic poet's quest for true identity. Moreover, the emergence of this aspiring autonomy may be traced to Kant. On the one hand, Kant's notion of freedom is based on obeying a law made by the rational self, in contrast to dependence on the will of others or on nature. But this rationality that creates law and demands obedience must be coupled with the Kantian idea that human nature is not simply given, but to be made over. A new problem emerges: According to what ideal or standard should this self-fashioned person be structured?

The obvious issue is Man's animal, instinctual, or, better yet, emotional nature. How was this aspect of his being to be combined satisfactorily with the rational faculty? Schopenhauer's pessimism was based on the assessment that the animal drives were unconquerable, and Kierkegaard's angst could only be overcome by relating oneself to God, thus stepping away from freedom as self-dependence. It was Nietzsche who grappled with the issue most dramatically. If freedom of self-dependence is ultimately empty, then "it risks ending in nihilism, that is, self-alienation through the rejection of all 'values'. . . . Only the will to power remains" (ibid., p. 563). By the end of the nineteenth century the untethered, alienated self – (originating in Marx's early philosophy

(1844, 1963) – was viewed from either a "positive" or a "negative" perspective. In the latter case, as most recently advanced by Foucault, little autonomy or freedom is allowed in the crucible of cultural forces that mold and manipulate both individual behavior and insight. In this essentially sociological analysis we are subject to the vicissitudes of our culture and time. The other response arises from the philosophical concern of how the world (viz. reality) is mediated (known), or, simply stated, from the concern in defining a post-Kantian subject–object relationship. And perhaps the question, more to the point, is, What is the post-Darwinian subject?

Fin-de-siècle Europe was keenly aware of the Darwinian ethos, or more precisely of the episteme of evolution. Schopenhauer's will was separated from Nietzsche's by *On the Origin of Species*. At least one astute critic, George Simmel, recognized that Nietzsche began where Schopenhauer had finished. The Schopenhaurian will, "because it can only grasp itself in a thousand disguises, is pushed forward from every point of rest on an endless path," yet there is no singular, unified *goal* in this universe (Simmel 1907, 1991, p. 5). Nietzsche discovered how evolution disallowed Schopenhaurian pessimism in the eternal recurrence. Schopenhauer's world of will and representation is valueless and devoid of meaning, but evolution offered Nietzsche an ethic in nature's seemingly ceaseless striving:

> Life is in itself, in its intimate and innermost essence, an increase, maximization, and growing concentration of the surrounding power of the universe in the subject. . . . [L]ife can become the goal of life. Thus, the question of a final goal beyond life's own natural process becomes moot. This image of life as a poetical-philosophical absolutization of the Darwinian idea of evolution . . . is ultimately decisive for every philosophy. The deep and necessary parting of the ways for Nietzsche and Schopenhauer lies here. (Ibid., p. 6)

Nietzsche found meaning not in something absolute and definitive, but in the Übermensch as evolution embodied. What is potential and germinal awakens to greater expansion and finer expression of its essence. With the Overman life becomes fuller and richer: "He is not a fixed goal which gives meaning to evolution, but only expresses the fact that there is no need for such a goal, that life in itself, in the process [of evolution] has its own value" (ibid., p. 7). This essentialist position, a *goalless* evolution, where only potentialities are recognized, allows for evolution to assume sui generis a *value* in itself. Characteristic and central to this view (echoed by others of the period, e.g., Bergson) is the faith in reliance on life to assume its own purposive character (in its

287

own evolutionary character) and to deny the need for external forces or deities. When such a philosophy assumes its expression in the self, there are jolting consequences.

Nietzsche ironically grasped eternity as the parameter of life's evolution continuing ad infinitum. In thus expanding the conception of life as radically incorporating change in its death–renewal cycle, he pronounced the ideal of life. No longer a "given," Man was seen as groping toward the ideal. But this ideality is not a definable goal. In this sense, the ideal represents, alone and unqualified, evolutionary movement. Changes, adjustment, improvement are the responses of life to its challenges, both external and from within. In the sequence of evolutionary movement every phenomenon is an irreplaceable and unrepeatable step. Each increment is also unique and nonreiterable, both in time and in its consequence. Such is freedom in Nietzsche's world, and this ethos serves as the foundation of his new morality. But we need not be Nietzscheans to recognize how this philosophy articulates Man in the embrace of a post-Darwinian world. There is no stable norm, no given, no static entity. The very character of evolution disallows permanence in any sense, and in each instant the organism must encounter its environment and itself to evolve, that is, respond. Life is dialogical – an ongoing and ceaseless dialogue between organism and its world. In a world of flux, relativity, and indeterminism, evolution replaces rigid hierarchy and ordered cosmic forces. Probability has replaced determinism, and cause arises from disparate and indefinite sources of likelihood. If we truly are Darwinians, then we remain unhinged in a world of change, bereft of the predictability of an ordered world of deterministic causality, as Darwin himself lamented: "I cannot think that the world is the result of chance; yet I cannot look at each separate thing as the result of Design. . . . I am, and shall ever remain, in a hopeless muddle" (letter from Darwin to Asa Gray, quoted in Birch 1988, p. 75).

Darwin expressed an ancient vexation that transcends the polarization of chance versus design. For despite constant change, an entity *has* an identity. How do we discern that character and what is it? Depending on the philosophy considered, the problem becomes to establish the criteria of *identity* – to perceive the governing Form (Aristotle's *telos*) or to discern the hidden Idea (e.g., Platonic). Aristotle clearly addressed the perplexities of defining a changing entity in his *Metaphysics:*

> The changing, when it is changing, does not exist. Yet it is after all disputable; for that which is losing a quality has something of that which is being lost, and of that which is coming to be, something must already be. And in general if a thing is perishing, there will be present something

288

that exists; and if a thing is coming to be, there must be something from which it comes to be and something by which it is generated, and this process cannot be *ad infinitum.* (Barnes 1984, p. 1595)

Aristotle sought essences, but in the post-Darwinian world *change itself is the essence.* The problem of seeking the essence of entities undergoing constant change was clearly articulated in ancient Greek philosophy. There is no need to explore these matters here; I only note that the difference between Nietzsche and the Ancients in this regard resides in the post-Darwinian acceptance of change as *integral* to the essence of the organism, constituting the very core of its being. Contrast this conception with Aristotle's description of how the Idealist position developed:

> The supporters of the ideal theory were led to it because they were persuaded of the truth of the Heraclitean doctrine that all sensible things are ever passing away, so that if knowledge or thought is to have an object, there must be some other and permanent entities, apart from those which are sensible; for there can be no knowledge of things which are in a state of flux. But it was natural that Socrates should seek the essence. For he was seeking to deduce, and the essence is the starting-point of deductions. For two things may be fairly ascribed by Socrates – inductive arguments and universal definition, both of which are concerned with the starting-point of science. But Socrates did not make the universals or the definitions exist apart; his successors, however, gave them separate existence, and this was the kind of thing they called Ideas. (Ibid., p. 1705)

A resting point never exists and never arrives. The norm is simply the most numerous. The ideal, the possible, the elusive potential has replaced our sense of finitude of a world with boundaries. Awash in this uncertain cosmos is the self, whose own sense gathers tenuously within elusive boundaries and pliable structure, much like Metchnikoff's phagocyte, which gropes in its "self-actualization" by sending out pseudopodia, feelers to touch its world, seeking to know friend and foe. This is the nether world in which Metchnikoff proposed a scientific theory of selfhood.

Perhaps Metchnikoff naively rejected Schopenhauer's pessimism in the promise of scientific discovery (Metchnikoff 1903; 1907), but more fundamentally he too was caught in the optimism of an evolutionary episteme that posited some ideality. Retrogressive, degenerate evolution was not integrated in this vision. Evolution offered promise not in the value-based notion of Man as the pinnacle of life's progression (a view no doubt held in high esteem by many social Darwinians of the time), but in the deeper sense of an assertion of life's promise to expand and

289

improve toward some undeclared yet undisguised ideal. It was this deep sense of identity as a metaphysical construct that lay at the core of Metchnikoff's thinking of the organism. Immunology was a vehicle to define what was essentially – and most profoundly in *all* particulars – a developmental process. I maintain that this conceptual foundation remains in place at the deepest strata of modern immunology.

In the final analysis, after much philosophical juggling, "the 'answer' to the question, what is the Self? is the anonymous language of the very question" (Corngold 1985). We are ultimately left with the self as a provocative question or as a generative concept, as when Nietzsche asks what created will and Zarathustra responds, "[T]he creative self" (Nietzsche 1885, 1959, p. 147). Nietzsche's attempt to come to terms with the questionable ego places him as a pivotal figure in modern philosophy. No matter how much Man is dismantled and reconstructed, however much autonomy is espoused as "self-overcoming" (as opposed to a more traditional "self-realization"), revealing his true identity remains *the* issue to pursue – and ultimately to assert. Fundamentally and barely hidden in Nietzsche's tantrums (as well as in the postmodernist thought that traces itself to a Nietzschean assertion) is the essential modernist quest for freedom of the self, captured by a philosophical construct firmly lodged within that problematic. This issue is crucial for assessing the philosophical origins of Metchnikoff's theory and the later revival of his project. Perhaps Kant, as the consummate architect of the modern self as an autonomous and self-determining subject, is a key inspiration of our own articulation.

We are still somewhat uncomfortable with the shifting boundaries of a Darwinian construct. We seek firmer definition, and we do so within a modernist tradition, as did Nietzsche. The striving for freedom as self-aggrandizement is clearly articulated by Nietzsche, the modernist, where ideals still allow for autonomy and self-actualization. As elusive as such an identity might be, it still represents an attainable ideal. In this reading of Nietzsche, modernity's essential failing was that "it had not been modern enough, that the restless, perpetually self-transforming, anomic, transient spirit of modernism had to be affirmed much more honestly and consistently" (Pippin 1991, p. 7). The argument for such a radicalized modern venture, to truly free the self, assumed its most exuberant expression in Nietzsche, but can be traced as well from Baudelaire through the antibourgeois sentiment that fueled much of fin-de-siècle's emotive energy. The self as independent and self-determining, self-conscious, always in action both as moral agent and ultimately in its knowledge of the world, constitutes in a strong sense the end result of the modernist quest that began with Kant. This pursuit may well be the true foundation of Metchnikoff's concept of the organism and of his notion of immune identity.

Viewed in this context, the Metchnikovian challenge is a disguised expression and assertion of the modernist ideal, where the very notion of the self comes to possess a pervasive and profound centrality in the metaphysical construction of immunology's endeavor. If immunity becomes the process that defines the self or if immunology assumes the scientific mantle (along with the neurosciences) to describe those processes that must discriminate the self from nonself, then we have articulated a beguiling problem. No longer is immunology content with simply discerning mechanisms for achieving and maintaining integrity; it seeks the deeper approbation of selfhood, that is, those functions that declare and legislate in truly a Kantian manner: identity.

I have argued that Burnet was a worthy heir to the Metchnikovian vision of immune function. Despite the elusiveness of Burnet's metaphor, his theory seems to fulfill the key modernist criteria of selfhood, namely, there was an appeal to such an entity as the self, which functions similarly to a Kantian category. Analogous to the apperception of the transcendental ego, Burnet's self is based upon a notion of unity – a coherent whole – to account for the inner, fundamentally integrated organism. An added feature, not fully appreciated at the time, later to become a central focus of immune theory, was that in defending the self, the immune system also defined it. The final element – again, not explicitly declared – was the bewildered search for the source of immune function. In other words, what constituted organismic identity? This issue must be recognized as the truest modernist aspiration, albeit a formulation in this earliest presentation of the self that remains hidden or perhaps implicitly understood. Thus, the underlying structure of Burnet's theory consisted of three elements: (1) the basic notion that there was such an entity as the self, (2) that the self was functionally identified by the immune system, and (3) that in some unspecified fashion it represented the source of immune activity. These are different levels of organization, each of which, in the modernist orientation, corresponds to an implicit understanding that a self exists. The self, never articulated by him in anything resembling a cogent scientific definition, remained a functional construct.

There is a clear scientific tradition that built on this version of selfhood, namely, the effort to define the self genetically. A molecular definition of selfhood emerged from transplant biology, where the very limits of successfully transplanting tissue from one individual to another demanded exploration of the factors that identified a transplanted organ as self or foreign. Spectacular success has been achieved in defining the molecular signature of the self in genetic terms. Defined by such means, the self would gain a certain finitude. In its simplest form, the issue is whether the self-markers originally hypothesized by Burnet are programmed by the genetic constitution of the host and whether they are operationally active in immune function. If these markers were in fact coded for in the genome, they would then be given, *a priori*. Defined through a given set of genes, the immune self would then take on a specific character, one given by the particular sequence of an individual's genetic endowment. That query has been remarkably successful. In exquisite detail, we have the structural and genetic definition of the protein complex that serves as the major identification system of the immune self. This so-called mixed histocompatibility complex (MHC) binds foreign substances, and thus becomes altered from its original state so that immune cells recognize the foreign element. Note that the foreign is perceived in the context of altered self. The important upshot is that immunologists who seek a ready definition of the self can now point to the genome and cite a genetic signature which regulates immune recognition mechanisms. I suggest that this line of research and pursuit represents the modernist background assumption that there is a self. *Because* the immune self is now definable in molecular terms at both the genetic and protein levels, a degree of exactitude has satisfied the epistemological and materialistic yearnings of the field, confirming that our commonsensical notions of identity can at some "basic" level be successfully applied to the problem of immune identity.

But this research tradition has been challenged. In Chapter 5 I suggested that the self should be defined as immune *process,* that "recognition" is a cognitive event and thus dependent on its contextural history (experience) and universe of known particulars. It is the unknown, the foreign that is reacted upon, and as the system closes around the antigen, that perception of nonidentity triggers immune recognition in its fully systemic cognition. Again, Burnet was a key architect of this competing formulation, for he originally perceived the cognitive potential of the self metaphor. From the very inception of immunology, the presence of the idea of *recognition* was evident. The identification of the foreign implicitly requires that something be doing the recognizing through a cognitive apparatus. But it was the richness of the self metaphor that allowed a cognitive model to be developed for immune theory and thereby fully expose the potential of the *self.* By invoking our deeply intuitive understanding of selfhood, Burnet articulated an organismic approach to immune identity, one that mirrored philosophical and psychological conceptions of our very selves. But what is raised immediately by such a parallel fashioning of immunology's model, gleaned from its metaphoric use of the self, is the accompanying critique of philosophy's unresolved debate over the proper testament of such a formulation. As James and Husserl left the self to be declared in its phenomenological experience, so I have deemed that posture relevant in the attempt to establish the philosophical foundation of current immunological theory.

This brings me to the last point of this summation, namely, the phenomenological orientation adopted in this discussion. This critique was stimulated in the attempt to redefine the basis of what might constitute the knowing and experiencing subject. The issue with James and Husserl revolves around identifying *cogito,* consciousness, and thereby the ego. The challenge posed by these phenomenological issues was considered in Chapter 6, and we need only summarily recall that reality is selected and formed by an active (intentional) self that, although not ultimately definable as an *entity,* emerges in its activity. How might I best describe the resonance of this approach for my argument? I believe it best espoused under the rubric of authenticity, which is the philosophical counterpart of a key element in current immune theory. Specifically, I am referring to alterity, the other. As already detailed, the immune system sees the foreign as a result of antigen binding to the MHC. The self is thus altered, and the foreign is "seen" because it modifies the natural state of the self's signature. The immune response ensues. This is a fascinating finding because of its parallel to phenomenological theory. Alterity is a major focus of how phenomenologists, beginning with Husserl, define selfhood. Briefly, the self, to the extent that it can be actualized, is, in varying views, defined by the other (e.g., Theunissen 1984; Taylor 1987). Obviously different critics see such a project with different degrees of optimism, but at least the problem is shared, and when the issue of alterity is discussed, we know roughly what is being debated. The question revolves around whether, and how, in response to an encounter, the self articulates itself or is altered as a consequence of that engagement. Also considered is how the engaged self might alter its object and their shared world. In short, the phenomenological approach explores how the self lives in its world, essentially in a universe of others. Either the self alone is miserably alienated or it actively engages the world and thereby becomes actualized. The debate revolves on the contingency of this process and its problematic opportunities for success. But by and large the parties agree that the potential

293

for self-aggrandizement must be realized *in* the world, and the self must ultimately actualize itself in the encounter with the other. I have found no explicit admission of immune selfhood as mirrored by these prominent philosophical discussions, but there is such a close correspondence in the manner in which the science has been explicated that I find it difficult to see them as two totally separate phenomena.

There are those immunologists who would resist defining immune activity in the context of the other. They are wedded to a genetic notion of a self entity, neatly defined and entailed by its own "selfness." This self is guarded by the immune system, a subsystem similarly (i.e., firmly) defined by a genetic prescription, the MHC. I believe this is a simple mechanical conception, where well-integrated parts function in line, like Descartes' clock: If we only better understood its workings, we would perceive its mechanical order and causal relationships. This is a rigid modernist aspiration, one that in my view is limited in its explanatory power. The phenomenologist dismisses this formulation, and replaces it with one centered on defining the self as emerging from its encounter with the world in *process*. In immune theory this view is represented by the cognitive paradigm.

In closing, I note that immunologists may well shrug off this essay as "mere philosophy," and that philosophers might agree with Heidegger (1927, 1962) that the self as self is not even a relevant subject of inquiry. In this sense the central general concern with the problem of selfhood has been deflected. Yet it is immunology that actively employs self as a fundamental concept, and my argument has been an attempt to explicate what that term might mean. To do so required identifying its entry into the scientific discourse, tracing its development, expounding on its metaphorical meanings, and attempting to reconstruct the significance of that etymology. To have remained within the boundaries of immunology would have limited the discussion to simply selecting and advocating one particular view amongst various understandings of the immune self. In this essay I have made a different gambit: In order to explore the implications of how to understand or constitute the self, we have marched through three centuries of philosophical debate, finding that its meaning has always been obscure and yet tantalizingly within our grasp. The self's immediacy and importance are poignant because of its obvious centrality to our understanding of ourselves. We cannot refuse or evade the term's use, but we are left with "mere philosophy" when we explicate its meaning critically.

There is a startling insight and a profound irony that accompany the coöption of *self* from philosophical discourse. The term harbors a series of problems: the question of personal identity, the position of the subject in his world, and the epistemological nature of the self. The debate on the issue of selfhood, originating in the post-Renaissance and extending uninterrupted as a crucial focus of modern philosophy, remains unresolved. At the end of the twentieth century philosophy has yet to solve the "I" question. One view is to insist that "any particular being in the world is simply not the true referent of 'I' at all, that the 'I' is the subject of the centreless view, and that this or that body, mind (or whatever) is merely one of the contents of the world correlated with the 'I.' The other is to deny the intelligibility of the centreless view of the world" (Janaway 1989, p. 316). The lesson that philosophers most cogently may offer immunologists is the fact that such an entity entails restrictions and ambiguities.

I accept the conundrum facing the modernist–postmodernist dispute regarding the self's existence and suggest we abandon attempts to pinpoint and define the self, allowing its definition by expression in its phenomenological address: The subject becomes defined in the process of its encounter with the world. Perhaps the search for *the* self is an anomaly of our subject–predicate language structure (Whorf 1958). In our everyday language only a subject can act. But recall how Husserl formulated the ego as regarding reality in terms of verbs, rather than of nouns, so that the subject became a predicate of action – never an object (see p. 219). Similarly the immune self is neither subject nor object, but is actualized in action; the self becomes, on this view, a subject-less verb (see pp. 224ff). Asking essentially the same questions concerning identity and cognition to explicate "the knower," some immune theorists have reached toward this phenomenological articulation. I have maintained that the self metaphor has directed their inquiry. Irrespective of whether the scientific approach is successful or not, this case reveals the rich interplay of science and its supporting philosophy. The *self* resonates with an unexpected rich vibrancy.

What the immune self truly represents is the drive to decipher the source of immunity, to uncover the ontological basis of identity. This is a worthy scientific undertaking and remains a profoundly perplexing issue. To define the self has become immunology's primary mission, the ultimate puzzle for the science that is attempting to identify the organism. To specify that endeavor, however, the discipline has borrowed a philosophical term to approximate a language that is inadequate to that task. It is in this sense that *self* is used metaphorically. Immunology's dipping into such language for a metaphoric approximation to its central problem again reveals how theory gropes for its footing in common experience. In the vast array of philosophical and historical contingencies that shape our thinking and behavior, it serves us well to reexamine the assumptions of our science and to recognize that the laboratory can hardly define its endeavors without borrowing from its culture. But the admission that immunology has been using a metaphor yields significant insights into its workings. It is necessary first to apply or pose, tentatively and indistinctly articulated concepts. Recognizing how *self* functions in the language of immunity is to recognize the dull point of our rapier. More saliently perhaps, it is to recognize the nature of our inquiry. To move into the strata where metaphors abound and govern the discourse is not only to acknowledge our ignorance, but to exhalt that a new vision is beckoning. From metaphor to theory to metaphor to theory – thus we evolve. And it is the notion of nondirectional, yet idealized adaptation, a distinctly post-Darwinian principle, that governs our modern quest, as well as our very identity.

Notes

ACKNOWLEDGMENTS

1. Funding from the Neutrophil Research Fund and a grant from the National Institutes of Health, HG00912, partially supported this research.
2. Permission to reprint excerpts and modifications from previously published papers was obtained from the University of Chicago Press, Chicago: Tauber, A. I., The immunological Self: A centenary perspective, *Perspectives in Biology and Medicine* 35:74–86, 1991; from Academic Press, San Diego: Tauber, A. I., The birth of immunology: III. The fate of the phagocytosis theory, *Cellular Immunology* 139:505–30, 1992; and from Kluwer Academic Publishers, Dordrecht: Tauber, A. I., A typology of Nietzsche's biology, *Philosophy and Biology* 9:25–44, 1994.

CHAPTER 1

1. The issue of Darwinian competition between cell lineages was reopened by Buss (1987) and has provoked spirited critiques (Falk and Sarkar 1992; Gilbert 1992; Chernyak and Tauber, 1992).
2. A much more celebrated case of similar structure was that advocated by Sigmund Freud. The essential characteristics of a disharmonious state composed, in this instance, of competing drives, striving for an unattainable ideal harmony is closely parallel to Metchnikoff's underlying thesis of the organism. With a model of appropriate psychosexual development providing the standard for the ideal state of mental health, Freudian psychoanalysis directs itself toward the same orthobiosis advocated by Metchnikoff, albeit in a different pathological context. In both cases health becomes the problematic, elusive ideal. As Freud stated, a normal ego "is like normality in general, an ideal fiction" (quoted in Boorse 1976). Each was heavily committed to the Darwinian paradigm, and thus we might expect to find resemblance in their respective conceptions of biological organiza-

297

tion. The influence of Darwinism on Metchnikoff has been detailed by Tauber and Chernyak (1991), and Freud's indebtedness has been discussed by Ritvo (1990). Sulloway, in an earlier study of Freud's general biological orientation, makes the following summary: "Darwin's influence upon Freud's scientific generation had been so extensive Freud himself probably never knew just how much he really owed to this one intellectual source. Darwinian assumptions 1) pervaded the whole nascent discipline of child psychology from which Freud drew . . .; 2) reinformed the immense importance of sexuality . . .; 3) alerted Freud . . . to the manifold potentials of historical reductionism . . .; 4) under[lay] Freud's fundamental conceptions of human psychosexual stages, and of the archaic nature of the unconscious; 5) contributed a number of major psychical concepts like those of fixation and regression – to Freud's overall theory of psychopathology. . . . Freud toward the end of his life, recommended that the study of evolution be included in every prospective psychoanalyst's program of training" (Sulloway 1979, 1983, p. 275–6).

3. Metchnikoff's most celebrated effort was to replace colonic bacteria with lactobaccilus; he thus advocated eating yogurt, championing what has become one of France's major dietary contributions on the basis of an unproven hypothesis. Believing that bacterial toxins of resident colonic bacteria seeped into the body, causing damage and thereby an immune response by phagocytes, Metchnikoff offered a remedy: He suggested that the more benign lactobacillus should be ingested to replace the deleterious microorganisms, whereby damage would be limited (Metchnikoff 1906; Stillwell 1991).

4. Metchnikoff's polarized position among his colleagues did not detract from his broad popularity with the general public. For instance, his concept of orthobiosis was translated to practical nutritive matters, as discussed in note 3. He also did not shy away from promoting notions of social order derived from his understanding of biology. His musings on marriage, child rearing, the human races, and social organization each reflect his general notions of biological harmony/disharmony, which he unabashedly applied to any arena he deemed needing reform (Tauber and Chernyak 1991).

5. As detailed elsewhere (Tauber 1992a), George Sternberg putatively exemplified a competing position in regard to Metchnikoff's phagocytosis theory. But simply to suggest a mechanism that was similar to Metchnikoff's argument concerning phagocytic destruction of bacteria did not warrant joint authorship of the same *phagocytosis theory*. Sternberg did not develop phagocyte biology as a comprehensive theory of immunity, and in fact he eventually embraced the humoral theory as a natural consequence of his fundamental failure to understand Metchnikoff's concepts of immunity.

6. Many of the crucial papers of this period have been translated with commentary by Bibel (1988). See also the comments of Silverstein (1989), Tauber and Chernyak (1991), and Moulin (1991a).

7. The archives of the Royal Academy of Sciences and its Nobel Committees, as a result of a rule instituted in 1974, permit access to its records for prize decisions dating back at least fifty years. Crawford (1984) has provided a careful study of the selection process for the Physics and Chemistry prizes between 1901 and 1915. Material from the Nobel Archives (Nobel Archives, 1900–8) was kindly provided by the Nobel Committee for Physiology or Medicine. The records consist of letters of nomination, reports on the candidate, and minutes of the Nobel Committee and the Academy (both section and plenary sessions); the minutes record only the votes taken, not the discussions, although this particular file contains three Committee members' written explanation of their respective votes. The Nobel Archives on Metchnikoff contain fifty letters of recommendations and four secret Swedish reports to the Committee revealing that Metchnikoff was nominated with two letters in the first year of the competition (1901). Support for Metchnikoff, however, increased steadily; by 1907 there are nine letters, and in 1908, fourteen. Not surprisingly virtually all the support came from the French, who championed their Pasteur Institute colleague, although there are other scattered letters of endorsement (see Tauber 1992a for details). As expected, the correspondence of largely Committee-solicited opinions are laudatory.

8. There are four confidential reports: (1) Carl Sundberg and Ernst Almquist, June 10, 1902; (2) Carl Sundberg, August 12, 1907; (3) Carl Sundberg, July 20, 1908; (4) Alfred Pettersson, July 7, 1908 (Nobel Archives 1900–8). Each is quite detailed (ranging from approximately three hundred to seventy-five hundred words) and offers a rich record of how Metchnikoff's theory was critically regarded by knowledgeable scientists of the period. Almquist and Pettersson were active investigators in immunology, and Sundberg was a physician; his reports, however, reveal a detailed knowledge of immunology. (See Tauber 1992a for further details.)

9. For forty years prior to Metchnikoff's phagocytosis theory it was known that leukocytes contained bacteria, but the significance was not understood, most believing the leukocyte served as a convenient transport vehicle or harbor for the microbe (Rather 1972).

10. Biochemical studies of living cells began with the demonstration of oxygen consumption by leukocytes (Graefe 1911), their increased oxidative metabolism with phagocytosis (Baldridge and Gerard 1933), and the production of large quantities of the oxygen-derived toxic species, hydrogen peroxide (Iyer, Islam, and Quastel 1961) and superoxide (Babior, Kipnes, and Curnutte 1973), which have been shown to be potent mediators of biological damage.

11. Aschoff (1924) originally classified "professional" phagocytes as part of the reticuloendothelial system (RES) on the basis of their shared *intra vitam* staining (by lithium carmine, trypan blue, and so on), their common function of producing reticulum, and their anatomic distribution of lining

sinusoid blood and lymph spaces. In 1924 splenic stroma, lymphatic tissue, endothelium of diverse organs (e.g., lymphatic, splenic, and adrenal tissue) were combined as one complex but ill-defined immune system through these nineteenth-century pathological criteria of histochemical staining and cell association. But Aschoff also included such diverse functions as blood cell production and destruction, general metabolism of proteins and fats, and bile acid metabolism. The voracious phagocytic appetite of some cells included in the RES was not equally shared by most components of Aschoff's classification, and the migrating histiocytes of various kinds were therefore uniquely designated.

CHAPTER 2

1. In rejecting atomism, Kant argued that the metaphysical-dynamical hypothesis accounted for impenetrability in terms of more basic repulsive forces, there being no point past which an object cannot be compressed (Brittan 1978). (The reductionists later used the polar force argument for their own purposes.) The antiatomism of *naturphilosophie* continued the movement of resistance to a mechanized model of biological function, which originated in the seventeenth century and was by 1740 already in retreat (Guerrini 1985; Hankins 1985).

2. Bernard's positivism as a deliberately adopted and self-consciously applied philosophical view, for which he was indebted to Comte, has been discussed in detail. See Canguilhem (1966, 1989, pp. 69–80) and Holmes (1974, pp. 403–6, 454–5); for the later history of positivism, see Frank (1941).

3. It is important to note that the mechanical philosophy of the seventeenth century was reductionist (that is, physical laws and facts may be explained by, "reduced to," local interactions of matter) but was ultimately discarded (Stein 1958). In the nineteenth century the success of explaining optics by classical electromagnetism and the laws of thermodynamics by the principles of statistical mechanics buoyed those enthusiasts who wished to apply the same strategy to other scientific and social endeavors (Sarkar 1992a).

4. It is of interest to note Darwin's notion of continuity of mind, in *The Descent of Man*, with Metchnikoff's argument of an evolutionary history of the psyche that might be traced to the phagocyte's behavior. The issue of a primitive consciousness was further explored during the fin-de-siècle period by Jennings (1906, 1976). The philosophical question of how the mental might be construed beyond our particular anthropocentric experience (and prejudice) is provocatively explored by Nagel (1986, pp. 22ff).

5. Consciousness, and the psyche in general, as a manifestation of *entelechy* is, however, very important to Driesch's explication and in fact is used to illustrate his vitalistic notions. See, for example, his "direct justification of entelechy" given in his 1908 Gifford Lectures (1908, vol. 2, pp. 277–86).

6. Fleck carefully shows how the Wassermann diagnostic test evolved to define syphilis (1935, pp. 70–6). This development constitutes a most revealing portrait of scientific practice in general, and of immunology in particular, during this crucial foundational period. More generally Fleck eloquently recognized that "the organism can no longer be construed as a self-contained, independent unit with fixed boundaries" (ibid., p. 60) and understood the implications of this expanded definition of selfhood for immunology.

7. *Immunochemistry* was coined as a term by Arrhenius in 1904 (Arrhenius 1907). Note that, as late as the early 1930s, whether antibodies were definite chemical substances was still in debate (Heidelberger 1932–3), and the question was not firmly resolved until their identification as gamma globulin – thus, the name *immunoglobulin* (Tiselius and Kabat 1939).

8. It seems fitting then that the major figure in establishing the mechanism of G.O.D., Susumu Tonegawa, still in 1990, when I interviewed him with Ilana Löwy, did not view himself as an immunologist! In that interview he explained that he arrived at the Basel Institute for Immunology (BII) in 1971 as a molecular biologist, an identity he seemingly still prefers. As a result of visa problems, he required a position outside the United States, and this was arranged by his supervisor, Renato Dulbecco, who had considerable influence at the BII. Tonegawa initially worked on viral mutants and had minimal contact with Jerne or any other immunologists. Sometime late in the first year George Steinberg, a geneticist and one of the few other non-immunologists at the Institute, alerted Tonegawa to the Dreyer–Bennett paper (1965). Tonegawa believed this hypothesis was not widely believed and only served as a model that required evidence because the concept was "so unconventional." Tonegawa claimed that, despite having a copy, he did not read this paper, and beyond being skeptical of the hypothesis, he was not initially interested enough even to test it experimentally. In any case, Tonegawa claimed that as a member of the Institute he felt compelled to find an "interesting" immunological problem that might be addressed by molecular biological techniques. He thus turned to the problem of immunoglobulin diversity, which he recognized resided in the notion of gene rearrangement, that is, how constant regions might combine with various variable regions. Tonegawa opposed what he referred to as the "germ liners," who included not only Dreyer and Bennett, but most notably Phil Leder, Alan Williamson, and Bernard Mach, who were the main competitors in the formative 1974–5 period. The germ line theory argued that for each chain of the immunoglobulin molecule there existed a germ line gene. Tonegawa argued that the number of germ line V kappa genes was too small to account for the sequence diversity (Tonegawa 1976). The crucial insight was to recognize that antibody variability arose from somatic rearrangement of various genetic components to be assembled within the variable region. These variable regions were not preprogrammed into distinctly encoded genomic units; they were newly assembled ensembles from

301

genetic sequences from the V and J "libraries." For elucidation of this mechanism, Tonegawa was awarded the Nobel Prize in 1987.

9. Clarence Little proposed that tumor susceptibility was controlled by several genes, any one of which could cause rejection, so that the number of such tumor-prone individuals would be $(3/4)n$ (n = number of genes); so if, for instance, there were fifteen genes (3/4 15) then only 1 percent of the F2s would be susceptible (Little 1914). When the experiment was conducted with many animals, in fact, 1.6 percent of the F2 mice were tumor transplantable (Little and Tyzzer 1916). Little extended this observation using normal tissue (spleen) transplanted into two crossed inbred strains (Little and Johnson 1921).

10. Snell's so-called congenic strains, which now number over three hundred, became the single most significant research resource for cellular immunology, or more precisely, lymphocyte immunology. Gorer, Lyman, and Snell (1948) collaborated to show, by linkage studies, that the resistance factor of a congenic line was an allele at the locus coding for the antigen II, and they agreed, following Snell's proposal that the tumor resistance factors be designated histocompatibility (H) genes, to call this first identification histocompatibility-2 (later abbreviated to H_2 and finally to H-2). The corresponding complex in humans was discovered in 1958 and later designated HLA (human leukocyte antigen); MHC has been identified in the frog, the chicken, and fifteen mammalian species (Klein 1986; Shreffler 1988).

11. In the 1960s observations were made in outbred mice (by Hugh McDevitt and Michael Sela) and guinea pigs (by Baruj Benacerref) that indicated marked quantitative variability with respect to their ability to develop humoral and cellular immune responses to certain antigens (McDevitt and Sela 1965; Benacerraf, Green, and Paul 1967). Subsequent studies of inbred animals and synthetic antigens showed that these responses were under the genetic control of a single, autosomal dominant gene located within the MHC, the so-called immune response (Ir) genes (McDevitt and Benacerraf 1969; Bluestein, Green, and Benacerraf 1971; McDevitt et al. 1972). In fact, a large number of antigens are under Ir control that exhibit T cell dependency and are required for development of cellular immunity, as well as antibody production.

12. The MHC class I alpha and beta polypeptides associate and appear on all nucleated cells, and the MHC class II alpha and beta proteins (similarly associate as heterodimers) are expressed in high quantity on B lymphocytes and activated or virus-infected T lymphocytes, and are inducible by interferon in APC (for example, macrophage and dendritic cells).

CHAPTER 3

1. For the substantiation of the thesis argued here, I am much indebted to my research assistant, Scott Podolsky, for his erudite examination of Burnet's 1925–55 writings.

2. Tangential to my concerns but perhaps important for the history of phage research, this quote shows that Burnet's supposedly prophetic view of the phage expressed between 1929 and 1932 was not retained permanently, and hence its claim to prophecy must be debated. Sexton (1991, p. 70) erroneously attributes this position to Burnet as a permanent viewpoint.

3. Note that Burnet's notion that "by some modification perhaps of its surface, the bacterium on which the amoeba . . . feeds becomes resistant to the digesting effect" (Burnet 1940, pp. 30–1) is from his actual benchwork and selective theorizing.

4. *Syngeneic* refers to genetically identical individuals (who have natural tolerance for each other's cells), *xenogeneic* refers to grafting across species (the classical Greek monster!), and *autogeneic* is grafting within the same individual (as might occur in skin transplantation for burn victims).

5. Many of the virus papers of this time were being incorporated into such a framework; viruses were changing the enzymes of human red cells, and such changes were rendered into analogues for modification of cells. See Burnet, McCrea, and Stone (1946) and Burnet and Fenner (1949, pp. 96–7) for an explicit delineation of this analogue.

6. "A well-defined mechanism has been evolved to allow a differentiation of the response of macrophage and similar cells to damaged or effete body constituents on the one hand, and to foreign organic matter (always microorganismal under natural conditions) on the other. . . . It is therefore necessary to endow each scavenging or antibody-producing cell with a capacity to 'recognise' as foreign any potential antigenic determinants. . . . Unless we are prepared to admit an extraordinarily complex 'code' by which such recognition of foreign structure is made, we are driven to the conclusion that all body cells carry some or all of a relatively small number of marker components [in *Antibodies* five to ten markers are attributed to the red blood cell (Burnet and Fenner 1949, p. 100)] whose specific character is determined by a correspondingly small number of genes" (Burnet and Fenner 1948, pp. 317–18).

7. That Burnet was not at ease with molecular biology was attested by Joshua Lederberg, who worked with Burnet in Australia when clonal selection theory was being formulated a decade later. Lederberg observed: "Our discussion became intense, although somewhat clouded by Burnet's tendency to resist the 'simplistic' mechanisms of DNA-based molecular genetics that are today's foundation stone. I would receive his exciting ideas, and then have to translate them into a contemporary idiom to get the full benefit of his marvelous biological intuition" (Lederberg 1988).

303

8. Another version offered by Burnet written earlier: "It gradually dawned on me that Jerne's selection theory would make real sense if cells produced a characteristic pattern of globulin for genetic reasons and were stimulated to proliferate by contact with the corresponding antigenic determinant. This would demand a receptor on the cell with the same pattern as antibody and a signal resulting from contact of antigenic determinant and receptor that would initiate mitosis or other cellular reaction. Once that central concept was clear, the other implications followed more or less automatically and were published in September, 1957. Just before that was written Talmage's (1957) review was seen, in which a brief suggestion of the same type was made. His subsequent development of the idea moved a long way away from mine. The next step was taken when Lederberg was a guest in my laboratory in November–December 1957. He had come to work on influenza virus genetics but he found more interest in the newly born Clonal Selection Theory – the name, by the way, dates from the first preliminary paper. It was in fact the first significant attempt to bring genetic ideas into immunology and we were both excited about the possibilities" (Burnet 1967).

9. The intellectual and scientific polemics surrounding Thorndike's theory are volatile and complex, but they are well summarized by Dennett. Dennett's own bias is reflected by his wry closing comment: "Finally, I cannot resist passing on a wonderful bit of incidental intelligence reported by Hadamard [Jacques Hadamard, *The Psychology of Inventing in the Mathematical Field,* Princeton, 1949]: The Latin verb *cogito* is derived, as St. Augustine tell[s] us, from Latin words meaning *to shake together,* while the verb *intelligo* means *to select among.* The Romans, it seems, knew what they were talking about" (1981, p. 89).

10. Although I have focused on the conceptual issues, a comprehensive history of post–World War II immunology would have to account for important technological and methodological advances of this period. See, for example, Rasmussen (1993).

CHAPTER 4

1. There is evidence that persistent antigen stimulation is required for memory, and models have been proposed to account for such findings (Antia, Levin, and Williamson 1991.)

2. "In a network arrangement, the idiotype of antibody 1 can bind to the variable region of the immunoglobulin on the surface of B-cell 2, thereby stimulating the cell to secrete antibody 2. In this way, an idiotype produced during the immune response to a particular antigen can stimulate the production of a corresponding antiidiotype. Antibody 2, a kind of autoantibody, is called an autoantiidiotype. Antibody 3 (an anti-antiidiotype) has also been found during the immune response, but there

304

the circuit usually ends because of the resemblance of antibody 3 to anti-body 1. In a self-contained immune network, the designation 'idiotype' or 'antiidiotype' is arbitrary; any idiotype is just as likely to be an antiidio-type" (Burdette and Schwartz 1987, p. 220).

3. These figures were obtained from a Medline search of immunology articles written in English between 1966 and 1992, containing *self* in either title or abstract. Admittedly this is neither a comprehensive nor a well-defined sur-vey, but at least by gross parameters the word usage is approximated as showing rapid growth in this period.

4. The issues concerning the changing views of organism and an expanded, biological understanding of immunology during the 1930s are analyzed by Löwy (1991).

5. Notwithstanding Flew's tirade, in the same year, Gilbert Ryle published his widely read and influential *The Concept of Mind* (1949), where the elusive "I" is discussed in several logical and body/mind contexts. One section is simply titled "The Self," and he employs the term liberally. *Self* is even used by the logical positivist A. J. Ayer in *Language, Truth and Logic* (1936) for a chapter title, "The Self and the Common World." As a freestanding noun, *self*, as used by Jones, was commonly employed by William James in *The Principles of Psychology* (1890) (discussed in detail in Chapter 6) and is widely found in nineteenth-century philosophical and psychological works (*Oxford English Dictionary* 1971, p. 2715 [original edition, pp. 410–11]).

6. In *Alcibiades* Socrates explains that to make or mend a shoe, one must understand its nature and purpose; similarly we cannot self-improve unless we first understand what we ourselves are. Therefore, our first duty is to obey the Delphic command, *"Gnothi se auton."* Knowledge depends on the distinction between the user and what he or she uses; for example, the shoe-maker is distinct from his tools. Generalizing, the body is but the instru-ment of a man to fulfill his purposes. Man therefore cannot be the body, but essentially exists as psyche, which uses and controls the body (Jowett 1953.)

7. English and German philosophers have commented extensively on the difficulty of translating *self* or *Selbst* into French. It is interesting that even in English discourse, *self* was viewed as having its province in philosophy, not in everyday language, as late as 1949 (Flew 1949); the observation was made that *self* only occurred as a suffix to pronouns, for example, *myself* or *herself.* Given the broad use today of *self* in psychological and sociological discussions, this change in language over such a short period is all the more remarkable. Williams makes the interesting observation that the French would have to employ *"le même"* to approximate our use of the nonreflex-ive self (Williams 1989, p. 74). For related discussion, consult Toulmin (1977) and Ricoeur (1992, pp. 1–3).

8. Consciousness includes both memory and awareness, but memory presup-poses – and does not constitute – personal identity (Flew 1951), and this issue leads to various inconsistencies in Locke's argument, which we will not dwell upon. (See Penelhum 1967).

9. Entwined in Locke's epistemological definition, we find his legal foundation, for the individual so defined becomes the unit of government, divided between its freedom and the rights of the majority. *Self* becomes a "forensic term" to which the law is applicable. Locke's person by this criterion is the moral agent, reflecting his ideal of responsible agency. Individualism, or what MacPherson (1962) calls "possessive individualism," is thus celebrated and, moreover, assured as established by an epistemological system from which an independent ethical unity consistently arises. We are heirs of this seventeenth-century liberalism. The foundations of our political and public moral fiber define the free rational individual as the irreducible unit of the good society. The assumption of this position is that one becomes fully human to the extent that one is free from dependence on the wills of others, which further argues that interrelations must be voluntarily engaged and that the individual is essentially the proprietor of his or her own person and capacities, and the products of his or her labor, that is, property.

10. Hume was effectively influential in showing that memory alone cannot constitute personal identity; it is only one of many causal links between earlier and later psychological states of a person (see, e.g., Hirsch 1982; Unger 1990). And he is the primary author of a reductionist tradition in discussing personal identity as exemplified by the logical empiricists of our own century. A short but excellent critique of Hume's position is given by Noonan (1989, pp. 77–103), who regards Parfit (1984) as the direct (and most influential?) modern heir of the Humean argument. The topic is endlessly complex but well surveyed by Rorty (1976).

11. We might note in passing how antithetical this position is to that of Nietzsche, who demanded a confrontational, self-creative encounter in which the self is challenged and responds by adaptation or absorption (an argument discussed in Chapter 7). More important, Rousseau's self is a changing entity of untethered vicissitudes, with no control mechanisms. Again, it is unlike Nietzsche's self of overcoming, which has an ethos defining its evolving vector.

CHAPTER 5

1. Burnet acknowledged a long fascination with the possible application of mind functions to immunity: "Believe it or not, in 1923 I sketched out some ideas on how Semon's mnemic theory of memory, now forgotten, might be applied to antibody function. I showed them to my chief, C. H. Kellaway, who gave me John Hunter's advice to Jenner: 'Don't think. Do the experiments'" (Burnet 1972).

2. The issue of determinancy is posed by Eco as "i) a triadic model, where between A and B there is an unpredictable and potentially infinite series of Cs, and ii) a diadic model where A provokes B without any mediation.

C is a space of choice and of supposed indeterminacy, while the non-space between A and B is a space of blind necessity and of unavoidable determination. . . . Perhaps someday science will demonstrate that the C space is only a figment like ether, presupposed in order to fill up an 'empty' interval where deterministic phenomena – that escape our present knowledge – take place. But until that moment we have to deal with C spaces. In any case, we know that in the C space a phenomenon takes place which is semiotically detectable: communicative contexts" (Eco 1988).

3. In considering the best genetic marker of the self, it is of interest to note that MHC (along with immunoglobulin and the T-cell receptor) is evolutionarily part of the supergene family that includes recognition proteins crucial for cell–cell interactions in development, serving as primitive regulators of morphogenesis (Edelman 1987; Hunkapiller and Hood 1989). Whether MHC directly functions as a general plasma anchorage site of organogenesis-directing proteins (Ohno 1977) has been questioned by gene depletion studies (Koller et al. 1990; Zijlstra et al. 1990), but its evolutionary heritage suggests such a function, and the degree of participation in vertebrate developmental processes remains to be elucidated.

4. A likely culprit is cyclosporin, an immunosuppressive drug often used following bone marrow transplantation. Although a precise mechanism has not been established, at least in those cases linked to this treatment GVHD seems related to (1) elimination or alteration of clonal deletion mechanisms in the thymus, and (2) the failure to regenerate the lymphocyte autoregulatory system. The peripheral autoregulatory system is eliminated by total body irradiation (preparative procedure for transplantation), giving rise to a similar phenomenon. In each case, these elements appear to alter the delicate balance of autoregulatory and autoreactive lymphocytes, leading to autoaggression (Hess 1990). I have already considered the case that autoimmunity is natural, while autoimmune *disease* is the imbalance of regulatory components (Cohen 1989; Cohen and Young 1991).

CHAPTER 6

1. Edie (1987, p. 37) notes that the approach of James is essentially the one that Sartre later employed, although Sartre makes no reference to the earlier source. "The *me* being an object, it is evident that I shall never be able to say: *my* consciousness, that is, the consciousness of my *me*. . . . The ego is not the owner of consciousness; it is the object of consciousness. To be sure, we constitute spontaneously our states and actions as productions of the ego. But our states and actions are also objects. We never have a direct intuition of the spontaneity of an instantaneous consciousness as produced by the ego. . . ." (Sartre 1937, 1957, pp. 96–7).

2. James built his theory of consciousness on this foundation of visceral perception. He might well have cited the concept of *cénesthésia* that was ad-

vanced in 1871 by Moritz Schiff, whom James referred to as "indefatigable" (James 1890, 1983, p. 105), to account for the sensory information emanating from the body. More broadly it constituted all the visceral and extracorporeal sensations perceived by consciousness at any given moment. In Schiff's early view, consciousness of self was an unstable modality of *cénesthésie* and represented a specific case of general consciousness. The concept was soon expanded to include sensory-inspired thought: As a single sensation sets in motion a chain of sensations and sensory images, these would initiate a series of thoughts when joined with other primitive sensations. "Cénesthésia, thus viewed, no longer represents a simple segment (the visual or muscular segment, with its blurred and deaf character) of sensory experience, but the totality of the sensory and intellectual events occurring at a given moment: it is the sum, at each different instant, of the 'peripheral' sensations and the 'central sensations.' Having renounced distinguishing internal and external sensations, striving to prove the *constructed* character of the discrimination between the interior world and the exterior world, Schiff can only save the concept of cénesthésie by making it coincide, paradoxically, with the entire *psychic life*. In proposing this new definition, Schiff had as his goal, above all, to criticize the traditional philosophic conception which makes of the *self*, of the *consciousness*, a separate instance, an observer of the life of the body, receiver of sensory experience, guarantor of the unity of the individual. Our self, our consciousness are not distinct from sensory experience: they are identical to these. They are not assured any permanence. The *identity* which perceives one individual is but the result of the sensations at each instant that revive former sensation." (Starobinski 1977, p. 8). This view is close to James's position, which is discussed in this chapter. (I appreciate Roger Shattuck for alerting me to *cénesthésie* and to Anne Dubitzky's translation.)

3. "It [consciousness] is, in the language of later phenomenologists, a 'prereflexive' or 'pre-thematic' awareness *that* – which is the experienced condition of all reflexive awareness of *what*" (Edie 1987, p. 43). Edie again quotes Sartre as following the same line of reasoning: "But it must be remembered that all writers who have described the *Cogito* have dealt with it as a reflexive operation . . . a consciousness which takes consciousness as an object. . . . But the fact remains that we are in the presence of two consciousnesses, one of which is conscious *of* the other. . . . Thus the consciousness which says I *think* is precisely not the consciousness which thinks" (Sartre 1937, 1957, pp. 44–5).

4. This discussion of Husserl is largely based on discussions I have enjoyed with my colleague Erazim Kohák and on his treatment of the subject in Kohák (1978).

5. Wilshire (1968) has clearly documented how James developed his notions of intentionality, which were much in line with Husserl's later work. Brentano distinguished conscious processes as "having objects," but James went beyond him by emphasizing their active and selective properties, and de-

scribing how the "sense of the same" is "the keel of the mind" (James, 1890, 1983, p. 434).

6. In Husserl's later philosophy transcendental intentionality as a function of the ego becomes "productive" and "creative" – ascribed to Fink (Eugen 1933, 1970) by Ricoeur (1967, p. 27).

7. Husserl uses the nomenclature of *noema* and *noesis* (Husserl 1913a, 1982, pp. 210–27). Noema is experience that is meaningful or constituted as meaningful in the act; objects as noemata present themselves in the context of experience. Noema perform, within phenomenological brackets, a function analogous to that performed by the natural object from the natural standpoint (Kohák 1978, p. 127). Noesis is experience as meaning – giving, that is, constituting, its data as meaningful.

CHAPTER 7

1. *The Will to Power,* the notebooks published by Nietzsche's sister in 1904, are collectively referred to as the disputed *Nachlass.* It has been drawn upon in this discussion, for I agree with Warren (1988) that it may be employed when there is no contradiction with the ideas in Nietzsche's published works; the material quoted from *Will to Power* has been chosen on the basis that it particularly elaborates what may less clearly be discerned in the published corpus. For comprehensive cross-references to these sources, see Moles (1990), and for an interpretative argument, see Behler (1991, pp. 13–15).

2. *Apperception* is the perception of one's own consciousness, which is the mental state that refers to the external world. *Empirical apperception* is the ego's awareness of actual and changing states of consciousness, a function performed by the *empirical ego,* in contrast to the *pure ego,* which produced a *transcendental apperception* (also referred to as the transcendental unity of apperception). The transcendental apperception is that aspect of consciousness that endures as a unity throughout time and thus affixes the fundamental, unchanging unity of consciousness. This pure ego, or self, precedes (is transcendent to) the content of our perceptions and makes possible their experienced order and meaning. This structured unity consists of (1) the intuitions of time and space, which are modes by which we perceive and are not objects of our perception, and (2) the categories of understanding. Transcendental apperception is thus the necessary condition for synthesizing experience into meaning (Angeles 1981, p.16).

3. See Allison (1983, pp. 72–82) and Guyer (1992) for distinctions of judgment, analytic and synthetic, in Kant's apperception, and its role in attaining objectivity in the transcendental deduction of the categories. The categories as necessary conditions of self-consciousness itself ("apperception") are also employed, a fortiori, as conditions for the representations of any object. The risk that these categories are but necessary conditions

for objective knowledge as opposed to actually applying to these objects, endangers the project of knowing the world objectively. How Kant deals with this problem is well discussed by Guyer (1992).

4. A modern restatement of the self as a transcendental construct is offered by Gunter Stent (1975). It is a particularly interesting view, since it is written from the perspective of a leading molecular biologist who has surveyed the neurosciences with some dismay. "The meaning of 'self' is intuitively obvious. It is another Kantian transcendental concept, one which we bring a priori to man, just as we bring the concepts of space, time, and causality to nature. The concept of self can serve the student of man as long as he does not probe too deeply. However, when it comes to explaining the innermost workings of the mind – the deep structure of structuralism – then this attempt to increase the range of understanding raises, in Bohr's terms, 'questions as to the sufficiency of concepts and ideas incorporated in daily language'. Thus the image of man as a Russian doll, with the outer body encasing an incorporeal inner man, is evidently a presupposition hidden in the rational linguistic use of the term 'self', and the attempt to eliminate the inner man from the picture only denatures that intuitive concept beyond the point of psychological utility."

5. "What is known (*Erkenntnis*) is, for a German-speaker, only a variation on what is familiar ('das Bekannte'). In English, the relationship between knowledge and familiarity is entirely semantic, but in German it is phonological and morphological as well, suggesting a stronger and deeper affiliation. The verb *erkennen,* from which is formed the term German philosophers use for 'knowledge' (*Erkenntnis*), has in its everyday usage the sense of 'recognize'" (Koelb 1990, p. 159).

6. Karl Löwith demonstrates the genealogy of the Übermensch as originating in the philosophical circle of Max Stirner, despite the absence of Nietzsche's citations. "The superman was originally the God-man and Christ-man; after Feuerbach's anthropological use, it changes its meaning: in relationship to the universally human, the superman becomes on the one hand *in*human, and on the other *more* than merely human. M. Hess employed the words 'superman' (*Übermensch*) and 'brute' (*Unmensch*) in this sense, applying the former to Bauer and the latter to Stirner. Bauer's thesis, that in the Christian relation man worships 'brutality' as his essential nature, corresponds to Stirner's thesis that so long as Christ is the superman, man can not be an 'I.' Therefore, the surmounting of Christianity is identical with man's surmounting of himself. This connection between the *God-man* Christ, *man* of Christian anthropology, and the individual's personal 'I,' which is a 'brute' in contrast to the former, corresponds to Nietzsche's no less consistent connection between the death of God and man's surmounting himself for the goal of the superman, who vanquishes both God and nothingness. By recognizing the full significance for the humanity of man in the 'great event' of God's death, Nietzsche also saw that for the

individual man, possessed of a will, the death of God is a 'freedom for death'" (Löwith 1964, p. 187–8).

7. Heidegger, in the context of discussing Nietzsche's views of racial breeding, offhandedly remarked: "Nietzsche's thought of will to power was ontological rather than biological" (Heidegger 1961, 1979, vol. 3, p. 231). In this fashion, Heidegger would characterize Nietzsche's racial thought as "metaphysical rather than biological in meaning" (ibid.). Although the will is clearly philosophically ontological, its character is conceived as organic, and thus the two aspects (contra Heidegger) cannot be regarded as contrasting. This issue is obviously complex, convoluted, and controversial and leads us well beyond our project, but to address the implicit concern as to the political and moral implications of Nietzsche's philosophy in the context of his biologicism, see Tauber (1994a; 1995b).

8. This passage of *Zarathustra* may be read as a critique of Hegel, discussed later in this chapter. "Rather than making body (self) a posit of absolute spirit, Nietzsche says that the ego is a projection of body with the consequences that only development of the body, rather than development of the spirit, could improve human beings. . . . This is, of course, not a very interesting differential critique of Hegel, since it would apply to all idealisms and doesn't recognize any version of Hegelism that sees spirit as immanent to the world, requiring only elaboration to be discussed" (Ackermann 1990, p. 50).

9. Hegel's specific thinking about biology is found in the second part of his *Encyclopedia, The Philosophy of Nature,* vol. 3, Organics (1970). In the general evolutionary scheme adopted throughout the *Encyclopedia,* tracing the progression from the simple to the complex, Hegel did not understand *evolution* in the sense expounded by Darwin. He was not aware of the potential implications of important developments in the comparative anatomy (Cuvier, Bichet) and the embryology (Pander, von Baer) of the 1820s, and thus remained a firm believer in the fixity of species. (See interpretative notes by Petry in Hegel 1970, vol. 3, pp. 229ff, pp. 370ff.) He was very much influenced by the prevailing notion of his day that there existed stable qualitative differences, arranged in an orderly hierarchy involving ascending degrees of complexity. His system was an attempt to define such an order to include physics, biology, Man, and ultimately spirit. The dialectic originated in the quest to establish the relationship between these levels of complexities.

10. "Becoming is ultimate and it is more real than Being, which is nothing but an abstraction, a static snapshot of becoming artificially arrested by a fictitious, instantaneous cut made by our perception and thought, at least in its dynamic aspects" (Čapek 1984, p. 117).

11. Čapek assigns the simple first person to this fragment, which Kahn (1979) translates with considerably less poetic simplicity: "One cannot step twice into the same river, nor can one grasp any mortal substance in a stable

condition, but it scatters and again gathers; it forms and dissolves, and approaches and departs" (#51, p. 53.).

CHAPTER 8

1. Note that this assessment is opposite to that of Bernauer, who views Foucault's critical process as the basis, on its own, of an ethics; that is, ethics becomes the practice of intellectual freedom (Bernauer 1990). A similarly sympathetic reading is offered by Rajchman (1991). Scott's analysis is more sensitive to the underlying basis of why Foucault dismisses ethics: In the dissolution of the self, quite simply an ethics is not possible (Scott 1990). Foucault's clearest statement (albeit in an unfinished format) regarding ethics was given shortly before his death (Foucault 1983b; 1984b) and critiqued by Taylor (1986) and Rorty (1991). I deal with this matter in detail elsewhere (Tauber, 1994a).

2. "Capital is thus the *power of command* over labour and its products" (Marx 1844, 1963, p. 85). Marx's critique of labor is leveled at the alienation of the self from nature, of the self's world as separation between the subject and its object. The focus becomes the object of labor: "What is embodied in the product of his labour is no longer his own. The greater this product is, therefore, the more he is diminished" (Ibid., p. 122). Thus, the affinity between Foucault and Marx resides fundamentally in this alienation, where the self is lost to itself. The difference between them, of course, resides in Marx's proposed social solution and more profoundly in his self-willed optimism of seeking Man's freedom. Foucault allowed no such choice.

3. Yet there is no consensus of what "postmodernism" *is*. The term *postmodern* was probably used for the first time in 1934 to describe a reaction within modernism, by the Spanish author Frederico De Onis (Rose 1991, p. 171). The term has assumed many connotations as it has been applied to art, literature, film, architecture, political theory, philosophy, psychology, sociology, science, and religion, and therefore it is notoriously difficult to offer a coherent definition (see, e.g., Rose 1991; Jencks 1992).

4. "Who was Félix Fénéon? Ask the average well-educated person today, and the name would probably ring no bell. . . . Yet he was one of those cultural personages who leave indelible marks on their epochs without writing an important book or poem, or painting a significant picture" (Thaw 1989). Fénéon, a major art critic, devised the system of artists' contracts with galleries that still prevails today, and he was instrumental in establishing neo-impressionism as a significant art movement. Politically active in fin-de-siècle France as an anarchist, Fénéon's fascinating career has been recently documented by Halperin (1988).

References

Abrams, M. H. 1953. *The Mirror and the Lamp: Romantic Theory and the Critical Tradition.* Oxford: Oxford University Press.

Ackerman, R. J. 1990. *Nietzsche: A Frenzied Look.* Amherst: The University of Massachusetts Press.

Ada, G. L. 1989. The conception and birth of Burnet's clonal selection theory. In *Immunology 1930–1980: Essays on the History of Immunology,* edited by P. M. H. Mazumdar, pp. 33–40. Toronto: Wall and Thompson.

Ada, G. L., and Nossal, G. 1987. The clonal-selection theory. *Scientific American* 257:62–9.

Ader, R., Felten, D. L., and Cohen, N. (eds.). 1991. *Psychoneuroimmunology,* 2d ed. San Diego: Academic.

Alexander, J. 1931. Some intracellular aspects of life and disease. *Protoplasma* 14:296–306.

Allen, G. 1978. *Life Sciences in the Twentieth Century.* Cambridge University Press.

Allison, H. E. 1983. *Kant's Transcendental Idealism.* New Haven: Yale University Press.

Amrine, F., Zucker, F. J., and Wheeler, H. 1987. *Goethe and the Sciences: A Reappraisal.* Dordrechet: Reidel.

Angeles, P. A. 1981. *Dictionary of Philosophy.* New York: Harper Collins.

Antia, R., Levin, B., and Williamson, P. 1991. A quantitative model suggests immune memory involves the colocalization of B and Th cells. *Journal of Theoretical Biology* 153:371–84.

Arrhenius S. 1907. *Immunochemistry: The Application of the Principles of Physical Chemistry to the Study of Biological Antibodies.* New York: Macmillan. [*Immunochemie.* Leipzig: Akademische Verlagsgesellschaft, 1907.]

Aschoff L. 1924. Reticulo-endothelial system (Janeway Lecture, New York). In *Lectures on Pathology.* pp. 1–33. New York: Hoeber.

Atlan, H. 1989. Noise, complexity and meaning in cognitive systems. *Revue Internationale de Systemique* 3:237–49.

Atlan, H. 1994. Intentionality in nature. Against an all-encompassing evolu-

313

References

tionary paradigm: Evolutionary and cognitive processes are not instances of the same process. *Journal for the Theory of Social Behaviour* 24:67–87.

Atlan, H., and Cohen, I. R. 1989. *Theories of Immune Networks.* Berlin: Springer-Verlag.

Avrameas, S. 1986. Natural autoreactive B cells and autoantibodies: The "know thyself" of the immune system. *Annales de l'Institut Pasteur* 137D:150–6.

———. 1991. Natural antibodies: From "horror autotoxicus" to "gnothi seauton." *Immunology Today* 12:154–9.

Ayer, A. J. [1936] n.d. *Language, Truth and Logic.* Reprint, New York: Dover.

Babior, B. M., Kipnes, R. S., and Curnutte, J. T. 1973. Biological defense mechanisms: The production by leukocytes of superoxide, a potential bactericidal agent. *Journal of Clinical Investigation* 52:741–4.

Bail, O., and Tsuda, K. 1909. Versuche über bakteriolytische Immunkörper mit besonderer Berücksichtigung des normalen Rinderserums. *Zeitschrift für Immunitätsforschung* 1:546–612.

Bakhtin, M. M. 1986. *Speech Genes and Other Late Essays.* Translated and edited by V. W. McGree, C. Emerson, and M. Holquist. Austin: University of Texas Press.

Baldridge, C. W., and Gerald, R. W. 1933. The extra respiration of phagocytosis. *American Journal of Physiology* 103:235–6.

Barnes, J., ed. 1984. *The Complete Works of Aristotle.* Princeton: Princeton University Press.

Barzun, J. 1983. *A Stroll with William James.* Chicago: The University of Chicago Press.

Baumler, E. 1984. *Paul Ehrlich. Scientist for Life.* New York, Holmes and Meier.

Beck, C. J. 1965. *The Metaphysics of Descartes.* Oxford: Clarendon Press.

Beck, L. 1978. Towards a Meta-Critique of Pure Reason. In *Essays on Kant and Hume.* New Haven: Yale University Press.

Behler, E. 1991. *Confrontations: Derrida, Heidegger, Nietzsche.* Translated by S. Taubeneck. Stanford: Stanford University Press.

Behring, E. von. [1890] 1961. Untersuchvenger über das Zustandekommen der Diptherie-Immunität ber Theiren. *Deutsche medezinische Wochenschrift* 16:1145–8. English translation (Studies on the mechanism of immunity to diphtheria in animals) in *Milestones in Microbiology,* edited by T. D. Brock, pp. 141–4. Englewood Cliffs, NJ: Prentice-Hall.

Benacerraf, B., Green, I., and Paul, W. E. 1967. The immune response of guinea pigs to hapten-poly-L-lysine conjugates as an example of the genetic control of the recognition of antigenicity. *Cold Spring Harbor Symposium on Quantitative Biology* 32:569–75.

Benner, R., van Oudenaren, A., Bjroklund, M., Ivars, F., and Holmberg, D. 1982. "Background" immunoglobulin production measurement, biological significance and regulation. *Immunology Today* 3:243–9.

Benton, E. 1974. Vitalism in nineteenth-century scientific thought: A typology and reassessment. *Studies in History and Philosophy of Science* 5:17–48.

References

Bergson, H. [1907] 1911. *Creative Evolution*. Translated by A. Mitchell. New York: Holt.

1946. *The Creative Mind: An Introduction to Metaphysics*. New York: Philosophical Library.

Bernard, C. [1865] n.d. *An Introduction to the Study of Experimental Medicine*. Translated by H. C. Green. (Reissue of 1927 English edition, New York: Dover.)

Bernauer, J. W. 1990. *Michel Foucault's Force of Flight: Toward an Ethics for Thought*. Atlantic Highlands, NJ: Humanities.

Bernstein, F. 1924. Ergebnisse einer biostatistischen zusammenfassenden Betrachtung über die erblichen Blutstrukturen des Menschen. *Klinische Wochenschrift* 3:1495–7.

Bibel, D. J. 1988. *Milestones in Immunology: A Historical Exploration*. Madison: Science Tech.

Billingham, R. E., and Brent, T. 1957. Tolerance. A simple method for inducing tolerance of skin homografts in mice. *Transplantation Bulletin* 4:67–71.

Billingham, R. E., Brent, L., and Medawar, P. B. 1953. "Actively acquired tolerance" of foreign cells. *Nature* 172:603–6.

1954. Quantitative studies on tissue transplantation immunity: I. The survival times of skin homografts exchanged between members of different inbred strains of mice. *Proceedings of the Royal Society of London*, Series B 143:43–80.

1955. Acquired tolerance of skin homografts. *Annals of the New York Academy of Sciences* 59:409–16.

Binns, R. M., Feinstein, A., Gurner, B. W., and Coombs, R. R. A. 1972. Immunolgobulin determinants on lymphocytes of adult, neonatal and foetal pigs. *Nature* 239:114–16.

Birch, C. 1988. The postmodern challenge to biology. In *The Reenchantment of Science: Postmodern Proposals,* edited by D. R. Griffin, pp. 69–78. Albany: State University of New York Press.

Black, M. 1978. How metaphors work: A reply to Donald Davidson. In *On Metaphor,* edited by S. Sacks, pp. 181–92. Chicago: The University of Chicago Press.

1982. *Models and Metaphors*. Ithaca: Cornell University Press.

Blondel, E. 1991. *Nietzsche: The Body and Culture*. Translated by S. Hand. Stanford: Stanford University Press.

Bluestein, H. G.; Green, I.; and Benacerraf, B. 1971. Specific immune response genes of the guinea pig: II. Relationship between the poly-L-lysine gene and the genes controlling immune responsiveness to copolymers of L-glutamic acid and l-tyrosine in random-bred Hartley guinea pigs. *Journal of Experimental Medicine* 134:471–81.

Bona, C. A. 1987. *Regulatory Idiotypes*. New York: Wiley.

Bona, C. A., and Kohler, H., eds. 1983. *Immune Networks. Annals of the New York Academy Sciences,* vol. 418.

Bonner, J. 1963. Analogies in biology. *Synthese* 15:275–9.

References

Bordet, J. 1909. *Studies in Immunity.* Translated by F. P. Gay. London: Chapman and Hall.

Bordet, J., and Ciuca, M. 1921. Remarques sur l'historique des récherches concernmant la lyse microbienne transmissible. *Comptes Rendus Hebdomadaires des Seances de l'Académie des Sciences (Paris)* 1984:745–7.

Bowler, P. J. 1983. *The Eclipse of Darwinism.* Baltimore: Johns Hopkins University Press.

Breinl, F., and Haurowitz, F. 1930. Chemische Untersuchung des Präzipitates aus Hämoglobin und Anti-Hämoglobinserum und Bemerkungen über die Natur des Antikörper. *Zeitschrift für Physiologische Chemie* 192:45–57.

Brentano, F. 1874. *Psychologie von empirischen Standpunkt.* Leipzig: Duncker and Humblot.

Brittan, G. G. 1978. *Kant's Theory of Science.* Princeton: Princeton University Press.

Brown, A. J. 1902. Enzyme action. *Transactions of the Chemical Society* 8:373–88.

Brown P. K. 1902. A fatal case of acute primary infectious pharyngitis with extreme leukopenia. *American Medicine* 3:649–52.

Burdette, S., and Schwartz, R. S. 1987. Idiotypes and idiotypic networks. *New England Journal of Medicine* 317:219–24.

Burgio, C. R. 1990. The "biological ego": From Garrod's "chemical individuality" to Burnet's "Self." *Acta Biotheoretica* 38:143–59.

Burnet, F. M. 1925a. The nature of the acquired resistance to bacteriophage action. *Journal of Pathology and Bacteriology* 28:407–18.

1925b. The conditions governing the appearance of tâches vièrges in bacteriophage activity. *Journal of Pathology and Bacteriology* 28:419–25.

1925c. Hydrogen peroxide and bacterial growth. *Australian Journal of Experimental Biology and Medical Science* 2:65–76.

1929a. "Smooth–Rough" variation in bacteria in its relation to bacteriophage. *Journal of Pathology and Bacteriology* 32:15–42.

1929b. Bacteriophage in its clinical aspects. *The Medical Journal of Australia* 1:406–10.

1930. Bacteriophage activity and the antigenic structure of bacteria. *Journal of Pathology and Bacteriology* 33:647–64.

1932. Lysogenicity as a normal function of certain salmonella strains. *Journal of Pathology and Bacteriology* 35:851–63.

1933a. Recent work on the biological nature of bacteriophages. *Transactions of the Royal Society of Tropical Medicine and Hygiene* 26:409–16.

1933b. The classification of dysentery-coli bacteriophages: II. The serological classification of coli-dysentery phages. *Journal of Pathology and Bacteriology* 36:307–18.

1934. The bacteriophages. *Biological Reviews* 6:332–50.

1940. *Biological Aspects of Infectious Diseases.* Cambridge University Press.

1941. *The Production of Antibodies.* London and New York: Macmillan.

References

1948. The Edward Stirling Lectures: I. The basis of allergic disease. *Medical Journal of Australia* 1:29–35.

1956. *Enzyme Antigen and Virus: A Study of Macromolecular Pattern in Action.* Cambridge University Press.

1957. A modification of Jerne's theory of antibody production using the concept of clonal selection. *Australian Journal of Science* 20:67–9.

1959. *The Clonal Selection Theory of Acquired Immunity.* Cambridge University Press.

1962. *The Integrity of the Body: A Discussion of Modern Immunological Ideas.* Cambridge: Harvard University Press.

1965. The Darwinian approach to immunity. In *Molecular and Cellular Basis of Antibody Formation,* edited by J. Sterzl, pp. 17–20. New York: Academic.

1967. Impact on ideas of immunology. *Cold Spring Harbor Symposium on Quantitative Biology* 32:1–8.

1969a. *Cellular Immunology.* Victoria: Melbourne University Press and Cambridge University Press.

1969b. *Self and Not-Self: Cellular Immunology.* 1. Victoria: Melbourne University Press and Cambridge University Press.

1969c. *Changing Patterns: An Atypical Autobiography.* New York: America Elsevier.

1970. *Immunological Surveillance.* Sydney: Pergamon.

1972. Immunology as scholarly discipline. *Perspectives in Biology and Medicine* 16:1–10.

1984. Letter from Burnet to N. K. Jerne, October 18. Burnet papers (Melbourne). Cited in C. Sexton, *The Seeds of Time: The Life of Sir Macfarlane Burnet.* Oxford: Oxford University Press, 1991, p. 140.

Burnet, F. M., and Anderson, S. G. 1947. The "T" antigen of guinea-pig and human red cells. *Australian Journal of Experimental Biology and Medical Science* 25:213–17.

Burnet, F. M., and Fenner, F. 1948. Genetics and immunology. *Heredity* 2:289–324.

1949. *The Production of Antibodies,* 2d ed. London: Macmillan.

Burnet, F. M., Keogh, E. V., and Lush, D. 1937. The immunological reactions of the filtrable viruses. *Australian Journal of Experimental Biology and Medical Science* 15:231–368.

Burnet, F. M., McCrea, J. F., Stone, J. D. 1946. Modification of human red cells by virus action: I. The receptor gradient for virus action in human red cells. *British Journal of Experimental Pathology* 27:228–36.

Burnet, F M., and McKie, M. 1929. Observations on a permanently lysogenic strain of *B. Enteritidis Gaertner. Australian Journal of Experimental Biology and Medical Science* 6:277–84.

Bury, J. B. 1932. *The Idea of Progress: An Inquiry into Its Origin and Growth.* New York: Macmillan.

317

References

Buss, L. W. 1987. *The Evolution of Individuality.* Princeton: Princeton University Press.

Callebaut, W., and Pinxten, R. 1987. *Evolutionary Epistemology: A Multiparadigm Program.* Dordrecht: Reidel.

Canguilhem, G. 1963. The role of analogies and models in biological discovery. In *Scientific Change,* edited by A. C. Crombie, pp. 507–20. New York: Basic.

[1966] 1989. *The Normal and the Pathological.* Translated by Carolyn R. Fawcett. New York: ZONE.

Čapek, M. 1984. Hegel and the organic view of nature. In *Hegel and the Sciences,* edited by R. S. Cohen and M. W. Wartofsky, pp. 109–21. Dordrecht: Reidel.

Caron, J. A. 1988. "Biology" in the life sciences: A historiographical contribution. *History of Science* 26:223–68.

Cassirer, E. 1950. *The Problem of Knowledge.* New Haven: Yale University Press.

[1954] 1989. *The Question of Jean-Jacques Rousseau,* 2d ed. Edited and translated by P. Gay. New Haven: Yale University Press.

Celada, F. 1988. Does the human mind use a logic of signs developed by lymphocytes 10^8 years ago? In *The Semiotics of Cellular Communication in the Immune System.* NATO ASI Series, vol. H23, edited by E. E. Sercarz, F. Celada, N. A. Mitchison, and T. Tada, pp. 71–9. Berlin: Springer-Verlag.

Celada, F., and Seiden, P. E. 1992. A computer model of cellular interactions in the immune system. *Immunology Today* 13:56–62.

Changeux, J. -P. 1985. *Neuronal Man: The Biology of Mind.* New York: Pantheon.

Chernyak, L., and Tauber, A. I. 1990. The idea of immunity: Metchnikoff's metaphysics and science. *Journal of the History of Biology* 23:187–249.

1991. The dialectical Self: Immunology's contribution. In *Organism and the Origins of Self,* edited by A. I. Tauber, pp. 109–56. Dordrecht: Kluwer Academic.

1992. Concerning individuality. *Philosophy and Biology* 7:489–99.

Churchill, W. H., and Kurtz, S. R. 1988. *Transfusion Medicine.* Boston and Oxford: Blackwell Scientific.

Clayton M. 1992. *Leonardo da Vinci: The Anatomy of Man.* Houston: The Museum of Fine Arts and Boston: Little, Brown.

Clements, F. E. 1936. Nature and structure of the climax. *Journal of Ecology* 24:252–84.

Cohen, I. R. 1989. Natural id-anti-id networks and the immunological homunculus. In *Theories of Immune Networks,* edited by H. Atlan and I. R. Cohen, pp. 6–12. Berlin: Springer-Verlag.

1992a. The cognitive principle challenges clonal selection. *Immunology Today* 13:441–4.

1992b. The cognitive paradigm and the immunological homunculus. *Immunology Today* 13:490–4.

References

1994. Kadishman's tree, Escher's angels, and the immunological homunculus. In *Autoimmunity: Physiology and Disease*, edited by A. Coutinho and M. D. Kazatchkine, pp. 7–18, New York: Wiley-Liss.

Cohen, I. R., and Young, D. B. 1991. Autoimmunity, microbial immunity and the immunological homunculus. *Immunology Today* 12:105–10.

Cohen, R. S. 1970. Ernst Mach: Physics, perception and the philosophy of science. In *Ernst Mach: Physicist and Philosopher*, edited by R. S. Cohen and R. J. Seeger, pp. 126–64. Dordrecht: Reidel.

Cohn, M. 1966. The concept of functional idiotype network for immune regulation mocks all and comforts none. *Annales de l'Institut Pasteur/Immunologie (Paris)* 137:64–76.

Coleridge, S. T. [1817] 1983. *Biographica Literaria*. Princeton: Princeton University Press.

Coleman, W. 1977. *Biology in the Nineteenth Century: Problems of Form, Function and Transformation*. Cambridge University Press.

Corngold, S. 1985. The question of the self in Nietzsche during the axial period (1882–1888). In *Why Nietzsche Now?*, edited by D. O'Hara, pp. 55–98. Bloomington: Indiana University Press.

Corsi, P., and Weindling, P. J. 1985. Darwinism in German, France and Italy. In *The Darwinian Heritage*, edited by D. Kohn, pp. 683–729. Princeton: Princeton University Press.

Coutinho, A., Forni, L., Holmberg, D., Ivars, F., and Vaz, N. 1984. From an antigen-centered, clonal perspective of immune responses to an organism-centered, network perspective of autonomous activity in a self-referential immune system. *Immunological Reviews* 79:151–68.

Coutinho, A., and Stewart, J. 1991. A hundred years of immunology: Paradigms, paradoxes and perspectives. In *Immunology: Pasteur's Heritage*, edited by P. A. Cazenave and G. P. Talwar, pp. 175–99. New Delhi: Wiley Eastern.

Craddock, C. G. 1980. Defenses of the body: The initiators of defense, the ready reserves, and the scavengers. In *Blood, Pure and Eloquent: A Story of Discovery, of People, and of Ideas*, edited by M. M. Wintrobe, pp. 417–54. New York: McGraw-Hill.

Crawford, E. 1984. *The Beginnings of the Nobel Institution: The Science Prizes, 1901–1915*. Cambridge University Press.

Cunningham, A., and Jardine, N., eds. 1990. *Romanticism and the Sciences*. Cambridge University Press.

Dallmayr, F. R. 1984. Introduction to M. Theunissen, *The Other*, pp. ix–xxv. Cambridge: The MIT Press.

Davenport, H. W. 1897. *Doctor Dock: Teaching and Learning Medicine at the Turn of the Century*. New Brunswick: Rutgers University Press.

Davidson, D. 1978. What metaphors mean. In *On Metaphor*, edited by S. Sacks, pp. 29–45. Chicago: The University of Chicago Press.

Dawkins, R. 1976. *The Selfish Gene*. New York: Oxford University Press.

1986. *The Blind Watchmaker*. New York: Norton.

References

de Kruif, P. [1926] 1954. *The Microbe Hunters.* San Diego: Harcourt, Brace, Jovanovich.

Deleuze, G. 1983. *Nietzsche and Philosophy.* Translated by H. Tomilinson. New York: Columbia University Press.

Dennett, D. C. 1981. Why the Law of Effect will not go away. In *Brainstorms: Philosophical Essays on Mind and Psychology,* pp. 71–89. Cambridge: The MIT Press.

Desmond, A. and Moore, J. 1991. *Darwin.* New York: Warner Books.

d'Herelle, F. 1917. Sur un microbe invisible antagonistic des bacilles dysenteriques. *Comptes Rendus Hebdomadaires des Seances de l'Académie des Sciences (Paris)* 165:373–5.

——— [1921] 1922. *The Bacteriophage: Its Rôle in Immunity.* Translated by G. H. Smith. Baltimore: Williams and Wilkins.

——— 1926. *The Bacteriophage and Its Behavior.* Translated by G. H. Smith. Baltimore: Williams and Wilkins.

Diamond, L. K. 1980. A history of blood transfusion. In *Blood, Pure and Eloquent,* edited by M. M. Wintrobe, pp. 659–88. New York: McGraw-Hill.

Dowling, L. 1986. *Language and Decadence in the Victorian Fin de Siècle.* Princeton: Princeton University Press.

Dreyer, W. J., and Bennett, J. C. 1965. The molecular basis of antibody-formation: A paradox. *Proceedings of the National Academy of Sciences, (U.S.A.)* 54:864–8.

Dreyfus, H. L., ed. 1982. *Husserl, Intentionality and Cognitive Science.* Cambridge: The MIT Press.

Driesch, H. 1908. *The Science and Philosophy of the Organism: The Gifford Lectures.* London: Adam and Charles Black.

——— 1914. *The Problem of Individuality.* London: Macmillan.

Duckworth, D. H. 1987. History and basic properties of bacterial viruses. In *Phage Ecology,* edited by S. M. Goyal, C. P. Gerber, and G. Bitton, pp. 1–43. New York: Wiley.

Dutton, R. W., and Mishell, R. J. 1967. Cell populations and cell proliferation in the in vitro response of normal spleen to heterologous erythrocytes: Analysis by the hot pulse technique. *Journal of Experimental Medicine* 126:443–54.

Dwyer, J. M. 1988. *The Body at War: The Miracle of the Immune System.* New York: New American Library.

Eco, V. 1988. On semiotics and immunology. In *The Semiotics of Cellular Communication in the Immune System,* NATO ASI Series, vol. H23, edited by E. E. Sercarz, F. Celada, N. A. Mitchison, and T. Tada, pp. 3–15. Berlin: Springer-Verlag.

Edelman, G. M. 1987a. *Neural Darwinism: The Theory of Neuronal Group Selection.* New York: Basic.

——— 1987b. CAMs and Igs: Cell adhesion and the evolutionary origins of immunity. *Immunological Reviews* 100:11–45.

320

References

Edie, J. M. 1987. *William James and Phenomenology.* Bloomington: Indiana University Press.

Ehrlich, P. [1897] 1957. Die Werbemessung des Diphtherieheilserums und deren theoretische Grundlagen. *Klinische Jahrbuch* 6:299–326. English translation (The assay of the activity of diphtheria-curative serum and its theoretical basis) in *The Collected Papers of Paul Ehrlich,* vol. 2, compiled, translated, and edited by F. Himmelweit, M. Marquardt, and H. Dale, pp. 107–25. London: Pergamon.

——— [1908] 1960. Experimental researches on specific therapy: On immunity with special reference to the relationship between distribution and action of antigens. First and Second Harben Lectures. *The Harben Lectures for 1907 of the Royal Institute of Public Health.* London: Lewis. Reprinted in *The Collected Papers of Paul Ehrlich,* vol. 3, compiled, translated, and edited by F. Himmelweit, M. Marquardt, and H. Dale, pp. 106–17, 118–29. London: Pergamon.

Ehrlich, P., and Morgenroth, J. [1900] 1957a. Über Hämolysine III: Mitteilung. *Berliner Klinische Wochenschrift* 37:453–8, 681–7. English translation (On hemolysins: Third communication) in *The Collected Papers of Paul Ehrlich,* vol. 2, compiled, translated, and edited by F. Himmelweit, M. Marquadt, and H. Dale, pp. 205–12. London: Pergamon.

——— [1901] 1957b. Über Hämolysine V: Mitteilung. *Berliner Klinische Wochenschrift* 38:251–7. English translation (On Haemolysins: Fifth Communication) in *The Collected Papers of Paul Ehrlich,* vol. 2, compiled, translated, and edited by F. Himmelweit, M. Marquadt, and H. Dale, pp. 246–55. London: Pergamon.

Eisley, L. 1961. *Darwin's Century: Evolution and the Men Who Discovered It.* Garden City, NY: Anchor.

Ellenberger, H. F. 1970. *The Discovery of the Unconscious: The History and Evolution of Dynamic Psychiatry.* New York: Basic.

Ellman, R. 1987. *Oscar Wilde.* New York: Knopf.

Emery, W. D. 1909. *Immunity and Specific Therapy.* New York: Hoeber.

Ernst, H. C. 1903. *Modern Theories of Bacterial Immunity.* Boston: Journal of Medical Research.

Falk, R., and Sarkar, S. 1992. Harmony from discord. *Philosophy and Biology* 7:463–72.

Farber, M. 1967. *Phenomenology and Existence.* New York: Harper.

Farley, J. 1982. *Gametes and Spores: Ideas about Sexual Reproduction. 1750–1914.* Baltimore: The Johns Hopkins University Press.

Feuerbach, L. [1843] 1972. *Principles of the philosophy of the future.* In *The Fiery Brook: Selected Writings of Ludwig Feuerbach,* translated and edited by Z. Hanfi. Garden City: Doubleday.

Fink, E. [1933] 1970. Die phänomenologische Philosophie Edmund Husserl's in der gegenwärtigen Kritik. *Kantstudien* 38:31–83. English translation (The phenomenological philosophy of Edmund Husserl and contemporary criticism) in *The Phenomenology of Husserl,* translated and edited by R. O. Elveton, pp. 73–147. Chicago: Quadrangle Club.

Fischer, E. P., and Lipson, L. 1988. *Thinking About Science: Max Delbrück and the Origins of Molecular Biology.* New York: Norton.

Fleck, L. [1935] 1979. *Genesis and Development of a Scientific Fact.* Translated by F. Bradley and T. J. Trenn. Chicago: The University of Chicago Press.

Flew, A. G. N. 1949. Selves. *Mind* 58:355–8.

1951. Locke and the problem of personal identity. *Philosophy* 26:53–68.

Flexner, A. 1910. *Medical Education in the United States and Canada.* Carnegie Foundation for the Advancement of Teaching, Bulletin no. 4. New York: Carnegie Foundation for the Advancement of Teaching.

Flood, P. M. 1988. Modes of communication within the immune system: Action or reaction? In *The Semiotics of Cellular Communication in the Immune System,* NATO ASI Series, vol. H23, edited by E. E. Sercarz, F. Celada, N. A. Mitchison, and T. Tada, pp. 120–9. Berlin: Springer-Verlag.

Ford, W. L. 1980. The lymphocyte – its tranformation from a frustrating enigma to a model of cellular function. In *Blood, Pure and Eloquent,* edited by M. M. Wintrobe, pp. 457–508. New York: McGraw-Hill.

Foster, P. 1991. Directed mutation in *Escherichia coli:* Theory and mechanisms. In *Organism and the Origins of Self,* edited by A. I. Tauber, pp. 213–34. Dordrecht: Kluwer Academic.

Foucault, M. [1963] 1973. *The Birth of the Clinic: An Archaeology of Medical Perception.* New York: Random House.

[1966] 1970. *The Order of Things: An Archaeology of the Human Sciences.* New York: Vintage.

[1966] 1989. Introduction. In *The Normal and the Pathological,* by G. Canguilhem. New York: Zone Books, pp. 7–24.

[1967] 1986. Nietzsche, Freud, Marx. Translated by J. Anderson and G. Hentzi. *Critical Texts* 3:1–5.

[1971] 1984a. Nietzsche, genealogy, history. In *The Foucault Reader,* edited by P. Rabinow, pp. 76–100. New York: Pantheon.

[1975] 1977. *Discipline and Punish: The Birth of the Prison.* Translated by A. Sheridan. New York: Pantheon.

1983a. The subject and power. In *Michel Foucault: Beyond Structuralism and Hermeneutics,* 2 d. ed., edited by H. L. Dreyfus and P. Rabinow, pp. 208–26. Chicago: The University of Chicago Press.

1983b. On the genealogy of ethics: An overview of work in progress. In *Michel Foucault,* edited by H. L. Dreyfus and P. Rabinow, pp. 229–52. Chicago: The University of Chicago Press.

1984b. Politics and ethics: An interview. In *The Foucault Reader,* edited by R. Rabinow, pp, 373–80. New York: Pantheon.

Fouillée, A. 1901. Nietzsche and Darwinism. *International Monthly* 3:134–65.

Frank, P. [1941] 1961. *Modern Science and its Philosophy.* New York: Collier.

Fremont, D. H., Matsumura, M., Stura, E. A.; Peterson, P. A., and Wilson,

References

I. A. 1992. Crystal structures of two viral peptides in complex with murine MHC class I H-2K^b. *Science* 257:919–26.

Galaty, D. H. 1974. The philosophical basis of mid-nineteenth century German reductionism. *Journal of the History of Medicine and Allied Sciences* 29:295–316.

Gasking, E. B. 1967. *Investigations into Generation. 1651–1828.* Baltimore: The Johns Hopkins University Press.

Gibson, A. B. 1932. *The Philosophy of Descartes.* New York: Russell and Russell.

Gilbert, S. F. 1992. Cells in search of community: Critiques of Weismannism and selectable units in ontogeny. *Biology and Philosophy* 7:473–87.

Gill, J. H. 1991. *Merleau-Ponty and Metaphor.* Atlantic Highlands, NJ: Humanities.

Gilman, S. L. 1988. *Disease and Representation: Images of Illness from Madness to AIDS.* Ithaca: Cornell University Press.

Glenny, A. T., and Hopkins, B. E. 1923, 1924. Duration of passive immunity. *Journal of Hygiene (London)* 22:12–51; 208–21.

Gode-von Aesch, A. 1941. *Natural Science in German Romanticism.* New York: Columbia University Press.

Goethe, J. W. von. [1790] 1952. *On Morphology.* In *Goethe's Botanical Writings,* translated by B. Mueller. Honolulu: University Press of Hawaii. (Reprinted 1989, Oxbow Press, Woodbridge, CT.)

——— [1792] 1988. The experiment as mediator between object and subject. In *Goethe: Scientific Studies,* edited and translated by D. Miller, pp. 11–17. New York: Suhrkamp.

——— [1818] 1988. Doubt and resignation. In *Goethe: Scientific Studies,* edited and translated by D. Miller, pp. 33–4. New York: Suhrkamp.

Golub, E. S. 1988. Semiosis for the immune system but not the immune response, or What can be learned about language by studying the immune system? In *The Semiotics of Cellular Communication in the Immune System,* NATO ASI Series, vol. H23, edited by E. E. Sercarz, F. Celada, N. A. Mitchison, and T. Tada, pp 65–9. Berlin: Springer-Verlag.

Golub, E. S., and Green, D. R. 1991. *Immunology: A Synthesis.* 2d ed. Sunderland, MA: Sinauer.

Golumb, J. 1989. *Nietzsche's Enticing Psychology of Power.* Ames: Iowa State University Press.

Gorer, P. A. 1936a. Detection of hereditary antigenic difference in blood of mice by means of human group A serum. *Journal of Genetics* 32:17–31.

——— 1936b. Detection of antigenic differences in mouse erythrocytes by employment of immune sera. *British Journal of Experimental Pathology* 17:42–50.

——— 1937a. Further studies on antigenic differences in mouse erythrocytes. *British Journal of Experimental Pathology* 18:31–6.

——— 1937b. The genetic and antigenic basis of tumor transplantation. *Journal of Pathology and Bacteriology* 44:691–7.

References

1938. Antigenic basis of tumor transplantation. *Journal of Pathology and Bacteriology.* 47:231–52.

Gorer, P., Lyman, S., and Snell, G. D. 1948. Studies on the genetic and antigenetic basis of tumour transplantation: Linkage between a histocompatibility gene and "fused" in mice. *Proceeding of the Royal Society of London, Series B,* 135:499–505.

Gosselin, E. J., Wardwell, K., Rigby, W. F. C., and Guyre, P. M. 1993. Induction of MHC class II on human polymorphonuclear neutrophils by granulocyte/macrophage colony-stimulating factor, IFN-gamma and IL-3. *Journal of Immunology* 151:1482–90.

Gotesky, R. 1967. Rudolf Hermann Lotze. In *The Encyclopedia of Philosophy,* vol. 5, edited by P. Edwards, pp. 87–9. New York: Macmillan.

Grafe, E. 1911. Die Steigerung des Stoffwechsels bei chronischer Leukamie und ihre Ursachen. *Deutsches Archiv Klinische Medizin* 102:406–30.

Grebe, S. C., and Streilein, J. W. 1976. Graft-versus-host reactions: A review. *Advances in Immunology* 22:119–221.

Gregory, F. 1977. *Scientific Materialism in Nineteenth Century Germany.* Dordrecht: Reidel.

Griswold, C. L. 1986. *Self-knowledge in Plato's Phaedrus.* New Haven: Yale University Press.

Grover, J. Z. 1988. AIDS: Keywords. In *AIDS: Cultural Analysis, Cultural Activism,* edited by D. Crimp, pp. 17–30. Cambridge: The MIT Press.

Grünbaum, A. S. F. 1903. Theories of immunity and their clinical application. *Lancet* 1:775–9, 853–6, 943–7.

Guerrini, A. 1985. James Keill, George Cheyne, and Newtonian Physiology, 1690–1740. *Journal of the History of Biology* 18:247–66.

Guo, H. -C., Jardetsky, T. S., Garrett, T. P. J., Lane, W. S., Strominger, J. L., and Wiley, D. C. 1992. Different length peptides bind to HLA-Aw68 similarly at their ends but bulge out in the middle. *Nature* 360:364–6.

Guthrie, W. K. 1971. *Socrates.* Cambridge University Press.

Gutting, G. 1989. *Michel Foucault's Archaeology of Scientific Reason.* Cambridge University Press.

Guyer, P. 1992. The transcendental deduction of the categories. In *The Cambridge Companion to Kant,* edited by P. Guyer, pp. 123–60. Cambridge University Press.

Hahlweg, K., and Hooker, C. A. 1989. *Issues in Evolutionary Epistemology.* Albany: State University of New York Press.

Haldane, E., and Ross, G. R. T., trans. and eds. 1911. *The Philosophical Works of Descartes.* Vol. 1. Cambridge University Press.

Haldane, J. B. S. 1937. Biochemistry of the individual, In *Perspectives in Biochemistry,* edited by J. Needham and D. E. Green, pp. 1–10. Cambridge University Press.

1942. *New Paths in Genetics.* London: Allen and Unwin.

Halperin, J. U. 1988. *Félix Fénéon: Aesthete and Anarchist in Fin-de-Siècle Paris.* New Haven: Yale University Press.

References

Hamlyn, D. W. 1980. *Schopenhauer.* London: Routledge and Kegan Paul.

Hankins, T. L. 1985. *Science and the Enlightenment.* Cambridge University Press.

Haraway, D. 1989. The biopolitics of postmodern bodies: Determinations of self in immune system discourse. *Differences* 1:3–43.

Hare, P. H., and Chakrabarti, C. 1980. The development of William James's epistemological realism. In *History, Religion and Spiritual Democracy,* edited by J. A. Martin, Jr., pp. 231–45. New York: Columbia University Press.

Harms, E. 1962. Introduction: On the genesis of the psychology of the self. In *Fundamentals of Psychology: The Psychology of the Self,* edited by E. Harms. *Annals of the New York Academy of Sciences* 96:683–6.

Harvey, D. 1989. *The Contributions of Postmodernity.* Cambridge: Basil Blackwell.

Haurowitz, F. 1953. Protein synthesis and immunochemistry. In *Essays on the Use of Information Theory in Biology,* edited by H. Quastler, pp. 125–46. Urbana: University of Illinois Press.

Hayman, R. 1980. *Nietzsche: A Critical Life.* New York: Oxford University Press.

Hebert, J., Bernier, D., Boutin, Y., Jobin, M., and Mourad, W. 1990. Generation of anti-idiotypic and anti-anti-idiotypic monoclonal antibodies in the same fusion. *Journal of Immunology* 144:4256–61.

Hegel, G. W. F. 1970. *Philosophy of Nature.* Vol. 3. Translated and edited by M. J. Petry. New York: Humanities Press.

Heidegger, M. [1927] 1962. *Being and Time.* Translated by J. Macquarrie and E. Robinson. New York: Harper and Row.

——— [1961] 1979. *Nietzsche.* 4 vols. Translated by D. F. Krell. San Francisco: Harper.

Heidelberger, M. 1932–3. Contributions of chemistry to the knowledge of immune processes. *The Harvey Lectures,* pp. 184–201. Baltimore: Wilkins and Wilkins.

Heims, S. J. 1980. *John von Neumann and Norbert Weiner: From Mathematics to the Technologies of Life and Death.* Cambridge: MIT Press.

Henderson, L. J. 1928. *Blood: A Study in General Physiology.* New Haven: Yale University Press.

Henri V. 1903. *Lois Générales de l'Action des Diastases.* Paris: Hermann.

Hess, A. D. 1990. Syngeneic graft-vs.-host disease. In *Graft vs. Host Disease: Immunology, Pathophysiology and Treatment,* edited by S. J. Burakoff, H. J. Deeg, J. Ferrara, and K. Atkinson, pp. 95–107. New York: Dekker.

Hesse, M. 1966. *Models and Analogies in Science.* Notre Dame: University of Notre Dame Press.

Hirsch, E. 1982. *The Concept of Identity.* New York: Oxford University Press.

Hiss, P. H. 1908–9. Some problems in immunity and the treatment of infectious diseases. *The Harvey Lectures,* pp. 240–79. Philadelphia: Lippincott.

Hoffman, G. W., Levy, J. G., and Nepom, G. T., eds. 1986. *Paradoxes in Immunology.* Boca Raton: CRC.

References

Hoffmann, R. R. 1985. Some implications of metaphor for philosophy and psychology of science. In *The Ubiquity of Metaphor,* edited by W. Paprotte and R. Dirven, pp. 327–80. Amsterdam/Philadelphia: Benjamins.

Holmes, F. L. 1974. *Claude Bernard and Animal Chemistry.* Cambridge: Harvard University Press.

Hozumi, N., and Tonegawa, S. 1976. Evidence for somatic rearrangement of immunoglobulin genes coding for variable and constant regions. *Proceedings of the National Academy of Sciences (U.S.A.)* 73:3628–32.

Hull, D. L. 1988. *Science as a Process.* Chicago: The University of Chicago Press.

Hume, D. [1739] 1962. *A Treatise of Human Understanding.* Book 1, *Of the Understanding.* Edited by D. G. C. Macnabb. Cleveland: World.

Hunkapiller, T., and Hood, L. 1989. Diversity of the immunoglobulin gene superfamily. *Advances in Immunology* 44:1–63.

Husserl, E. [1913a] 1982. *Ideas Pertaining to a Pure Phenomenology and to a Phenomenological Philosophy: First book.* Translated by F. Kersten. Dordrecht: Kluwer Academic.

——— [1913b] 1950. Ideen zu einer reinen Phänomenologie und Phänomenologischen Philosophie, vol. 3, Book 1. Haag: Nijhoff.

——— [1948] 1973. *Experience and Judgement.* Translated by J. S. Churchill and K. Ameriks and edited by L. Landgrebe. Evanston: Northwestern University Press.

——— [1954] 1970. *The Crisis of European Sciences and Transcendental Phenomenology.* Evanston: Northwestern University Press.

Hynes, N. 1991. Cover design. *Immunology Today.* 12, no. 6.

Iyer, G. Y. N., Islam, M. F., and Quastel, J. H. 1961. Biochemical aspects of phagocytosis. *Nature* 192:535–41.

James, W. [1890] 1983. *The Principles of Psychology.* Cambridge: Harvard University Press.

——— [1897] 1979. *The Will to Believe and Other Essays in Popular Philosophy.* Cambridge: Harvard University Press.

——— [1902] 1987. *The Varieties of Religious Experience.* In *William James: Writings 1902–1910,* pp. 3–477. New York: Library of America.

——— [1904] 1987. Does consciousness exist? In *William James: Writings 1902–1910,* pp. 1203–14. New York: Library of America.

——— [1911] 1987. *Some Problems of Philosophy.* In *William James: Writings 1902–1910,* pp. 979–1106. New York: Library of America.

Jameson, F. 1972. *The Prison-House of Language: A Critical Account of Structuralism and Russian Formalism.* Princeton: Princeton University Press.

Janaway, C. 1989. *Self and World in Schopenhauer's Philosophy.* Oxford: Clarendon Press.

Jantsch, E. 1980. *The Self-organizing Universe.* Oxford: Pergamon.

Jaret, P. 1986. Our immune system: The wars within. *National Geographic* 169:702–35.

References

Jennings, H. S. [1906] 1976. *The Behavior of the Lower Organisms.* Blooming-ton: Indiana University Press.

Jensen, C. O. 1903. Experimentelle Untersuchungen über Krebs bei Mausen. Centralblatt für Bakteriologie Parasitenkunde und Infektionskrankheiten 34:28–122.

Jerne, N. K. 1955. The natural selection theory of antibody formation. *Proceedings of the National Academy of Sciences (U.S.A.)* 41:849–57.

—— 1956. The presence in normal serum of specific antibody against bacterio-phage T_4 and its increase during the earliest stages of immunization. *Journal of Immunology* 76:209–16.

—— 1960. Immunological speculations. *Annual Review of Microbiology.* 14:341–58.

—— 1966a. The natural selection theory of antibody formation; ten years later. In *Phage and the Origins of Molecular Biology,* edited by J. Cairns, G. S. Stent, and J. D. Watson, pp. 301–12. Cold Spring Harbor: Cold Spring Harbor Laboratory.

—— 1966b. Antibody formation and immunological memory. In *Macromolecules and Behavior,* edited by J. Gaito, pp. 151–7. New York: Appleton Century Crafts.

—— 1967a. Summary: Waiting for the End. *Cold Spring Harbor Symposium on Quantitative Biology* 32:591–603.

—— 1967b. Antibodies and learning: Selection versus instruction. In *Neurosciences,* edited by G. C. Quarton, T. Melnechuk, and F. O. Schmitt, pp. 200–205. New York: The Rockefeller University Press.

—— 1974. Towards a network theory of the immune system. *Annals de l'Institut Pasteur/Immunologie (Paris)* 125C:373–89.

—— 1984. Idiotypic networks and other preconceived ideas. *Immunological Reviews* 79:5–24.

—— 1985. The generative grammar of the immune system. *EMBO Journal* 4:847–52.

—— 1992. Letter to Edward Goldberg, February 23. Tufts University School of Medicine, Boston.

Jonas, H. 1982. *The Phenomenon of Life: Toward a Philosophical Biology.* Chicago: The University of Chicago Press.

Jones J. R. 1950. Selves. A reply to Mr. Flew. *Mind* 59:233–6.

Jowett B., trans. 1953. *The Dialogues of Plato,* 4th edition. Oxford: Clarendon Press.

Jung, C. G. [1934–9] 1988. *Nietzsche's Zarathustra.* 2 vol. Princeton: Princeton University Press.

Kahn, C. H. 1979. *The Art and Thought of Heraclitus.* Cambridge University Press.

Kalmus, H. 1950. A cybernetical aspect of genetics. *Journal of Heredity* 41:19–22.

Kant, I. [1781, 1787] 1966. *Critique of Pure Reason.* Translated by F. Muller. Garden City, NY: Anchor Books.

References

[1784] 1959. What is enlightenment? In *Foundations of the Metaphysics of Morals,* translated by C. W. Beck, pp. 85–92. New York: Bobbs-Merrill.

[1786] 1985. *Metaphysical Foundations of Natural Science.* Translated by P. Carus and revised by J. W. Ellington. Indianapolis. Hackett.

[1790] 1987. *Critique of Judgement.* Translated by W. S. Pluhar. Indianapolis: Hackett.

Karush, F. 1989. Metaphors in immunology. In *Immunology 1930–1980: Essays on the History of Immunology,* edited by P. M. H. Mazumdar, pp. 73–80. Toronto: Wall and Thompson.

Kauffman, S. A. 1993. *The Origins of Order: Self-organization and Selection in Evolution.* New York: Oxford University Press.

Kaufman, M.; Urbain, J.; and Thomas, R. 1985. Towards a logical analysis of the immune response. *Journal of Theoretical Biology* 114:527–61.

Kaufmann, W. 1974. *Nietzsche: Philosopher, Psychologist, Antichrist,* 4th ed. Princeton: Princeton University Press.

Kay, L. E. 1993. *The Molecular Vision of Life: Caltech, The Rockefeller Foundation and the Rise of the New Biology.* New York: Oxford University Press.

——. 1994. Who wrote the book of life? Information and the transformation of molecular biology, 1945–55. In *Experimentalsysteme in den Biologische-Medizinschen Wissenschaften: Objekt differenzen, Konjunkturen,* edited by M. Hagner and H.-J. Rheinberger. Berlin: Akademie Verlag.

Keating, P., and Ousman, A. 1991. The problem of natural antibodies, 1894–1905. *Journal of the History of Biology* 24:245–63.

Kelsoe, G., and Schulze, D. H. 1987. *Evolution and Vertebrate Immunity: The Antigen-Receptor and MHC Gene Families.* Austin: University of Texas Press.

Kent, T. 1991. Hermeneutics and genre: Bakhtin and the problem of communicative interaction. In *The Interpretive Turn: Philosophy, Science, Culture,* edited by D. R. Hiley, J. F. Bohman, and R. Shusterman, pp. 282–303. Ithaca: Cornell University Press.

Kern, S. 1983. *The Culture of Time and Space, 1880–1918.* Cambridge: Harvard University Press.

Kierkegaard, S. [1849] 1955. *The Sickness Unto Death.* Translated by W. Lowrie. Garden City, New York: Doubleday.

Kingsland, S. E. 1991. Defining ecology as a science. In *Foundations of Ecology: Classic Papers with Commentaries,* edited by L. A. Real and J. H. Brown, pp. 1–13. Chicago: Chicago University Press.

Kittay, F. F. 1987. *Metaphor: Its Cognitive Force and Linguistic Structure.* Oxford: Clarendon Press.

Klein, J. 1982. *Immunology: The Science of Self-Non-Self Discrimination.* New York: Wiley.

——. 1986. *Natural History of the Major Histocompatibility Complex.* New York: Wiley.

——. 1990. *Immunology.* Boston and Oxford: Blackwell Scientific.

References

Kockelmans, J. J. 1977. Husserl and Kant on the pure ego. In *Husserl, Expositions and Appraisals,* edited by F. Elliston and P. McCormick, pp. 269–85. Notre Dame: University of Notre Dame Press.

Kocks, C., and Rajewsky, K. 1989. Stable expression and somatic hypermutation of antibody V regions in B-cell developmental pathways. *Annual Review of Immunology* 7:537–9.

Koelb, C., ed. 1990. *Nietzsche as Postmodernist: Essays Pro and Con.* Albany: State University Press of New York.

Koestler, A., and Smythies, J. R. 1971. *Beyond Reductionism.* Boston: Beacon.

Kōgaku, A. 1991. The problem of the body in Nietzsche and *Dōgen.* In *Nietzsche and Asian Thought,* edited by G. Parkes, pp. 214–25. Chicago: The University of Chicago Press.

Kolakowski, L. 1985. *Bergson.* Oxford: Oxford University Press.

Koller, B. H., Marrack, P., Kappler, J. W., and Smithies, O. 1990. Normal development of mice deficient in beta-2, MHC class I proteins, and CD8$^+$ T Cells. *Science* 248:1227–30.

Koshland, D. E. 1990. The rational approach to the irrational. *Science* 250:189.

Kourilsky, P., and Claverie, J.-M. 1989. MHC-antigen interaction: What does the T cell receptor see? In *Advances in Immunology,* vol. 45, edited by F. J. Dixon, pp. 107–93. San Diego: Academic.

Kremer, R. L. 1990. *The Thermodynamics of Life and Experimental Physiology, 1770–1880.* New York: Garland.

Kunkel, H. G., Mannik, M., and Williams, R. C. 1963. Individual antigenic specificity of isolated antibodies. *Science* 140:1218–19.

Lakatos, I. 1978. *The Methodology of Scientific Research Programmes.* Cambridge University Press.

Lakoff, G., and Johnson, M. 1980. *Metaphors We Live By.* Chicago: The University of Chicago Press.

Landsteiner, K. 1900. Zur Kenntnis der antifermentativen, lytischen und agglutinierenden Wirkungen des Blutserums und der Lymphe. *Centralblatt für Bakteriologie* 27:357–62.

    1901. Über Agglutinationserscheinungen normalen menschlichen Blutes. *Wien Klinische Wochenschrift* 14:1132–4.

    1917. Über die Antigeneigenschaften von methyliertem Eiweiss: VII. Mitteilung über Antigene (The antigenic properties of methylated protein). *Zeitschrift für Immunitätsforschung* 26:122–33.

    [1933] 1945. *The Specificity of Serological Reaction,* 2d ed. Cambridge: Harvard University Press. (Reprinted 1962 by New York: Dover.)

Landsteiner, K., and Reich, M. 1905. Über Unterschiede zwischen normalen und durch Immunisierung entstandenen Stoffen des Blutserums. *Centralblatt für Bakteriologie* 39:712–17.

Lang, R. A., and Bishop, J. M. 1993. Macrophages are required for cell death and tissue remodeling in the developing mouse eye. *Cell* 74:453–62.

Langbaum, R. 1982. *The Mysteries of Identity: A Theme in Modern Literature.* Chicago: The University of Chicago Press.

References

Langman, R. E., and Cohn, M. 1986. The "complete" idiotype network is an absurd immune system. *Immunology Today* 7:100–1.

Leder, P. 1982. The genetics of antibody diversity. *Scientific American* 246(5):102–15.

Lederberg, J. 1959. Genes and antibodies. *Science* 129:1649–53.

1988. Ontogeny of the clonal selection theory of antibody formation. Reflections on Darwin and Ehrlich. *Annals of the New York Academy of Science* 546:175–82.

Lenoir, T. 1982. *The Strategy of Life: Teleology and Mechanism in Nineteenth Century German Biology*. Dordrecht: Reidel. (Reissued 1989 by Chicago: The University of Chicago Press.)

Levere, T. H. 1981. *Poetry Realized in Nature: Samuel Taylor Coleridge and Early Nineteenth-century Science*. Cambridge University Press.

Levin, D. M. 1985. *The Body's Recollection of Being: Phenomenological Psychology and the Deconstruction of Nihilism*. London: Routledge and Kegan Paul.

Levinas, E. [1963] 1973. *The Theory of Intuition in Husserl's Phenomenology*. Translated by A. Orianne. Evanston: Northwestern University Press.

Levins, R., and Lewontin, R. C. 1985. *The Dialectical Biologist*. Cambridge: Harvard University Press.

Lewontin, R. C., Rose, S., and Kamin, L. J. 1984. *Not in Our Genes: Biology, Ideology, and Human Nature*. New York: Pantheon.

Lister, J. 1891. Discussion of immunity: Section II. Bacteriology. *British Medical Journal* 2:378–80.

Litman, G. W., Amemiya, C. T., Hair, R. N., and Shamblott, M. J. 1990. Antibody and immunoglobulin diversity. *Bioscience* 40:751–7.

Little, C. C. 1914. A possible Mendelian explanation for a type of inheritance apparently non-Mendelian in nature. *Science* 40:904–6.

Little, C. C., and Johnson, B. W. 1922. The inheritance of susceptibility to implants of splenic tissue in mice: I. Japanese waltzing mice albinos, and their FI generation hybrids. *Proceedings of the Society of Experimental Biology and Medicine* 19:163–7.

Little, C. C., and Tyzzer, E. E. 1916. Further studies on inheritance of susceptibility to a transplantable tumor of Japanese waltzing mice. *Journal of Medical Research* 33:393–453.

Locke, J. [1689] 1939. An essay concerning human understanding. In *The English Philosophers from Bacon to Mill*, edited by E. A. Burtt, pp. 238–402. New York: The Modern Library.

Loeb, J. [1912] 1964. *The Mechanistic Conception of Life*, edited by D. Fleming. Cambridge: Harvard University Press.

Loeb, L. 1901. On transplantation of tumors. *Journal of Medical Research* 6:28–38.

1908. Über Entstehung einer Sarkoms nach Transplantation eines Adenocarcinoms eines japanischen Maus. *Zeitschrift für Krebsforschung Berlin* 7:80–110.

References

1937. The biological basis of individuality. *Science* 86:1–5.

1945. *The Biological Basis of Individuality.* Springfield: Thomas.

Lloyd, A. R., and Oppenheim, J. J. 1992. Poly's lament: the neglected role of the polymorphonuclear neutrophil in the afferent limb of the immune response. *Immunology Today* 13:169–72.

Longino, H. E. 1990. *Science as Social Knowledge.* Princeton: Princeton University Press.

Lovejoy, A. O. 1936. *The Great Chain of Being: A Study of the History of an Idea.* Cambridge: Harvard University Press.

Lowe, V. 1949. The influence of Bergson, James, and Alexander on Whitehead. *Journal of the History of Ideas* 10:267–96.

1990. *Alfred North Whitehead: The Man and His Work,* vol. 2, *1910–1947.* Baltimore: The Johns Hopkins University Press.

Löwith, K. 1964. *From Hegel to Nietzsche: The Revolution in Nineteenth Century Thought.* Translated by D. E. Green. New York: Columbia University Press.

Löwy, I. 1991. The immunological construction of the self. In *Organism and the Origins of Self,* edited by A. I. Tauber, pp. 43–75. Dordrecht: Kluwer Academic.

Lundkvist, I., Coutinho, A., Varela, F., and Holmberg, D. 1989. Evidence for a functional idiotypic network among natural antibodies in normal mice. *Proceedings of the National Academy of Sciences (U.S.A.)* 86:5074–8.

Lyotard, J.-F. [1980] 1991. *Phenomenology.* Translated by B. Beakley. Albany: State University of New York.

MacKay, A. L. 1975. A metaphor for molecular evolution. *Journal of Theoretical Biology* 54:399–401.

Mach, E. [1883] 1893. *Die Mechanik in iher Entwicklung historisch-kritisch dargestellte.* Translated by T. J. McCormack as *The Science of Mechanics.* Chicago: 1893. (Reissue 1960 by LaSalle: Open Court.)

[1886] 1914. *Beiträge zur Analyse der Empfindungen.* Jena: Fischer. (Enlarged as *Die Analyse der Empfindungen,* 5th ed., Jena: Fischer, 1906.) Translated by C. M. Williams and S. Waterlow as *The Analysis of Sensations.* Chicago.

Mackenzie, J. N. 1896. The production of the so-called "rose cold" by means of an artificial rose. *American Journal of Medical Science* 91:45–57.

MacPherson, C. B. 1962. *The Political Theory of Possessive Individualism: Hobbes to Locke.* Oxford: Clarendon Press.

Magee, B. 1983. *The Philosophy of Schopenhauer.* Oxford: Clarendon Press.

Manwaring, W. H. 1930a. Biochemical relativity. *Science* 72:23–7.

1930b. Renaissance of pre-Elrich immunology. *Journal of Immunology* 19:155–63.

Margulis, L. 1981. *Symbiosis in Cell Evolution.* San Francisco: Freeman.

Markus, H. R. 1991. Culture and Self: Implications for cognition, emotion and motivation. *Psychological Reviews* 98:244–53.

Marshall, M. S. 1959. The concept of immunity. *The Centennial Review* 3:95–113.

References

Martin, E. 1990. Toward an anthropology of immunology: The body as nation state. *Medical Anthropology Quarterly* 4:410–26.

1992. The end of the body? *American Ethnologist* 19:121–40.

Marx, K. [1844] 1963. *Economic and Philosophical Manuscripts.* Translated and edited by T. B. Bottomore. New York: McGraw-Hill.

Masaki, H., and Irimajiri, K. 1992. Generation of helper of T cells that recognize a cross-reactive idiotype through a network mechanism. *Microbiology and Immunology (Tokyo)* 36:279–95.

Masson, D. 1867. *Recent British Philosophy.* London.

Matsumara, M., Fremont, D. H., Peterson, P. A., and Wilson, I. A. 1992. Emerging principles for the recognition of peptide antigens by MHC class I molecules. *Science* 257:927–34.

Mayr, E. 1982. *The Growth of Biological Thought: Diversity, Evolution, and Inheritance.* Cambridge: Harvard University Press.

Mazumdar, P. M. H. 1972. Immunity in 1890. *Journal of the History of Medicine and Allied Sciences* 27:312–24.

1974. The antigen-antibody reaction and the physics and chemistry of life. *Bulletin of the History of Medicine* 48:1–21.

1976. Karl Landsteiner and the problem of species. Ph.D. diss. The Johns Hopkins University. Baltimore.

ed. 1989. *Immunology 1930–1980: Essays on the History of Immunology.* Tornoto: Wall and Thompson.

McAlister, L. 1982. *The Development of Franz Brentano's Ethics.* Amsterdam: Rodopi.

McClintock, B. 1987. *The Discovery and Characterization of Transposable Elements.* New York: Garland.

McDevitt, H. O., and Benacerraf, B. 1969. Genetic control of specific immune responses. *Advances in Immunology* 11:31–74.

McDevitt, H. O., Deak, B. D., Shreffler, D. L., Klein, J.; Stimpfling, J. H., and Snell, G. D. 1972. Genetic control of the immune response. Mapping of the Ir-1 locus. *Journal of Experimental Medicine* 135:1259–78.

McDevitt, H. O., and Sela, M. 1965. Genetic control of the antibody response: I. Demonstration of determinant-specific differences in response to synthetic polypeptide antigens in two strains of inbred mice. *Journal of Experimental Medicine* 122:517–31.

McFarland, J. D. 1970. *Kant's Concept of Teleology.* University of Edinburgh Press.

Medawar, P. B. 1944. The behavior and fate of skin autografts and skin homografts in rabbits. *Journal of Anatomy* 78:176–99.

Megill, A. 1985. *Prophets of Extremity: Nietzsche, Heidegger, Foucault, Derrida.* Berkeley: University of California Press.

Mendelsohn, E., Weingart, P., and Whitley, R., eds. 1977. *The Social Production of Scientific Knowledge.* Dordrecht: Reidel.

Merleau-Ponty, M. [1945] 1962. *Phenomenology of Perception.* Translated by C. Smith. London: Routledge.

References

1964. *The Primacy of Perception.* Evanston: Northwestern University Press.

Merz, J. T. [1896] 1965. *A History of European Thought in the Nineteenth Century.* 4 vols. New York: Dover.

Metal'nikov, S., and Chorine, V. 1926. Rôle des reflexes conditionels dans l'immunité. *Annales de l'Institut Pasteur* 40:893–900.

Metcalf, D. *The Molecular Control of Blood Cells.* Cambridge: Harvard University Press.

Metchnikoff E. [1892] 1893. *Lectures on the Comparative Pathology of Inflammation.* Translated by F. A. Starling and E. H. Starling. London: Kegan, Pauls Trench Trubner. (Reprinted 1968 by New York: Dover.)

1892. Études sur l'immunité, 5eme memoire: Immunité des lepins vaccines contre le microbe du Hogcholéra. *Annales de l'Institut Pasteur* 6:289–320.

[1901] 1905. *Immunity in Infective Diseases.* Translated by F. G. Binnie. Cambridge University Press.

1903. *The Nature of Man: Studies in Optimistic Philosophy.* Translated by P. C. Mitchell. New York: Putnam.

1906. *The New Hygiene: Three Lectures on the Prevention of Infectious Disease.* Translated by E. R. Lankester. London: Heinemann, and Chicago: Kenner.

1907. *The Prolongation of Life: Optimistic Studies.* Translated by P. C. Mitchell. London: Heinemann. (Also published New York: Putnam, 1910.)

1907. Nobel Lecture, Dec. 11, 1908. In *Nobel Lectures: Physiology or Medicine 1901–1921,* pp. 281–300. Nobelstifelsen, Amsterdam: Elsevier.

Metchnikoff, O. 1924. *Life of Elie Metchnikoff 1845–1916.* Translated by E. R. Lankester. London: Constable, and Boston: Houghton Mifflin.

Michaelis, L., and Menten, M. L. 1913. Zur Kinetik der Inventinwirkung. *Biochimische Zeitschrift* 49:333–69.

Mitchison, N. A. Passive transfer of transplantation immunity. *Proceedings of the Royal Society of London,* Series B, 142:72–87.

Mittenthal, J. E., and Baskin, A. B. 1992. *The Principles of Organization in Organisms.* Reading: Addison-Wesley.

Moles, A. 1990. *Nietzsche's Philosophy of Nature and Cosmology.* New York: Lang.

Monod, J. 1972. *Chance and Necessity.* Translated by A. Wainhouse. New York: Vintage.

Mörner K. A. H. [1908] 1967. Presentation speech, Nobel Prize for Physiology or Medicine. In *Nobel Lectures: Physiology or Medicine 1901–1921,* Nobelstiftelsen, pp. 277–80. Amsterdam: Elsevier.

Morrissey, R. J. 1988. Jean Starobinski and Otherness. In J. Starobinski, *Jean-Jacques Rousseau: Transparency and Obstruction,* pp. xiii-xxxviii. Chicago: The University of Chicago Press.

Mostert, P. 1970. Nietzsche's reception of Darwinism. *Bijdragen tot de Dierkunde* 49:235–46.

Moulin, A. M. 1989. Immunology old and new: The beginning and the end. In

References

Immunology 1930–1980: Essays on the History of Immunology, edited by P. M. H. Mazumdar, pp. 291–8. Toronto: Wall and Thompson.

1991a. *Le Dernier Langage de la Médicine: Histoire de l'immunologie de Pasteur au Sida.* Paris: Presses Universitaires de France.

1991b. Death and resurrection of immunology at the Pasteur Institute (1917–1940). In *Immunology: Pasteur's Heritage,* edited by P. A. Cazenave and G. P. Talwar, pp. 53–69. New Dehli: Wiley Eastern.

Mourant, A. E. 1971. Transduction and skeletal evolution. *Nature* 231:466–7.

Mudd, S., 1932. A hypothetical mechanism of antibody formation. *Journal of Immunology* 23:423–7.

Murphy, J. B. 1926. The lymphocyte in resistance to tissue grafting, malignant disease and tuberculosis infection. [1912–24]. *Monographs of the Rockefeller Institute of Medical Research* 21:1–168.

Myers, G. E. 1986. *William James: His Life and Thought.* New Haven: Yale University Press.

Nagel, T. 1986. *The View from Nowhere.* New York: Oxford University Press.

Navarre, M. 1988. Fighting the victim label. In *AIDS: Cultural Analysis, Cultural Activism,* edited by D. Crimp, pp. 143–6. Cambridge: The MIT Press.

Nehamas, A. 1985. *Nietzsche: Life as Literature.* Cambridge: Harvard University Press.

Nelson, D. S., ed. 1989. *Natural Immunity.* San. Diego: Academic Press.

Nietzsche, F. [1872] 1956. *The Birth of Tragedy.* Translated by F. Golffing. Garden City, NY: Doubleday Anchor.

[1874] 1983. Schopenhauer as educator. In *Untimely Meditations,* translated by R. J. Hollingdale, pp. 125–94. Cambridge University Press.

[1878] 1986. *Human, All Too Human.* Translated by R. J. Hollingdale. Cambridge University Press.

[1881] 1982. *Daybreak.* Translated by R. J. Hollingdale. Cambridge University Press.

[1882] 1974. *The Gay Science.* Translated by W. Kaufmann. New York: Vintage.

[1885] 1959. *Thus Spoke Zarathustra.* In *The Portable Nietzsche,* edited and translated by W. Kaufmann, pp. 112–439. New York: Penguin.

[1886] 1966. *Beyond Good and Evil.* Translated by W. Kaufmann. New York: Vintage.

[1887] 1967. *On the Genealogy of Morals.* Translated by W. Kaufmann and R. J. Hollingdale. New York: Vintage.

[1888a] 1959. *The Antichrist.* In *The Portable Nietzsche,* edited and translated by W. Kaufmann, pp. 568–656. New York: Penguin.

[1888b] 1959. *Twilight of the Idols.* In *The Portable Nietzsche,* edited and translated by W. Kaufmann, pp. 463–563. New York: Penguin.

[1904] 1967. *The Will to Power.* Translated by W. Kaufmann and R. J. Hollingdale. New York: Vintage.

References

Nilsson, L. 1987. *The Body Victorious: The Illustrated Story of Our Immune System and Other Defenses of the Human Body.* New York: Delacorte.

Nisbet, R. 1980. *History of the Idea of Progress.* New York: Basic.

Nisonoff, A. 1991. Idiotypes: Concepts and applications. *Journal of Immunology* 147:2429–38.

Nobel Archives. 1900–8. Nobel Archives related to the Nobel Prize for Physiology or Medicine. File concerning Elie Metchnikoff. Karolinska Institutet.

Noonan, H. 1989. *Personal Identity.* London and New York: Routledge.

Nossal, G. J. V. 1989. The coming of age of the clonal selection theory. In *Immunology 1930–1980: Essays on the History of Immunology,* edited by P. M. H. Mazumdar, pp. 41–48. Toronto: Wall and Thompson.

Nossal, G. J. V., and Lederberg, J. 1958. Antibody production by single cells. *Nature* 181:1419–20.

Nuttall, G. H. F. [1888] 1988. Experimente über die bacterienfeindliches Einflüsse des thierischen Körpers. *Z. Hyg.* 4:353–94. English translation (Experiments on the antibacterial influence of animal substances) in D. J. Bibel, *Milestones in Immunology,* pp. 161–66. Madison: Science Tech.

Ofek, I., and Sharon, N. 1988. Lectinophagocytosis: A molecular mechanism of recognition between cell surface sugars and lectins in the phagocytosis of bacteria. *Infection and Immunology* 56:539–47.

Ohno, S. 1977. The original function of MHC antigens as the general plasma membrane anchorage site of organogenesis-directing proteins. *Immunological Reviews* 33:59–69.

Ollman, B. 1976. *Alienation: Marx's Conception of Man in Capitalist Society,* 2d. ed. Cambridge University Press.

Onians, R. B. 1951. *The Origins of European Thought About the Body, the Mind, the Soul, the World, Time and Fate.* Cambridge University Press.

Orgony, A., ed. 1979. *Metaphor and Thought.* Cambridge University Press.

Oudin, J., and Michel, M. 1963. Une nouvelle forme d'allotypie des globulines gamma du serum de lapin apparemment liée à la fonction et à la spécificitie des anti-corps. *Comptes Rendues de l'Académie Sciences (Paris)* 257:805–8.

Owen, R. D. 1945. Immunogenetic consequences of vascular anastomoses between bovine twins. *Science* 102:400–1.

Oxford English Dictionary. 1971. Vol. 2. Oxford: Clarendon Press.

Parfit, D. 1984. *Reasons and Persons.* Oxford: Clarendon Press.

Parisi, G. 1990. A simple model for the immune network. *Proceedings of the National Academy of Sciences (U.S.A.)* 87:429–33.

Paternek, M. A. 1987. Norms and normalization: Michel Foucault's overextended panoptic machine. *Human Studies* 10:97–121.

Paton, H. J. 1951. *In Defense of Reason.* New York: Hutchinson's University Library.

Paton, R. C. 1992. Towards a metaphorical biology. *Biology and Philosophy* 7:279–94.

References

Pauling, L. 1940. A theory of the structure and process of formation of antibodies. *Journal of the American Chemical Society* 62:2643–57.

1970. Fifty years of progress in structural chemistry and molecular biology. *Daedalus* 99:909–10.

Pauling, L., and Campbell, D. H. 1942a. Manufacture of antibodies in vitro. *Journal of Experimental Medicine* 76:211–20.

1942b. The production of antibodies in vitro. *Science* 95:440–1.

Pauling, L., and Delbrück, M. 1940. The nature of the intermolecular forces operative in biological processes. *Science* 92:77–9.

Penelhum, T. 1967. Personal identity. In *The Encyclopedia of Philosophy*, vol. 6, edited by P. Edwards, pp. 95–107. New York: Macmillan.

Pepper, S. C. 1942. *World Hypotheses*. Berkeley: University of California Press.

Pereira, P., Bandeira, A., Coutinho, A., Marcos, M.-A., Toribio, M., and Martinez, A.-C. 1989. V-region connectivity in T cell repertoires. *Annual Review of Immunology* 7:209–49.

Perelson, A. S., ed. 1987. *Theoretical Immunology (Part Two)*, vol. 3. Reading, MA: Addison-Wesley.

1989. Immune network theory. *Immonogical Reviews* 110:5–36.

Petry, M. J. 1970. Introduction. In *Hegel's Philosophy of Nature*, vol. 1, edited by M. J. Petry, pp. 11–177. New York: Humanities.

Pettersson, A. 1905–6. Über die Bedeutung der Leukocyten dei der intraperitonealen Infektion des Meerschweinchens mit Typhusbacillen. *Centralblatt für Bakteriologie, Parasitenkunde und Infektionskrankheiten* Abt. 1, vol. 40.

Pfeiffer, R. F. J. [1894] 1988. Weitere Untersuchungen über das Wesen der Cholera immunität und über specifische bactericide process. *Zeitung für Hygiene* 18:1–16. English translation (Further investigation on the nature of immunity to cholera and specific bactericidal processes) in D. J. Bibel, *Milestones in Immunology*, pp. 197–200. Madison: Science Tech.

Phillips, G. N. Jr., Fillers, J. P., and Cohen, C. 1980. Motions of tropomyosin crystal as metaphor. *Biophysics Journal* 32:485–502.

Phillips, J. 1935. Succession, development, the climax and the complex organism: An analysis of concepts. Part III. *Journal of Ecology* 23:488–508.

Piaget, J. 1952. *The Origins of Intelligence in Children*. New York: Norton.

Piatelli-Palmerini, M. 1984. The rise of selective theories: A case study and some lessons from immunology. In *Language, Learnability and Concept Acquisition*, edited by A. Marras and W. Demopoulos, pp. 117–30. Norwood, NJ: Ablex.

1991a. Evolution, selection and cognition: From learning to parameter-fixation in biology and in the study of mind. In *Immunology: Pasteur's Heritage*, edited by P. A. Cazenave and G. P. Talwar, pp. 95–158. New Delhi: Wiley Eastern.

1991b. The rise of selective theories: A case study and some lessons from immunology. In *Immunology: Pasteur's Heritage*, edited by P. A. Cazenave and G. P. Talwar, pp. 159–72. New Dehli: Wiley Eastern.

References

Pick, E. P. 1912. *Handbuch der pathogenen Mikroorganismen,* vol. 1, 2d ed. Edited by W. Kolle and A. von Wasserman. Jena: Fischer.

Pippin, R. A. 1989. *Hegel's Idealism: The Satisfactions of Self-Consciousness.* Cambridge University Press.

1991. *Modernism as a Philosophical Problem.* Oxford and Cambridge: Basil Blackwell.

Pittendrigh, C. S. 1958. Adaptation, natural selection, and behavior. In *Behavior and Evolution,* edited by A. Roe and G. G. Simpson, pp. 390–416. New Haven: Yale University Press.

Pletsch, C. 1991. *Young Nietzsche: Becoming a Genius.* New York: The Free Press.

Playfair, J. H. L. 1984. *Immunology at a Glance,* 3rd edition. Oxford: Blackwell Scientific.

Plum, P. 1935. Agranulocytosis due to aminopyrine; experimental and clinical study of 7 new cases. *Lancet* 1:15–18.

Polanyi, M. 1969. Life's irreducible structure. In *Knowing and Being,* edited by M. Grene, pp. 225–39. Chicago: University of Chicago Press.

Potes, M. H. 1979. The functional analogy as a concealed metaphor: A critical examination of the use of biological analogy in biopolitics. Ph.D. diss., Kent State University.

Prodi, G. 1988. Signs and codes in immunology. In *The Semiotics of Cellular Communication in the Immune System,* NATO ASI series, vol. H23, edited by E. E. Sercarz, F. Celada, N. A. Mitchison and T. Tada, pp. 53–64. Berlin: Springer-Verlag.

Ptashne, M. 1992. *A Genetic Switch,* 2d ed. Cambridge: Cell Press and Blackwell Scientific.

Quastler, H. 1953. *Essays on the Use of Information Theory in Biology.* Urbana: University of Illinois Press.

Radnitzky, G., and Bartley, W. W. 1987. *Evolutionary Epistemology, Rationality, and the Sociology of Knowledge.* LaSalle, IL: Open Court.

Raff, M. C., Feldmann, M., and de Petris, S. 1973. Monospecificity of bone marow derived lymphocytes. *Journal of Experimental Medicine* 137:1024–30.

Rajchman, J. 1991. *Truth and Eros: Foucault, Lacan and the Question of Ethics.* New York: Routledge.

Randall, J. H. Jr. 1962. *The Career of Philosophy: From the Middle Ages to the Enlightenment.* New York: Columbia University Press.

Rappeport, J. M. 1990. Syngeneic and autologous graft vs. host disease. In *Graft vs. Host Disease: Immunology, Pathophysiology and Treatment,* edited by S. J. Burakoff, H. J. Deeg, J. Ferrara, and K. Atkinson, pp. 455–66. New York: Dekker.

Rasmussen, N. 1993. Freund's adjuvant and the realization of questions in postwar immunology. *Historical Studies in the Physical and Biological Sciences* 23:337–66.

References

Rather, L. J. 1972. *Addison and the White Corpuscles.* London: Wellcome Institute of the History of Medicine.

——— 1982. On the source and development of metaphorical language in the history of Western medicine. In *A Celebration of Medical History,* edited by L. G. Stevenson, pp. 135–53. Baltimore: The Johns Hopkins University Press.

Ricketts, H. T. 1908. *Infection, Immunity and Serum Therapy.* Chicago: American Medical Association Press.

Ricoeur, P. 1967. *Husserl: An Analysis of His Phenomenology.* Translated by E. G. Ballard and C. E. Embree. Evanston: Northwestern University Press.

——— 1992. *Oneself as Another.* Translated by K. Blamey. Chicago: University of Chicago Press.

Ritvo, L. B. 1990. *Darwin's Influence on Freud: A Tale of Two Sciences.* New Haven: Yale University Press.

Roland, D. 1988. *In Search of Self in India and Japan.* Princeton: Princeton University Press.

Rorty, A. O. 1976. *The Identities of Persons.* Berkeley: University of California Press.

Rorty, R. 1979. *Philosophy and the Mirror of Nature.* Princeton: Princeton University Press.

Rorty, R. 1991. Moral identity and private autonomy: The case of Foucault. In *Essays on Heidegger and Others,* p. 193–8. Cambridge University Press.

Rose, M. A. 1991. *The Post-modern and the Post-industrial: A Critical Analysis.* Cambridge University Press.

Rosen, R. 1971. Some realizations of metaphorical paradigms for cellular activity systems and their interpretation. *Bulletin of Mathematics and Biophysics* 33:303–19.

Rousseau, J.-J. [1750] 1987. Discourse on the moral effects of the arts and the sciences. In *The Basic Political Writings,* translated by D. A. Cress, pp. 1–21. Indianapolis: Hackett.

——— [1755] 1987. Discourse on the origins of inequality. In *The Basic Political Writings,* translated by D. A. Cress, pp. 25–109. Indianapolis: Hackett.

——— [1762a] 1911. *Émile, or on Education* Translated by B. Foxley. New York: Dutton.

——— [1762b] 1987. *On the Social Contract.* In *The Basic Political Writings,* translated by D. A. Cress, pp. 139–227. Indianapolis: Hackett.

Roux, W. 1881. *Der Kampf des Theile um Organismus.* (The Struggle of Parts in the Organism). Leipzig: Engelmann.

——— [1894] 1986. The problems, methods, and scope of developmental mechanics. Translated by W. M. Wheeler. In *Defining Biology: Lectures from the 1890's,* edited by J. Maienschein, pp. 107–48. Cambridge: Harvard University Press. (Translated introduction to vol. 1 of *Archiv für Entwicklungsmechanik der Organismen* [1894–1895].)

References

1895. *Die Selbstregulation: Gesammelte Abhandlungen über Entwicklungs-mechanik der Organismen*, vol. 2. Leipzig: Engelmann.

Rowley D. A. 1991. Paul R. Cannon. In *Remembering the University of Chicago, Teachers, Scientists, and Scholars*, edited by E. A. Shils, pp. 15–31. University of Chicago Press.

Rubin, W. 1989. Picasso and Braque: An introduction. In *Picasso and Braque: Pioneering Cubism*, pp. 15–62. New York: The Museum of Modern Art.

Ruse, M. 1988. *But Is It Science?* Buffalo: Prometheus Books.

Ruskin, J. 1901. *Modern Painters*. London: Dent.

Ryan, J. 1991. *The Vanishing Subject: Early Psychology and Literary Modernism*. Chicago: The University of Chicago Press.

Ryle, G. 1949. *The Concept of Mind*. Harmondsworth: Penguin.

Sacker, R. A., ed. 1992. IVIG [intravenous immunoglobulin]: Current role in bone marrow transplant, malignancy, and immune hematologic disorders. *Seminars in Hematology*, vol. 29, pp. 1–133.

Salk, J. 1975. *Metaphores Biologiques*. Paris: Calmann-Levy (Libertés de l'Espirit).

Sallis, E. 1981. *Merleau-Ponty: Perception, Structure, Language. A Collection of Essays*. Atlantic Highlands, NJ: Humanities.

Salmon, D. E., and Smith, T. 1884. On a new method of producing immunity from contagious disease. *Proceedings of the Biology Society of Washington* 3:29–33.

Salthe, S. N. 1985. *Evolving Hierarchical Systems*. New York: Columbia University Press.

Saper, M. A., Bjorkman, P. J., and Wiley, D. C. 1991. Refined structure of the human histocompatibility antigen HLA-A2 at 2×6 Å resolution. *Journal of Molecular Biology* 219:277–319.

Sapp, J. 1987. *Beyond the Gene: Cytoplasmic Inheritance and the Struggle for Authority in Genetics*. New York and Oxford: Oxford University Press.

1994. *Evolution by Association: A History of Symbiosis*. New York and Oxford: Oxford University Press.

Sarkar, S. 1991a. Lamarck *contra* Darwin, Reduction *versus* statistics: Conceptual issues in the controversy over directed mutagenesis in bacteria. In *Organism and the Origins of Self*, edited by A. I. Tauber, pp. 235–71. Dordrecht: Kluwer Academic.

1991b. Reductionism and functional explanation in molecular biology. *Uroboros* 1:67–94.

1992a. Models of reduction and categories of reductionism. *Synthese* 91:167–94.

1992b. Haldane as biochemist: The Cambridge decade, 1923–1932. In *The Founders of Evolutionary Genetics*, edited by S. Sarkar, pp. 53–81. Dordrecht: Kluwer Academic.

Sartre, J. -P. [1937] 1957. *The Transcendence of the Ego: An Existentialist Theory of Consciousness*. Translated by F. Williams and R. Kirkpatrick. New York: Hill and Wang.

References

Sastry, K. N., and Ezekowitz, R. A. 1993. Collectins: Pattern recognition molecules involved in first line host defense. *Current Opinion in Immunology* 5:59–66.

Sauerbeck, E. 1904. *Die Krise in der Immunitätsforschung*. Leipzig: Klinkhardt.

Saussure, F. de. [1916] 1983. *Course in General Linguistics*. Translated by R. Harris. LaSalle, IL: Open Court.

Schaffner, K. F. 1967. Approaches to reduction. *Philosophy of Science* 34:137–47.

1974. The peripherality of reductionism in the development of molecular biology. *Journal of the History of Biology* 7:111–29.

1992. Theory change in immunology: Part II. The clonal selection theory. *Theoretical Medicine* 13:191–216.

Schneider, S. H., and Boston, P. J. 1991. *Scientists on Gaia*. Cambridge: The MIT Press.

Schofield, R. E. 1970. *Mechanism and Materialism: British Natural Philosophy in an Age of Reason*. Princeton: Princeton University Press.

Schopenhauer, A. [1819] 1969. *The World as Will and Representation*. 2 vols. Translated by E. F. J. Payne. New York: Dover.

Schorske, C. E. 1981. *Fin-de-Siècle Vienna: Politics and Culture*. New York: Vintage.

Schultz, E. A. 1990. *Dialogue at the Margins: Whorf, Bakhtin and Linguistic Relativity*. Madison: University of Wisconsin Press.

Schultz, W. 1922. Über eignenartige Halserkrankungen. *Deutsche Medizinische Wochenschrift* 48:1495–8.

Schwartz, R. S. 1993. Autoimmunity and autoimmune diseases. In *Fundamental Immunology*, 3d ed., edited by W. E. Paul, pp. 1033–97. New York: Raven Press.

Scott, C. E. 1990. *The Question of Ethics: Nietzsche, Foucault, Heidegger*. Bloomington: Indiana University Press.

Scruton, R. 1982. *From Descartes to Wittgenstein: A Short History of Modern Philosophy*. New York: Harper Colophon.

Seigfried, C. H. 1984. Extending the Darwinian model: James's struggle with Royce and Spencer. *Idealistic Studies* 14:259–72.

1990. *William James's Radical Reconstruction of Philosophy*. Albany: State University of New York Press.

Sercarz, E. E., Celada, F., Mitchison, N. A., and Tada, T., eds. 1988. *The Semiotics of Cellular Communication in the Immune System*, NATO ASI Series, vol. H23. Berlin: Springer-Verlag.

Sexton, C. 1991. *The Seeds of Time: The Life of Sir Macfarlane Burnet*. Oxford University Press Australia.

Shapiro, J. A. 1988. Bacteria as multicellular organisms. *Scientific American* 256(6): 82–9.

Shattuck, R. 1969. *The Banquet Years: The Origins of the Avant-Garde in France 1885 to World War I*. Salem: Ayer.

References

Shaw, G. B. [1906] 1985. Doctor's Dilemma. In *Three Plays by G. B. Shaw,* pp. 330–424. New York: New American Library.

Shibles, W. A. 1971. *Metaphor: An Annotated Bibliography and History.* Whitewater, WI: The Language Press.

Shimony, A. 1987. The methodology of synthesis: Parts and wholes in low energy physics. In *Kelvin's Baltimore Lectures and Modern Theoretical Physics,* edited by R. Kargo and P. Achinstein, pp. 379–95. Cambridge: The MIT Press.

Shreffler, D. C. 1988. Seventy-five years of immunology: The view from the MHC. *Journal of Immunology* 141:1791–8.

Silver, M. L., Guo H. -C., Strominger, J. L., and Wiley, D. C. 1992. Atomic structure of a human MHC molecule presenting an influenza virus peptide. *Nature* 360:367–9.

Silverstein, A. M. 1989. *A History of Immunology.* San Diego: Academic.

Simmel, G. [1907] 1991. *Schopenhauer and Nietzsche.* Translated by H. Loiskandl, D. Weinstein, and M. Weinstein. Urbana: University of Illinois Press.

Simon, W. 1973. Positivism in Europe to 1900. In *Dictionary of the History of Ideas,* vol. 3, edited by P. P. Wiener, pp. 532–9. New York; Scribner's.

Simonsen, M. 1957. The impact on the developing embryo and newborn animal of adult homologous cells. *Acta Pathologica et Microbiologica Scandinavica* 40:480–500.

Singer, J. A., Jennings, L. K., Jackson, C. W., Dockter, M. E., Morrison, M., and Walker, W. S. 1986. Erthrocyte homeostasis: Antibody-mediated recognition of the senescent state by macrophages. *Proceedings of the National Academy of Sciencies (U.S.A.)* 83:5498–501.

Siskind, G. W. 1984. Immunological tolerance. In *Fundamental Immunology,* edited by W. E. Paul, pp. 537–58. New York: Raven.

Smith, C. U. M. 1986. Friedrich Nietzsche's biological epistemics. *Journal of the Society of Biological Structure* 9:375–88.

——— 1987. "Clever beasts who invented knowing": Nietzsche's evolutionary biology of knowledge. *Biology and Philosophy* 2:65–91.

Snell, G. D. 1948. Methods for the study of histocompatibility genes. *Journal of Genetics* 49:87–108.

Snell, G. D., Smith, P., and Gabrielson, F. 1953. Analysis of the histocompatibility-2 locus in the mouse. *Journal of the National Cancer Institute* 14:457–80.

Söderqvist, T. 1994. "Ich Fühle mich wie ein Englischer Schlüssel": The selection theory of antibody formation as the confession of its maker. *Journal of the History of Biology.* In press.

Sonea, S., and Panisset, M. 1983. *A New Bacteriology.* Boston and Portola Valley, California: Jones and Bartlett.

Sontag, S. 1989. *AIDS and Its Metaphors.* New York: Farrar, Straus & Giroux.

References

Starobinski, J. [1971] 1988. *Jean-Jacques Rousseau: Transparency and Obstruction.* Translated by A. Goldhammer. Chicago: The University of Chicago Press.

—— 1977. Le concept de cénesthésie et les idées neuropsychologiques de Moritz Schiff. *Gesnerus* 34:2–20.

Stein, H. 1958. Some philosophical aspects of natural science. Ph.D. diss., University of Chicago.

Steiner, G. 1989. *Real Presences.* Chicago: The University of Chicago Press.

Stent, G. S. 1975. Limits to the scientific understanding of man. *Science* 187:1052–7.

Stewart, J., and Varela, F. J. 1989. Exploring the meaning of connectivity in the immune network. *Immunological Reviews* 110:37–61.

Stewart, J., Varela, F. J., and Coutinho, A. 1989. The relationship between connectivity and tolerance as revealed by computer simulation of the immune network: Some lessons for an understanding of autoimmunity. *Journal of Autoimmunity* 2 (Suppl.):15–23.

Stillwell, C. R. 1991. The wisdom of cells: The integrity of Elie Metchnikoff's ideas in biology and pathology. Ph.D. diss., Notre Dame University.

Strong, T. 1988. Nietzsche's political aesthetics. In *Nietzsche's New Seas: Explorations in Philosophy, Aesthetics and Politics,* edited by T. Strong and M. Gillespie, pp. 153–74. Chicago: University of Chicago Press.

Sturtevant, A. H. 1944. Can specific mutations be induced by serological methods? *Proceedings of the National Academy of Sciences (U.S.A.)* 30:176–8.

Sulloway, F. J. [1979] 1983. *Freud, Biologist of the Mind.* New York: Basic.

Sulzberger, M. D. 1930. Arsphenamine hypersensitiveness in guinea pigs: II. Experiments demonstrating the role of the skin, both as originator and as site of the hypersensitiveness. *Archives of Dermatology and Syphilology* 22:839–49.

Talmage, D. W. 1957. Allergy and immunology. *Annual Review of Medicine* 8:239–56.

—— 1959. Immunological specificity. *Science* 129:1643–8.

—— 1986. The acceptance and rejection of immunological concepts. *Annual Review of Immunology* 4:1–11.

Tauber, A. I. 1990. Metchnikoff, the modern immunologist. *Journal of Leukocyte Biology* 47:560–6.

——, ed. 1991a. *Organism and the Origins of Self.* Dordrecht: Kluwer Academic.

—— 1991b. Speculations concerning the origins of Self. In *Organism and the Origins of Self,* edited by A. I. Tauber, pp. 1–39. Dordrecht: Kluwer Academic.

—— 1991c. The immunological self: A centenary perspective. *Perspectives in Biology and Medicine* 35:74–86.

—— 1991d. A case of defense: Metchnikoff at the Pasteur Institute. In *Immunology: Pasteur's Heritage,* edited by P. A. Cazenave and G. P. Talwar, pp. 21–36. New Delhi: Wiley Eastern.

342

References

1992a. The birth of immunology: III. The fate of phagocytosis theory. *Cellular Immunology* 139:505–30.

1992b. The organismal self: Its philosophical context. In *Selves, People, and Persons*, edited by L. Rouner, Vol. 13. Boston University Studies in Philosophy and Religion, pp. 149–67. Notre Dame: Notre Dame University Press.

1992c. The two faces of medical education/Flexner and Osler revisited. *Journal of the Royal Society of Medicine* 85:598–602.

1993. Goethe's philosophy of science: Modern resonances. *Perspectives in Biology and Medicine* 36:244–57.

1994a. Nietzsche and Foucault: On the transvaluation of values. In *Papers in Honor of Robert Cohen*, edited by M. Wartovsky. Dordrecht: Kluwer Academic.

1994b. A typology of Nietzsche's biology. *Biology and Philosophy* 9:25–44.

1995a. From Descartes' dream to Husserl's nightmare. In *The Elusive Synthesis: Aesthetics and Science*, edited by A. I. Tauber. Dordrecht: Kluwer Academic.

1995b. From the self to the other: Building a philosophy of medicine. In *Meta-Medical Ethics: Foundations of American Bioethics*, edited by M. Grodin. Dordrecht: Kluwer Academic.

Tauber, A. I., and Chernyak, L. 1989. The birth of immunology: II. Metchnikoff and his critics. *Cellular Immunology* 121:447–73.

1991. *Metchnikoff and the Origins of Immunology: From Metaphor to Theory.* New York and Oxford: Oxford University Press.

Tauber, A. I., and Sarkar, S. 1992. The Human Genome Project and the limitations of reductionism. *Perspectives in Biology and Medicine* 35:220–35.

1993. The ideology of the Human Genome Project. *Journal of the Royal Society of Medicine* 86:537–40.

Taylor, C. 1975. *Hegel.* Cambridge University Press.

1986. Foucault on freedom and truth. In *Foucault: A Critical Reader*, edited by D. C. Hoy, pp. 69–102. Oxford: Basil Blackwell.

1989. *Sources of the Self: The Making of the Modern Identity.* Cambridge: Harvard University Press.

1991. The dialogical self. In *The Interpretive Turn: Philosophy, Science, Culture*, edited by D. R. Hiley, J. F. Bohman, and R. Shusterman, pp. 304–14. Ithaca: Cornell University Press.

Taylor, M. C. 1987. *Alterity.* Chicago: The University of Chicago Press.

Temkin, O. 1949. Metaphors of human biology. In *Science and Civilization*, edited by R. C. Stauffer, pp. 169–94. Madison: University of Wisconsin Press.

Thaw, E. V. 1989. The painter's saint. *New Republic*, Feb. 6, pp. 40–1.

The I Ching, or Book of Changes, 3d ed. 1967. Translated into German by R. Wilhelm and into English by C. F. Baynes. Princeton: Princeton University Press.

References

Thiele, L. P. 1990. *Friedrich Nietzsche and the Politics of the Soul.* Princeton: Princeton University Press.

Theunissen, M. [1977] 1984. *The Other: Studies in the Social Ontology of Husserl, Heidegger, Sartre, and Buber.* Translated by C. Macann. Cambridge: The MIT Press.

Thomas, K. 1983. *Man and the Natural World: A History of the Modern Sensibility.* New York: Pantheon.

Thorndike, E. L. 1911. *Animal Intelligence.* New York: Macmillan.

Tiselius, A., and Kabat, E. 1939. An electrophoretic study of immune sera and purified antibodies preparation. *Journal of Experimental Medicine* 69:119–31.

Tonegawa, S. 1976. Reiteration frequency of immunoglobulin light chain genes: Further evidence for somatic generation of antibody diversity. *Proceedings of the National Academy of Sciences (U.S.A.)* 73:203–7.

———. 1985. The molecules of the immune system. *Scientific American* 253(4):122–31.

Topley, W. W. C. 1930. Role of spleen in production of antibodies. *Journal of Pathology and Bacteriology* 33:339–51.

Toulmin, S. E. 1977. Self-knowledge and knowledge of the 'self.' In *Self: Psychological and Philosophical Issues,* edited by T. Mischel, pp. 291–317. Oxford: Basil Blackwell.

Traub, E. 1936. The epidemiology of lymphocytic choriomeningitis in white mice. *Journal of Experimental Medicine* 64:183–200.

———. 1938. Factors influencing the persistence of choriomeningitis virus in the blood of mice after clinical recovery. *Journal of Experimental Medicine* 68:229–50.

Treichler, P. A. 1988. AIDS, homophobia, and biomedical discourse: An epidemic of signification. In *AIDS: Cultural Analysis Cultural Activism,* edited by D. Crimp, pp. 31–70. Cambridge: The MIT Press.

Trentin, J. J. 1956. Mortality and skin transplantability in X-irradiated mice receiving isologous, homologous or heterologous bone marrow. *Proceedings of the Society of Experimental Biology and Medicine* 92:688–93.

Trowsdale, J. 1993. Genomic structure and function in the MHC. *Trends in Genetics* 9:117–22.

Twort, F. W. 1915. An investigation on the nature of ultra-microscopic viruses. *Lancet* 2:1241–3.

———. 1949. The discovery of the bacteriophage. *Science News* 14:33–4.

Tyzzer, E. E. 1909. A study of inheritance in mice with reference to their susceptibility transplantable tumors. *Journal of Medical Research* 21:519–73.

Unanue, E. 1981. The regulatory role of macrophages in antigenic stimulation. Part Two: Symbiotic relationship between lymphocytes and macrophages. *Advances in Immunology* 31:1–136.

Unger, P. 1990. *Identity, Consciousness and Value.* New York: Oxford University Press.

References

van Noppen, J. P., de Knop, S., and Jongen, R., eds. 1978. *Metaphor: A Bibliography of Post-1970 Publications.* Amsterdam and Philadelphia: Benjamins.

Varela, F. J. 1979. *Principles of Biological Autonomy.* New York: Elsevier-North Holland.

———. 1988. Structural coupling and the origin of meaning in a simple cellular automation. In *The Semiotics of Cellular Communication in the Immune System,* NATO ASI Series, vol. H23, edited by E. E. Sercarz, F. Celada, N. A. Mitchison, and T. Tada, pp. 151–61. Berlin: Springer-Verlag.

Verela, F. J., and Coutinho, A. 1991. Second generation immune networks. *Immunology Today* 12:159–66.

Varela, F. J., Coutinho, A., Dupire, B., and Vaz, N. N. 1988. Cognitive networks: Immune, neural, and otherwise. In *Theoretical Immunology,* part 2, edited by A. S. Perelson, pp. 359–75. Reading, MA: Addison-Wesley.

Vaughan, V. C. 1915. *Infection and Immunity.* Chicago: American Medical Association.

Vaz, N. M., and Varela, F. J. 1978. Self and nonself: An organism-centered approach to immunology. *Medical Hypotheses* 4:231–67.

Violi, P. 1988. A nonrestrictive semiotics of the immune system. In *The Semiotics of Cellular Communication in the Immune System,* NATO ASI Series, vol. H23, edited by E. E. Sercarz, F. Celada, N. A. Mitchison, and T. Tada, pp. 17–23. Berlin: Springer-Verlag.

von Dungern, E., and Hirschfeld, Z. 1910. Über Verebung gruppen-spezifischer Struckturen des Blutes. *Zeitschrift für Immunitätsforschung und experimentelle Therapie* 6:284–92.

Ward, M. 1986. The rhetoric of independence and innovation. In *The New Painting: Impressionism 1874–1886,* edited by C. S. Moffett, pp. 421–39. San Francisco: The Fine Arts Museums of San Francisco.

Warren, M. 1988. *Nietzsche and Political Thought.* Cambridge: The MIT Press.

Wartofsky, M. W. 1979. *Models, Representation and the Scientific Understanding.* Dordrecht: Reidel.

Watson, J. D., Gilman, M., Witkowski, J., and Zoller, M. 1992. *Recombinant DNA,* 2d ed. New York: Freeman.

Weber, E. 1986. *France, Fin de Siècle.* Cambridge: Harvard University Press.

Weismann, A. 1895. Germinal selection. *The Monist* 6:250–93.

Weiss, P. 1947. The problem of specificity in growth and development. *Yale Journal of Biology and Medicine* 19:235–78.

Wells, H. G.; Huxley, J. S.; and Wells, G. P. [1929] 1934. *The Science of Life.* New York: The Literary Guild.

Wells, H. G., and Osborne, T. B. 1911. The biological reactions of the vegetable proteins: I. Anaphylaxis. *Journal of Infectious Disease* 8:66–124.

White, G. [1798] 1985. *The Natural History of Selborne.* In *The Essential Gilbert White of Selborne,* edited by J. J. Massingham, pp. 1–232. Boston: Godine.

Whitehead, A. N. 1925. *Science and the Modern World.* New York: Macmillan.

References

Whorf, B. L. 1956. *Language, Thought and Reality,* edited by J. B. Carroll. Cambridge: The MIT Press.

Williams, C. J. F. 1989. *What Is Identity?* Oxford: Clarendon Press.

Williams, W. D. 1952. *Nietzsche and the French.* Oxford: Basil Blackwell.

Wilshire, B. 1988. *William James and Phenomenology: A Study of "The Principles of Psychology."* Bloomington: Indiana University Press.

Wilson, M. D. 1978. *Descartes.* London: Routledge and Kegan Paul.

Wimsatt, W. C. 1976. Reductive explanation: A functional account. In *Proceedings of the 1974 Biennial Meeting of the Philosophy of Science Association,* edited by R. S. Cohen, C. A. Hooker, A. C. Michalos, and J. van Evra, pp. 671–710. Dordrecht: Kluwer Academic.

Winchester, G., Sunshine, G. H., Nardi, N., and Mitchison, N. A. 1984. Antigen-presenting cells do not discriminate between self and non self. *Immunogenetics* 19:487–91.

Wittgenstein, L. 1961. *Notebooks 1914–16.* Oxford: Blackwell.

Wright, A. E. 1910. *Studies on Immunisation: Diagnosis and Treatment of Bacterial Infections.* New York: Wood.

Wright, A. E., and Douglas, S. R. 1903. An experimental investigation of the role of the blood fluids in connection with phagocytosis. *Proceedings of the Royal Society of London* 72:357–70.

Wright, A. E., and Reid, S. T. 1905–6. On spontaneous phagocytosis, and on the phagocytosis which is obtained with the heated serum of patients who have responded to tubercular infection, or as the case may be, to the inoculation of tubercle vaccine. *Proceedings of the Royal Society of London,* Series B 77:211–25.

Yates, F. E. 1987. *Self-Organizing Systems: The Emergence of Order.* New York: Plenum.

Young, J. 1992. *Nietzsche's Philosophy of Art.* Cambridge University Press.

Young, R. M. 1971. Darwin's metaphor: Does nature select? *The Monist* 55:442–503.

Zaner, R. M. 1975. Context and reflexivity: The genealogy of self. In *Evaluation and Explanation in the Biomedical Sciences,* edited by H. T. Engelhardt, Jr., and S. F. Spicker, pp. 153–74. Dordrecht: Reidel.

Zijlstra, M., Bix, M., Simister, N. E., Loring, J. M., Raulet, D. H., and Jaenisch, R. 1990. Beta-2-microglobulin deficient mice lack $CD4^-8^+$ cytolytic T cells. *Nature* 344:742–6.

Zirkle, C. 1941. Natural selection before *The Origin of Species. Proceedings of the American Philosophical Society* 84:71–123.

Index

adaptive enzymes, 102, 108
AIDS, 186–90
Alexander, Jerome, 72
Almquist, Ernst, 39, 299n8
antibody, 28, 31–4, 77, 87, 104, 109, 128,
    173
    antibody–antigen kinetics, 117
    antitoxin, 25, 29–31, 34–5, 84, 117
    isoagglutinins, 116
    natural, 74, 116–17, 119, 122, 173
    variable region, 75–6
antibody production, 31, 70, 102–3, 109–11,
    122, 125, 162–5, 302n11
    generation of [antibody] diversity
    (G.O.D.), 75–6, 127, 191, 200
    instruction theories, 70–3, 102–3, 110–
    11, 119; see also Burnet; Jerne
antibody receptor, 127
antibody repertoire, 71
antigen, 30, 71, 74–6, 78–9, 104, 109–11,
    124, 127–9, 168, 170, 179, 191, 194,
    304n1
antigen presenting cell (APC), 79, 124–6,
    192, 228
anti-idiotypic antibodies, 129–30, 167; see
    also idiotypes; idiotypic network
Aristotle, 1, 137, 288–9
Arrhenius, Svante, 31, 68, 301n7
Aschoff, Ludwig, 41, 299–300n11
Atlan, Henri, 130, 199
autoimmunity, 6, 36, 69, 76, 98, 120, 158,
    173, 175–7, 180–1, 185, 194–5
autonomous network theory (ANT), 173–5,
    196–7
Avrameas, Stratis, 185–6
Ayer, A. J., 305n5

Bacon, Francis, 51, 143
bacteriophage, 84–92, 116
Baer, Karl von, 48, 311n9
Bail, Oscar, 72
Baudelaire, Charles, 290
Baumgarten, Paul, 26
Behring, Emil von, 27–9, 34, 36, 37, 71
Benacerref, Baruj, 302n11
Beneke, Friedrich Edward, 204
Bennett, J. Claude, 75, 301n8
Bergson, Henry, 53, 65, 283, 287
Bernard, Claude, 6, 23, 51–2, 67,
    300n2
Bernstein, Felix, 78
Besredka, Alexander, 32, 40
Billingham, Rupert, 78–9
biochemistry, 67, 175, 200
biological ego, 132; see also self
Black, Max, 137
blood group antigens, 77–8
body, 178, 182–3, 211, 242, 249, 251–2,
    258, 260, 273–5, 277
Bona, Constantin, 129–30, 169, 171
Bordet, Jules, 28, 31, 77, 85–90
Bordet–d'Herelle debate, 89, 192
Bouchard, Charles, 28
Breinl, Friedrich, 72
Brent, Leslie, 78, 79
Brentano, Franz, 202, 204–5, 207, 213, 216,
    219, 285
Brown, A. J., 67
Buchner, Edward, 67
Buchner, Hans, 33, 71
Buffon, Comte de, 262
Burnet, Sir Frank Macfarlane, 7, 14, 31, 74,
    81 passim, 303n1–3, 5–7, 304n8

347

Index

Index

Ehrlich, Paul, 28–31, 35, 39, 42, 61, 69, 71, 73, 76–7, 100, 113, 116, 122–3, 126
  horror autotoxicus, 32, 76, 113, 185, 195
  side chain theory, 30, 70, 72, 117, 123, 194
embryology, 18, 20, 64, 66; *see also* developmental biology
Emery, d'Este, 40
enzyme, 67, 103–4, 109
Escher, Maurits, 179
evolution, 2, 6, 13, 21, 23, 43, 87, 253, 260, 262, 288
evolutionary episteme, 5, 261 *passim*
evolutionary epistemology, 121

Fénéon, Felix, 282, 284, 312n4
Fenner, Frank, 81, 99, 106, 108, 112, 157, 162
Feuerbach, Ludwig, 198
Fichte, Johann, 232, 283
fin-de-siècle, 15, 45, 201, 257, 270, 278–9, 300n4
Fink, Eugen, 309n6
Fisher, Emil, 30, 73
Fleck, Ludwig, 68, 301n6
Flew, Antony, 139, 305nn5, 8
Flexner Report, 68
Fluegge, Carl, 27
Foucault, Michel, 147, 227, 252, 270–78, 287, 312n1,2; *see also* Nietzsche; postmodernism; power
Freud, Sigmund, 15, 96, 243, 260, 283, 297–8n2
  ego, 142, 147, 204, 260, 297n2
  id, 142, 243, 260
  see also ego
Fries, Jakob, 47

Gaia hypothesis, 57
Gajdusek, Charleton, 120
Gallo, Robert, 187
gene, 9, 56, 70, 77–8, 167, 175, 180, 200
  C, J, and V genes, 75–6
  see also antibody production, generation of antibody diversity; Burnet; differentials; molecular biology; self as entity
genetics, 56–7, 74, 78, 107–8, 112, 123, 292
gnothi se auton, 142, 144, 185, 305n6
Goethe, Johann Wolfgang von, 47, 50, 51, 52, 57, 203, 209
Golub, Edward, 169–70
Gorer, Peter A., 78, 101, 302n10

graft rejection, 100
graft-versus-host (GVH) disease, 120, 194, 307n4
Griesinger, Wilhelm, 204
Grünbaum, A. S., 25

Haeckel, Ernst, 17, 65
Haldane, J. B. S., 73, 78, 106, 162–3
Haraway, Donna, 183–4
Haurowitz, Felix, 72, 110, 162–3
health, 23, 249
Heberden, William, 49
Hegel, Georg Wilhelm Friedrich, 47, 198, 258, 263–6, 280, 283, 311n8,9; *see also* Nietzsche
Heidegger, Martin, 135, 142, 169, 204, 259, 267, 283, 294, 311n7
Helmholtz, Hermann, 6, 49, 67, 284
Henderson, Lawrence, 54–5
Henri, Victor, 67
Heraclitus, 13, 261, 263, 265, 278, 289
Hershey, Alfred, 118
heterogeneous immunity, 87; *see also* d'Herelle
Hirshfeld, Ludwik, 77
His, Wilhem, 66
Hood, Leroy, 75
holism, 44, 48, 50, 62, 68, 96
host, 139, 141
human genome project, 56
human immunodeficiency virus (HIV), 186–9
Hume, David, 148–9, 207, 223, 236, 246, 306n10
humoral immunity, 126–7
humoral theory of immunity, 25, 27–8, 34–5, 43, 61, 97
humoralists, 6, 17, 34, 39, 42–4, 61, 62, 70
Husserl, Edmund, 11, 142, 197, 204, 213–26, 228, 270, 282, 285, 293, 295, 308n4,5, 309n6
Huxley, Julian, 94

I, 147, 150, 198, 212, 246–7, 249, 252, 260, 295; *see also* ego; identity; self
identity, 3, 7, 10–11, 13, 42, 80, 130, 132, 139–40, 145–6, 148–9, 175, 185–6, 189, 194, 197–9, 201, 252–3, 263, 279, 291–4
  Metchnikovian, 5–6, 20, 24, 26, 40, 43, 59, 62, 185, 190, 195
  see also integrity; self
idiom, 128

349

Index

idiotypes, 128–9, 167, 170
idiotypic network, 129–30, 167, 169, 171–2, 177–9
immune homunculus, 178–80, 197
immune memory, 127–8, 165
immune processing, 62, 127
immune recognition, 11–12, 41–2, 70, 75, 79
immune repertoire, 127–8, 172–3, 195
immune response, 4, 18, 25, 31, 33, 39, 78, 99, 103–4, 110, 130, 133
  active, 36
  natural, 33, 35, 126, 192
  passive, 36
immune self, 8, 76, 156, 189, 201, 228–9, 294; *see also* self
immune system, 3, 8, 58, 62, 96, 124–30, 165, 169, 173, 291
immunity, 4, 9, 11, 17–19, 20, 26, 31, 40, 86, 97
  acquired immunity, 33–5, 37, 42, 75, 127
  cell mediated, 120, 125–8
  natural, 124, 127, 133
immunization, 71
immunochemistry, 3, 8–9, 29, 31, 32, 62, 69, 71–3, 279, 301n7
immunodiagnostics, 68, 231n6
immunoglobulin, 9, 128–9, 168–9, 304n8, 304n2; *see also* antibody
immunological specificity, 70, 73, 116
individual, 100–1, 139, 146, 189, 198
infectious diseases, 5, 19, 105
inflammation, 19–21, 35, 41, 126
information, 161–5, 176, 197
integrity, 3, 6–7, 20, 62, 130, 230, 291
  Metchnikovian, 6–7, 20, 62
  see also identity
intentionality, 176, 205, 219
interleukins, 125, 127

Jahn, Ferdinand, 24
James, William, 11, 15, 202, 204–5, 207–15, 219, 231, 233, 269, 282, 284–5, 293, 305n5, 307n1, 308nn2,5
Janaway, Christopher, 241, 242, 243, 244, 247, 248, 249, 295
Jensen, Carl O., 77
Jerne, Niels, 31, 70, 74, 76, 103, 121, 128–9, 132, 161, 168, 172, 304n8
  immune network, 164–9
  natural selection theory, 114–19, 123, 158–9, 166
Johnson, Mark, 137–8, 182

Jones, J. R., 139
Jordan, Pascual, 73

Kant, Immanuel, 46–50, 142, 144, 203, 205, 232, 235–47, 252, 269, 280, 281–3, 286, 290–1, 300n1, 309n2, 309–10n3
  transcendental unity of apperception, 142, 220, 236–7, 239, 245, 309n2,3, 310n4; *see also* modernism
Kierkegaard, Soren, 116, 166, 168, 198, 278, 286
Kitasato, Shibasaburo, 27, 28, 36
Klein, Jan, 77–8, 100, 130, 134
Koch, Robert, 16–18
Kohák, Erazim, 216–20, 222, 308n4, 309n7
Kowalevsky, Alexander, 17
Kruif, Paul de, 17–18

Lakatos, Imre, 5
Lakoff, George, 137–8, 182
Lamarck, Jean Baptiste de, 46
Landsteiner, Karl, 68, 71–3, 77, 122
language, 10, 165, 167–72, 186
Lavoisier, Antoine, 44
learning, 109, 158, 161, 165–6
Leder, Philip, 75, 301n8
Lederberg, Joshua, 31, 74, 107, 114, 121, 123, 303n7, 304n8
Leibniz, Gottfried, 142
Leukart, Rudolf, 17
Levin, David, 257
Lindegren, Carl, 108
linguistics, 11, 168–9
Linnaeus, Carolus, 262
Little, Clarence, 302n9
Locke, John, 140, 145–9, 154, 209, 235, 305n8, 306n9
Loeb, Jacques, 55, 100
Loeb, Leo, 77, 100–1, 156
Lotze, Rudolf, 206
Löwith, Karl, 263, 264, 310n6
lymphocyte, 4, 40, 78, 81, 100, 120, 127, 158, 167, 178, 193–4, 224
  B lymphocyte, 79, 114, 125–9, 171, 179, 181, 302n12
  T lymphocyte, 76, 79, 114, 118, 125–6, 129, 171, 177, 179, 184, 186, 193, 302n11,12
  see also cell-mediated immunity

McDevitt, Hugh, 302n11
Mach, Bernard, 301n8
Mach, Ernst, 15, 202, 205–7, 227, 285
Mackenzie, J. N., 181

350

Index

macrophage, 42, 126, 192; *see also*
    phagocyte
Magee, Bryan, 241, 243, 246
major histocompatibility complex (MHC),
    76–9, 120, 125–7, 159, 173, 193,
    197, 292–4, 302nn10–12, 307n3
    antigen complex, 125, 133, 191–3
    restricted recognition, 79, 133–4, 177,
        224, 228
Margulis, Lynn, 58
Martin, Emily, 183–4, 186
Marx, Karl, 203, 275, 283, 286, 312n2
Maxwell, James, 284
Mayr, Ernst, 53, 261–2
meaning, 169–71, 179
mechanical, 63, 65
mechanistic biology, 24, 294
Medawar, Peter, 78–9, 121
memory, 127, 161, 165–6
Mendelian inheritance, 56–7
mental substance, 142
Menten, Maud, 67
Merleau-Ponty, Maurice, 211, 213, 216,
    226–7
Merz, John, 45, 206–7
mesoderm, 18, 20–1
Metal'nikov, Serge, 181
metaphor, 10, 136–42, 182–5, 189–90, 201,
    224, 228, 252, 279, 296
metaphysics, 12, 27, 42, 44, 45, 48, 131,
    135, 149, 206, 229–30, 234, 239,
    241, 246–7, 258, 265, 267, 276–8,
    283
Metchnikoff, Elie, 4–8, 12–14, 16–45, 58–
    63, 69, 77, 87–8, 96–8, 114, 123,
    126, 130, 155, 157, 209–10, 227,
    230–2, 234–5, 240, 250, 257, 261,
    263, 267–70, 279, 285, 289–91,
    297nn2–5, 299nn7–9, 300n4
    harmony/disharmony theory, 6, 21–4
    Immunity in Infective Diseases, 25, 35,
        38
    Lectures in the Comparative Pathology
        of Inflammation, 24, 37, 40
    Nobel Prize, 25, 28, 35–9, 64, 68–9, 119
    physiological inflamation, theory of,
        5–6, 21–3, 286
    teleology, 25, 33, 59–61
    theory of immunity, 4–8, 16–23, 25,
        34–5, 38, 58–62, 97, 126, 231,
        299n9
    see also cellularist–humoralist debate;
        identity; integrity; macrophage; neu-
        trophil; phagocyte

Michaelis, Leonor, 67
microbiology, 4, 18, 59, 61
mind, 1, 160, 169, 172, 177, 186, 282,
    300n4
mind–body dualism, 10, 49, 147, 168–9,
    225–6; *see also* Descartes
Mitchison, Avrion, 78
modernism, 13, 280–1, 291; *see also* Kant;
    Nietzsche; postmodernism
modernism–postmodernism, 12, 280–3,
    295; *see also* Nietzsche
molecular biology, 9, 72–3, 79, 200; *see also*
    DNA; gene
monad, 142
Monod, Jacques, 60–1, 199
Morgenroth, Julius, 77
Mörner, K. A. H., 36–8, 39
Mudd, Stuart, 72
Muller, Johannes, 48
Murphy, James B., 99, 100, 120

Nagel, Thomas, 202, 204, 300n4
natural selection, 6, 121, 122, 159, 253
natural selection theory (Jerne), 114,
    118
naturphilosophie, 46, 47, 50, 52, 300n1
nervous system, 9, 159–60, 165–7, 174,
    176–7
network, 125, 128–9, 134, 167, 173, 175,
    189, 228, 304n2
Neumann, John von, 162
neural functions, 9
neutrophil, 126–7, 192; *see also* phago-
    cyte
Newton, Isaac, 5, 49
Nietzsche, Friedrich, 12, 15, 65, 211, 231,
    243, 265–7, 269–70, 280–1, 283,
    286–90, 306n11, 309n1
    biologicism, 58, 249–61, 288
    eternal recurrence, 251, 265–6, 268
    Foucault and, 271–8
    Hegel and, 263–6, 311n8
    modernist vs. postmodernist, 13, 252,
        280–1, 283, 290
    nihilism, 250, 254, 268, 286
    Schopenhauer and, 245–9, 263–5, 283,
        287
    Ubermensch, 251, 255, 267, 274, 287,
        310n6
    will to power, 235, 248, 250–2, 254–5,
        265, 286, 309n1, 311n7
    see also Metchnikoff; self, philosophy of
nonself, 76, 99, 102, 134, 291; *see also* self
    and not self

351

Index

352

Index

as process, 58, 159, 168–9, 172, 177,
180, 185, 198, 225, 266–8, 294
punctual, 134–5, 155, 160, 170, 172,
186
see also autonomous network theory;
Burnet, views of the self; Descartes,
cogito; Cohen; Coutinho; clonal selection theory; ego; Freud; host; I;
identity; immune homunculus; immune self; individual; Metchnikoff;
organism; person; selfhood; subject
self and not self, 83, 98, 111, 132, 150–4,
164, 167, 173, 184, 187–9, 192,
194–6, 198, 203, 210, 228, 234, 263,
292
self, metaphor, 7, 10, 143–4, 156, 159–61,
164–8, 227–9, 252, 291, 292, 294–5
cognitive, 136, 161–80, 193, 196
embodied, 136, 181–90
ontological, 135, 156, 295
self, philosophy of
Descartes, René, 140, 144–5, 148–9,
216
Foucault, Michel, 271–8
Hegel, G. W. F., 266
Heidegger, Martin, 135, 142, 294
Hume, David, 148–9, 246
Husserl, Edmund, 219–23, 226
James, William, 211–15
Kant, Immanuel, 235–9, 246–7
Locke, John, 140, 145–9, 154
Marx, Karl, 203, 283, 286
Nietzsche, Friedrich, 232–4, 240, 248–9,
252, 256, 260, 263, 266–8, 286,
289–90, 306n11
Rousseau, Jean-Jacques, 149–54
Schopenhauer, Arthur, 239–45, 247–8
selfhood, 8–10, 26, 69, 90, 107, 201, 245,
252, 260, 278, 280, 294; *see also* self
semiotics, 169–71
serology, 3, 107
Seurat, Georges, 209, 284
Shattuck, Roger, 15, 308n2
Shaw, George Bernard, 33
Simmel, George, 241–4, 246–7, 264, 287
Simonsen, Morten, 120
Snell, George, 78, 302n10
Socrates, 142, 166, 289, 305n6
Söderqvist, Thomas, 115, 117
Sonneborn, Tracy, 108
Sontag, Susan, 186, 188
soul, 142, 219
Spiegelman, Sol, 108
Spinoza, Baruch, 48–9

Stahl, George, 49
Starobinski, Jean, 153–4
Steinberg, George, 301n8
Stent, Gunther, 118, 310n4
Sternberg, George, 298n5
Stewart, John, 172–3, 176, 195
stimulus–response reaction, 169–71
Stirner, Max, 310n6
Sturtevant, Alfred, 106
subject, 240, 244, 252, 273, 287, 295
Sundberg, Carl, 37, 39, 299n8
surveillance, 81
symbiosis, 58, 88, 112, 133
syphilis, 68, 301n6
system, 8, 126, 177

Talmage, David, 31, 74, 114, 121, 123, 158,
304n8
Taylor, Charles, 134–5, 144–5, 147, 152,
190, 198, 278, 283, 286, 292,
teleology, 25, 33, 44, 48–9, 59–61, 199, 243,
278, 288
teleomechanism, 46–9
teleonomy, 60, 199
Thorndike, Edward, 122, 304n9
tolerance, 43, 69, 75–6, 83, 93, 98–101,
104–5, 108–13, 123, 132, 173, 179,
197
Tonegawa, Susumu, 75, 301–2n8
Topley, William, 72
training, 84, 101, 104–5
transplantation, 43, 77–8, 100–1, 174, 291
Traub, Erich, 99, 112
Treichler, Paula, 187–9
Treviranus, Gottfried, 46
Twort, Frederick, 85–6
Tyzer, Ernest, 77

unconscious, 142, 260; *see also* Freud

Verela, Francisco, 134, 170, 172–6, 195–7,
200
Vaughn, Victor, 35, 39
Vaz, Nelson, 134, 172, 174–5, 195–7
Vienna Circle, 206
Violi, Patrizia, 170–1
Virchow, Rudolf, 16, 17, 19, 22, 24
vitalism, 19, 44, 46–50, 59, 64, 67, 69, 258,
300n5

Wassermann, August von, 301n6
Watson, James, 118, 162, 193

353